T0271449

## The Mechanics and Reliability of Films, Multilayers and Coatings

A wide variety of applications ranging from microelectronics to turbines for propulsion and power generation rely on films, coatings and multilayers to improve performance. As such, the ability to predict coating failure – such as delamination (debonding), mud-cracking, blistering, crack kinking and the like – is critical to component design and development. This work compiles and organizes decades of research that established the theoretical foundation for predicting such failure mechanisms and clearly outlines the methodology needed to predict performance. Detailed coverage of cracking in multilayers is provided with an emphasis on the role of differences in thermoelastic properties between the layers. The comprehensive theoretical foundation of the book is complemented by easy-to-use analysis codes designed to empower novices with the tools needed to simulate cracking; these codes enable not only precise quantitative reproduction of results presented graphically in the literature, but also the generation of new results for more complex multilayered systems.

**Professor Matthew R. Begley** is broadly recognized for seminal contributions in the mechanics of multilayered systems with an emphasis on computational aspects of the required analysis. His codes are employed in some industries to design experiments, assess current designs and evaluate novel multilayer systems for improved performance. Both authors are widely sought after for consulting work on the mechanics of thin films, coatings and multilayers by companies such as General Electric, Pratt & Whitney, Intel, Sunpower, Raytheon, Areva etc.

**Professor John W. Hutchinson** is a member of the US National Academies of Engineering and Sciences and a Foreign Member of the Royal Society of London. He is one of the leading experts in the mechanics of thin film systems, with a number of highly cited, seminal journal papers on the subject. Hutchinson is broadly credited with generating many of the conceptual developments in this field, as well as illustrations of those concepts to applications ranging from microelectronics to thermal barrier coatings, microfluidic devices, hypersonics etc.

# The Mechanics and Reliability of Films, Multilayers and Coatings

MATTHEW R. BEGLEY
University of California, Santa Barbara

JOHN W. HUTCHINSON
Harvard University

CAMBRIDGE
UNIVERSITY PRESS

Shaftesbury Road, Cambridge CB2 8EA, United Kingdom

One Liberty Plaza, 20th Floor, New York, NY 10006, USA

477 Williamstown Road, Port Melbourne, VIC 3207, Australia

314–321, 3rd Floor, Plot 3, Splendor Forum, Jasola District Centre, New Delhi – 110025, India

103 Penang Road, #05–06/07, Visioncrest Commercial, Singapore 238467

Cambridge University Press is part of Cambridge University Press & Assessment,
a department of the University of Cambridge.

We share the University's mission to contribute to society through the pursuit of
education, learning and research at the highest international levels of excellence.

www.cambridge.org
Information on this title: www.cambridge.org/9781107131866

First published 2017

*A catalogue record for this publication is available from the British Library*

*Library of Congress Cataloging-in-Publication data*
Names: Begley, Matthew R., 1969– author. | Hutchinson, John W., author.
Title: The mechanics and reliability of films, multilayers and coatings / Matthew R. Begley
    (University of California, Santa Barbara), John W. Hutchinson (Harvard University).
Description: Cambridge, United Kingdom ; New York, NY : Cambridge University Press, 2017. |
    Includes bibliographical references and index.
Identifiers: LCCN 2016036248 | ISBN 9781107131866 (hardback ; alk. paper) |
    ISBN 1107131863 (hardback ; alk. paper)
Subjects: LCSH: Thin films – Mechanical properties. | Protective coatings – Mechanical properties. |
    Fracture mechanics.
Classification: LCC TA418.9.T45 B44 2017 | DDC 667/.9–dc23 LC record available at
https://lccn.loc.gov/2016036248

ISBN    978-1-107-13186-6    Hardback

# Contents

# Acknowledgements

This book would not have been possible without the extensive contributions of others, on many levels.

At the most fundamental level, Tony Evans and Zhigang Suo were responsible for much of the scientific foundation utilized in this book, not to mention countless inspirational applications, examples and key insights. Simply put, this book would not exist without their pioneering efforts.

Those pioneering efforts were made possible to a significant degree by the National Science Foundation and Office of Naval Research, who deserve special recognition for supporting the underlying research of the material in this book – not only the authors themselves, but also the entire community working on thin films and coatings.

Cedric and Lily Xia deserve special recognition for the ancestral codes that inspired those in this book; while their original codes have been supplanted, they played in important role in forming the vision for the codes accompanying this book.

R. Wesley Jackson was an early adopter of the codes and made significant contributions to their utility, as well many, many insightful analyses on thermal barrier coatings. It would be difficult to overestimate the practical contributions of Ryan Latture, whose edits sped up the codes by orders of magnitude, and Foucault de Franqueville, who did critical work in validating the codes by conducting an estimated $10^4$ analyses. The authors are also deeply indebted to J. William Pro and Stephen Sehr for their help in data management and vetting the codes. Dr. Tyler Ray was a huge help with typesetting the book.

M. R. Begley would also like to acknowledge Carlos Levi, Bob McMeeking, Tresa Pollock and Frank Zok; my wonderful colleagues at UCSB provided encouragement at critical times and covered numerous obligations for me while I was 'in the bunker' writing the book. My appreciation for their direct and indirect support cannot be overstated.

# Notation

The list below refers to the most common usage:

| | |
|---|---|
| $\Lambda_U, \Lambda_B, \Lambda_o$ | strain energy per unit area in thin strips |
| $\alpha$ | coefficient of thermal expansion |
| $\epsilon_{ij}, \sigma_{ij}$ | strain and stress tensors, respectively |
| $E, v$ | elastic modulus and Poisson's ratio |
| $\bar{E} = E/(1 - v^2)$ | plane strain modulus |
| $E_*$ | effective interface modulus, eqn. (3.14) |
| $\eta_x, \eta_y, \eta_{xy}$ | large strain definitions |
| $\alpha_D$ | first Dundurs' parameter, eqn. (3.9) |
| $\beta_D$ | second Dundurs' parameter, eqn. (3.9) |
| $\epsilon$ | mismatch parameter when $\beta_D \neq 0$ |
| $\theta$ | stress-free misfit strain, such as thermal expansion |
| $\bar{c}$ | coefficient of misfit strain defined by geometry constraint |
| $u(x), w(x)$ | axial and transverse displacements, respectively |
| $\kappa$ | curvature (second derivative of transverse displacement) |
| $\epsilon_o$ | axial stretch of the reference axis |
| $\sigma_o, \sigma_c$ | residual stress and critical stress for buckling |
| $N, M$ | axial force and bending moment resultants, respectively |
| $a_{ij}, b_j$ | coefficients in multilayer equations to find $\epsilon_o$ and $\kappa$ |
| $G$ | energy release rate |
| $K_I, K_{II}$ | isotropic stress intensity factors |
| $K_1, K_2$ | interface stress intensity factors |
| $\Gamma_I, \Gamma_{II}$ | interface toughness in mode I and mode II |
| $\eta$ | dimensionless stress parameter for kinking/deflection |
| $\lambda$ | fitting parameter that dictates mode II toughness |
| $\psi$ | phase angle that defines mode-mix |
| $\omega$ | phase factor used to compute mode-mix |
| $\Delta T$ | temperature change from stress-free reference state |
| $q$ | heat flux |
| $k$ | thermal conductivity |

# 1 Introduction

A 'generic' system consisting of bonded layers of different materials and illustrations of various failure modes are shown in Figure 1.1. Arguably, the two dominant technological applications with this type of geometry are microelectronic devices and protective coatings in extreme environments (e.g., thermal barrier coatings). These applications involve layers with very disparate properties and are subjected to rather aggressive external stimuli. Cracking happens either between the layers (interface delamination or debonding) or within a layer (tunneling or channeling cracks). Delamination can occur regardless of whether the stresses are tensile or compressive, while buckling-driven delamination occurs only in layers experiencing compressive stress, and channeling or tunneling cracks require tensile stress in the layers.

Failure can be driven by a variety of factors, but it is probably fair to state that the integrity of the vast majority of multilayered devices is controlled by the layers' tendency to expand at different rates in response to thermal, mechanical or chemical stimuli. In essence, when left by their lonesome, the layers expand differently in response to temperature fields, mechanical loading and so forth. However, in the multilayer component they are not alone: they are constrained to experience conformal deformation where they are bonded. This constraint generates stresses (both tensile and compressive) and stored elastic energy that drive system failure. Crudely speaking, the layers would be happiest and in their lowest energy state as separate pieces, and they seek to return there – even if it means splitting themselves into pieces and leaving part of themselves stuck to another layer.

As with the development of any predictive framework, the initial challenge is to reduce the complexity of the actual system to produce a model that captures the salient features of the system, while ignoring those details that have little effect on the behavior of interest. This can be a critical step in the development of multilayered systems, because their geometrical and multimaterial complexity can make full numerical representations extremely costly. Make no mistake: this is an art.[1] Further, it can be highly problem specific. Nonetheless, there are some central idealizations that are widely applied to thin film systems that have served to generate considerable general insight regarding failure. These idealizations involve eliminating at least one, and very often

---

[1] As epitomized by the late Tony Evans, the Michaelangelo of knowing what to paint and what not to paint in a multilayer problem.

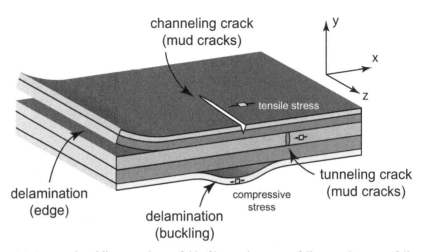

**Figure 1.1** A general multilayer made up of thin films and common failure modes; some failure modes occur regardless of the sign of the stresses in the layers (e.g., delamination), while others require either tensile stress in the layer (e.g., channeling and tunneling cracks) or compressive stress in the layer (e.g., buckling-driven delamination). In some failure modes (e.g., edge delamination or channeling cracks), the driving force for crack growth becomes independent of crack length for very long cracks and scales with the layer thickness: this is referred to as 'steady-state' cracking.

two (or even more), of the length scales associated with the failure modes shown in Figure 1.1.

Consider the edge delamination crack shown in Figure 1.1. The most common step is to eliminate from consideration the dimension in the $z$-direction, either because the structure is very thin in that direction (plane stress) or very wide (plane strain). The problem is now two-dimensional, but still with (at least) three length scales: the $(x, y)$ dimensions of the component and the crack length. We fully enter the realm of 'multi-layer analysis' (as it is commonly used) when we assume the $x$ dimension is much larger than either the $y$ dimension or the crack length. In such cases, the problem consists of 'thin' layers bonded together. If the crack length and $y$ dimensions are comparable, we are left with a two-dimensional problem with (at least) two characteristic length scales. Still, it is common to go a step further: suppose the crack length is much larger than the $y$ dimensions: we now have a two-dimensional problem dominated by $y$ dimensions (i.e., layer thickness).

This is arguably the central reference point in the multilayer failure universe: the semi-infinite crack in a semi-infinite plane of thin layers. Herein, this geometry is referred to as consisting of *blanket films*, since the only relevant dimensions are layer thickness. In this reduced problem space, many important aspects of the response – such as the stresses in the layers away from the crack tip and the energy released by crack advance – can be solved for analytically or with highly efficient one-dimensional numerical tools. (As will be discussed, a critical feature of the response – the behavior near the crack tip – will require a two-dimensional numerical analysis.) Cracking in this scenario is referred to as 'steady-state' because a change in the crack length does not

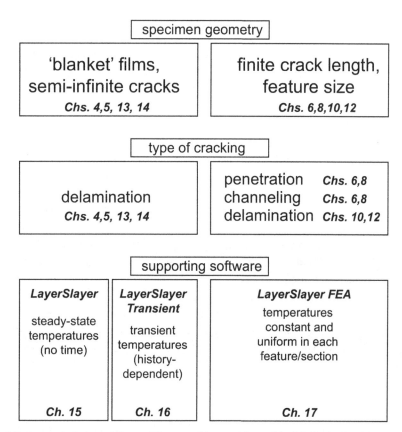

specimen geometry

| 'blanket' films, semi-infinite cracks *Chs. 4,5, 13, 14* | finite crack length, feature size *Chs. 6,8,10,12* |

type of cracking

| delamination *Chs. 4,5, 13, 14* | penetration *Chs. 6,8* channeling *Chs. 6,8* delamination *Chs. 10,12* |

supporting software

| *LayerSlayer* steady-state temperatures (no time) *Ch. 15* | *LayerSlayer Transient* transient temperatures (history-dependent) *Ch. 16* | *LayerSlayer FEA* temperatures constant and uniform in each feature/section *Ch. 17* |

**Figure 1.2** A snapshot view of the multilayer problems addressed in this book, the chapters covering each subject and the supporting LayerSlayer software that is provided to analyze various failure modes.

impact the problem: the geometry and solution are self-similar as the crack advances. In many instances, steady-state cracking results in the maximum possible driving force for cracking, which means this idealization plays a central role in conservative designs. In summary, the blanket film scenario is the easiest to analyze and most pessimistic (i.e., it is assumed that a large interface flaw exists) and therefore takes center stage in multilayer design.

Because this process of reducing the dimensionality of the problem is so central to analysis, it is worth repeating for a different failure mode. Consider the channeling crack shown in Figure 1.1. Here, we assume that the length of the crack in the $z$-direction is much larger than the dimensions in the $(x, y)$ plane, and that the total dimension of the multilayer in the $z$-direction is even larger still. Thus the device is 'infinite' in the $z$-direction while the crack is 'semifinite'. Since the problem is self-similar in the $z$-direction, the channel crack is at steady state with regard to propagation in this direction. It turns out that one can get the relevant quantities that control propagation of the crack in both the $z$ *and* $y$ directions (i.e., growth through the stack) by analyzing the $(x, y)$ planes ahead of the crack in the intact layer, and far behind the crack tip (where the crack is open). In some cases, the analysis can be analytical: in others, numerics are required. As you might expect, it is easiest to illustrate whether analytical analysis or numerical analysis is needed by presenting the assumptions associated with the analytical approach and then discussing when these assumptions are likely violated.

Once the geometry has been suitably simplified (if possible), the question then arises: what to compute? The answer in the simplest terms includes stresses, displacements and the strain energy associated with those quantities. The strain energy plays the central role, as it creates the driving force for failure because energy is released by crack advance. The rest of this book is basically just an illustration of what to compute, how to compute it and then how to use these results to predict whether or not something will fail. The associated theory is discussed principally in the context of examples involving blanket films, and then extended to more complicated geometries for which numerical analysis is required.

A powerful feature of this book is that software is provided to compute the parameters controlling failure modes such as those shown in Figure 1.1, complementing the text, which focuses largely on theory. Figure 1.2 provides an overview of the book's content, organized by geometric idealization, cracking mode and associated software – with indications of the chapters that provide the underlying theory. The chart in Figure 1.2 is intended as a handy reference for readers to identify the location of theory and software that addresses a specific type of problem.

# 2 Key Mechanics Concepts

This book relies heavily on concepts from strength of materials, continuum mechanics and the finite element method. Excellent books on these topics are abundant. The book on solid mechanics by Bower (2010) is a particularly complete reference that meshes well with much of the coverage here. This chapter is merely meant to provide a convenient reference for concepts used frequently in the rest of the book.

The majority of analytical solutions relevant to thin films and multilayers correspond to two-dimensional (planar) idealizations, which in one way or another represent a slice through a specific $(x, y)$ plane in Figure 1.1. Even more narrowly, the mechanics review presented here is focused on results used to analyze *blanket* thin films, in which the width of the layers in these slices (i.e., the dimension in the $x$-direction) is much greater than their thickness. In this scenario, the films behave as plates whose deformation is uniform in the $z$-direction. Often the term 'beam' is used with the understanding that the behavior in the $z$-direction may not correspond to plane stress.

In this chapter, isotropic linear elastic constitutive descriptions are reviewed first, which are used exclusively throughout the book. The reader is referred elsewhere for generalizations to orthotropic and/or nonlinear constitutive relationships, for example, Bower (2010). Then, the mechanics of beams and plates are reviewed; the majority of problems addressed in this book involve small deformations and the corresponding linear strain-displacement relationships. As there are a few important problems in coatings that require nonlinear strain-displacement relationships (e.g., buckling of coatings subject to compressive stresses), a brief introduction to moderate rotation beam/plate theory is provided. The extension of these results to multilayers, that is, the analysis of individual layers bonded together, is left for future chapters. Finally, this chapter concludes with a section on unidirectional heat transfer, which is invoked in later chapters to determine temperature distributions through multilayers, which generate stresses that drive failure.

## 2.1 Review of Linear Isotropic Elasticity

The foundation of much of the subject matter covered in this book rests on linear elasticity theory. Many solutions are two-dimensional (2D) idealizations of three-dimensional (3D) problems, and extensive use of plate and beam theory is made for modeling purposes. Throughout the book, almost without exception for the 2D problems, the

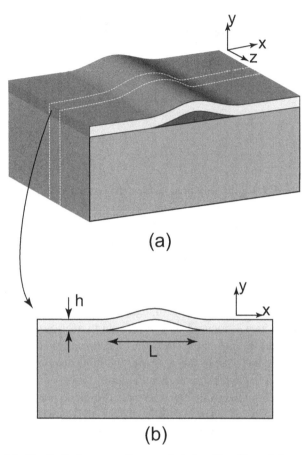

**Figure 2.1** (a) A thin film that bulges away from a substrate with uniform deformation in the $z$-direction can be analyzed using a one-dimensional idealization, as shown in (b), which depicts a representative slice from the interior of the film (away from the edges).

coordinate choice shown in Figures 1.1 and 2.1 will be employed with $x = x_1$ parallel to the layer in the direction of significant variation, $y = x_2$ normal to the film or layer, and $z = x_3$ in the out-of-plane direction. Especially for the 2D applications of elasticity considered in this book, the out-of-plane behavior is important. Specifically, all of the common idealizations regarding out-of-plane behavior will have to be considered, that is, plane strain, plane stress and generalized plane strain. The special case of generalized plane strain that is of most interest to us will be equi-biaxial in-plane straining, which is a common occurrence in films, coatings and multilayers. Thus, it is necessary to begin by considering the 2D behavior within the 3D setting.

Three-dimensional Hooke's law for an isotropic material is given in indicial notation by

$$\epsilon_{ij} = \frac{1}{E} \left[ (1 + v)\sigma_{ij} - v\sigma_{kk}\delta_{ij} \right] + \theta\delta_{ij} \tag{2.1}$$

where $E$ is the Young's modulus, $v$ is Poisson's ratio and $\delta_{ij}$ is the Kronecker delta, where $\delta_{ij} = 1$ for $i = j$ and $\delta_{ij} = 0$ for $i \neq j$. The isotropic misfit strain $\theta$ is any strain present under no stress. It can be a thermal strain defined as $\theta = \int_{T_o}^{T} \alpha(\tilde{T}) d\tilde{T}$ with $T_o$ as a reference temperature defined to have no strain, $T$ as the current temperature and $\alpha(T)$ as the coefficient of thermal expansion (expressed as a function of temperature). Alternatively, $\theta$ can represent a swelling or growth strain, or a strain introduced by processing. In thermal strain examples considered later in this book, the coefficient of thermal expansion $\alpha$ will be regarded as temperature-independent such that $\theta = \alpha (T - T_o)$, but the framework allows for a temperature-dependent $\alpha$.

For the 2D problems that we will consider, the only nonzero shear strain and stress component are $\epsilon_{12} = \epsilon_{xy}$ and $\sigma_{12} = \sigma_{xy}$. Depending on the problem under consideration, either indicial notation or $(x, y, z)$-component notation will be used. For the components of interest, the constitutive law (2.1) in component notation is

$$\epsilon_x = \frac{1}{E} \left( \sigma_x - v \left( \sigma_y + \sigma_z \right) \right) + \theta$$
$$\epsilon_y = \frac{1}{E} \left( \sigma_y - v \left( \sigma_x + \sigma_z \right) \right) + \theta \tag{2.2}$$
$$\epsilon_{xy} = \frac{1 + v}{E} \sigma_{xy}$$

The inversion of (2.1) is given by

$$\sigma_{ij} = \frac{E}{1 + v} \left( \epsilon_{ij} + \frac{v}{1 - 2v} \epsilon_{kk} \delta_{ij} \right) - \frac{E}{1 - 2v} \theta \delta_{ij} \tag{2.3}$$

For an elastically incompressible material with $v = 1/2$, the dilation is due entirely to the misfit strain, but the mean stress, $\sigma_{kk}$, becomes indeterminate. Numerical work on incompressible problems requires special consideration. Most of the analytic solutions given in this text will be valid in the limit of $v = 1/2$, but numerical computations for incompressible materials are often carried out using a value for $v$ slightly smaller than $1/2$ or using special techniques.

For plane stress, (2.3) applies with $\sigma_z = 0$, such that the vector-matrix notation of the 2D constitutive relation is

$$\begin{bmatrix} \epsilon_x \\ \epsilon_y \\ \gamma_{xy} \end{bmatrix} = \frac{1}{E} \begin{bmatrix} 1 & -v & 0 \\ -v & 1 & 0 \\ 0 & 0 & 2(1+v) \end{bmatrix} \begin{bmatrix} \sigma_x \\ \sigma_y \\ \sigma_{xy} \end{bmatrix} + \begin{bmatrix} \theta \\ \theta \\ 0 \end{bmatrix} \tag{2.4}$$

with $\gamma_{xy} = 2\epsilon_{xy}$ as the engineering shear strain and $\epsilon_z = -v \left( \sigma_x + \sigma_y \right) / E + \theta$. The inverse of (2.4) is

$$\begin{bmatrix} \sigma_x \\ \sigma_y \\ \sigma_{xy} \end{bmatrix} = \frac{E}{1 - v^2} \begin{bmatrix} 1 & v & 0 \\ v & 1 & 0 \\ 0 & 0 & \frac{1-v}{2} \end{bmatrix} \begin{bmatrix} \epsilon_x \\ \epsilon_y \\ \gamma_{xy} \end{bmatrix} + \frac{E}{1 - v} \begin{bmatrix} \theta \\ \theta \\ 0 \end{bmatrix} \tag{2.5}$$

For plane strain, (2.3) gives (with $\epsilon_z = 0$)

$$
\begin{bmatrix} \sigma_x \\ \sigma_y \\ \sigma_{xy} \end{bmatrix} = \begin{bmatrix} \frac{(1-v)E}{(1+v)(1-2v)} & \frac{vE}{(1+v)(1-2v)} & 0 \\ \frac{vE}{(1+v)(1-2v)} & \frac{(1-v)E}{(1+v)(1-2v)} & 0 \\ 0 & 0 & \frac{E}{2(1+v)} \end{bmatrix} \begin{bmatrix} \epsilon_x \\ \epsilon_y \\ \gamma_{xy} \end{bmatrix} + \frac{E}{1-2v} \begin{bmatrix} \theta \\ \theta \\ 0 \end{bmatrix} \tag{2.6}
$$

where $\sigma_z = [E/(1-2v)] \left[ v \left( \epsilon_x + \epsilon_y \right) / (1+v) - \theta \right]$. The inversion of (2.6) is

$$
\begin{bmatrix} \epsilon_x \\ \epsilon_y \\ \gamma_{xy} \end{bmatrix} = \frac{1-v^2}{E} \begin{bmatrix} 1 & -v & 0 \\ -v & 1 & 0 \\ 0 & 0 & \frac{2}{1-v} \end{bmatrix} \begin{bmatrix} \sigma_x \\ \sigma_y \\ \sigma_{xy} \end{bmatrix} + (1+v) \begin{bmatrix} \theta \\ \theta \\ 0 \end{bmatrix} \tag{2.7}
$$

For generalized plane strain, $\epsilon_z$ is independent of the coordinates and set by an overall constraint. The 2D stress-strain relations for generalized plane strain can be derived directly from (2.3) and have a form similar to (2.4) and (2.5), but with an extra term depending on $\epsilon_z$. The case of equi-biaxial straining, with $\epsilon_z = \epsilon_x$, features prominently in a number of examples in the book, and it will be discussed further below for specific constraints.

Denote the 2D stress-strain relation in vector-matrix notation by

$$
[\sigma] = [K][\epsilon] - [C][\theta] \tag{2.8}
$$

with $[\sigma] = \left[ \sigma_x, \sigma_y, \sigma_{xy} \right]^T$, $[\epsilon] = \left[ \epsilon_x, \epsilon_y, \gamma_{xy} \right]^T$ and $[\theta] = [\theta, \theta, 0]^T$ and with symmetric matrices $[K]$ and $[C]$ identified from (2.5) for plane stress or (2.6) for plane strain. The strain energy density for either plane stress or plane strain is

$$
\Pi = \frac{1}{2}\sigma_{ij}\epsilon_{ij}^e = \frac{1}{2}[\epsilon^e]^T [\sigma] = \frac{1}{2}\left( [\epsilon]^T - [\theta]^T \right) [\sigma] \tag{2.9}
$$

which, by (2.8), becomes

$$
\Pi = \frac{1}{2}[\epsilon]^T [K][\epsilon] - \frac{1}{2}[\epsilon]^T \left( [K] + [C] \right) [\theta] + \frac{1}{2}[\theta]^T [C][\theta] \tag{2.10}
$$

The potential energy of a body with volume $V$ loaded by prescribed tractions $\bar{T}_i$ on the portion of the surface denoted $S_T$ and prescribed displacements $\bar{u}_i$ on the remainder of the surface denoted $S_u$ is a function of the displacements

$$
PE(u_i) = \int_V \Pi dV - \int_{S_T} \bar{T}_i u_i dS \tag{2.11}
$$

where for linear elasticity $\epsilon_{ij} = \left( u_{i,j} + u_{j,i} \right) /2$, with commas indicating spatial derivatives. In the absence of a misfit strain (i.e., $[\theta] = 0$), (2.10) reduces to the conventional expression for the energy density. The middle term in (2.10) provides the interaction between the misfit strain and the total strain. This is the term that couples the total strains to the misfit strain when the energy of the system is minimized with respect to admissible displacements. The third term on the right (2.10) does not play a role in the minimization process.

## 2.2 An Overview of Thin Films and Coatings Modeled with Beam and Plate Theories

Plate and beam theories represent simplifications to full elasticity theory that are based on imposing a priori assumptions regarding the spatial distribution of displacements in the structure. Consider the illustration in Figure 2.1, which depicts a thin layer bonded to a substrate that has separated over a small region and subsequently bulges away from substrate in the $y$-direction. Both plate and beam theories impose simple mathematical relationships between the displacement in the $y$-direction ($u_y$) and the resulting strain(s) in the $x$- and $z$-directions; these relationships are described in the subsections of this chapter that follow. In this book, the term 'beam' simply refers to scenarios in which $u_y$ is an unknown function of only one direction (e.g., the $x$-direction in Figure 2.1), with displacements in the $z$-direction producing uniform strains. The term 'plate' refers to scenarios in which $u_y$ is a function of both the $x$- and $z$-directions.

Very often, thin layers that appear visually to be plates are analyzed with beam theory; in Figure 2.1, such a scenario would arise if the dimension in the $z$-direction were either very large or very small. If the dimension in the $z$-direction is small, it reasonable to assume plane stress conditions ($\sigma_z = \sigma_{yz} = \sigma_{xz} = 0$ ) because these values are zero on the two ($y, x$) faces which are close to one another. In this case, only the $u_x$ and $u_y$ displacements must be determined as a function of the $x$-coordinate. If the dimension in the $z$-direction is very large, and the variations in displacement near the ($y, x$) faces do not greatly influence the behavior near the center (i.e., the $z = 0$ plane). In this case, the $\epsilon_z$ strain component is approximately uniform away from the edges; one can either assume $\epsilon_z = 0$ (plane strain conditions) or $\epsilon_z = constant$ (generalized plane strain conditions). Either way, only the $u_x$ and $u_y$ displacements must be determined as a function of the $x$-coordinate, so again the problem would be described using beam theory.

Plate theory is simply an extension of beam theory to account for strains in the $z$-direction that cannot be approximated as constant. Broadly speaking, the expressions from beam theory that describe $\epsilon_x = f(u_x, u_y)$ are generalized to include $\epsilon_z = f(u_x, u_y, u_z)$, and all displacements are assumed to be functions of position in the ($x, z$)-plane. (Variations in displacement with the $y$-coordinate are defined a priori using plate kinematics.) Although conceptually beam and plate theories are the same, beam theories lead to ordinary differential equations (since all unknowns vary only in the $x$-direction) while plate theories lead to partial differential equations (since all unknowns vary in the $x$- and $z$-directions). In this book, the focus is on beam theories because they provide excellent approximations for many relevant scenarios, although important aspects of plate theory are covered to provide supporting material for certain special cases.

In most analyses, a linearized kinematic description that relates displacements and strains is sufficient, as displacements are typically small in comparison with the beam or plate thickness; however, for buckling problems, a nonlinear kinematic description is needed to capture coupling between in-plane and out-of-plane deformation. In the next section, we fully describe the nonlinear kinematic relations and then illustrate the

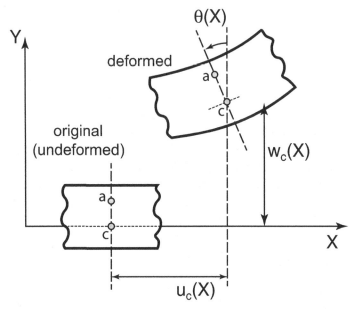

**Figure 2.2** Schematic illustration of the deformation of a beam segment (or plane-strain plate segment) assuming plane sections remain plane.

conditions under which the nonlinearities can be neglected. It should be understood that for the vast majority of the scenarios considered in this book, the simpler small deformation description is acceptable, with the exceptions related to buckling.

## 2.3     Beams: Strain-Displacement Relations and Constitutive Relations

In addition to the assumptions described above that reduce the dimensionality of the problem, plate and beam theories impose other assumptions regarding displacement distribution in the through-thickness direction. While a number of approximations are possible, this book assumes that planes initially perpendicular to the reference axis remain perpendicular to the tangent of the reference axis in the deformed state, as shown in Figure 2.2. This assumption is tantamount to saying that it is far easier to deform the beam through stretching and bending than through shearing, that is, shear strains are negligible in comparison to stretching and bending strains. This subsection describes beams with the corresponding plate theory left for later in this chapter.

The assumption that 'plane sections remain plane' can be used to derive two common kinematic relationships for beams. The first and most common is the linear strain-displacement relationship described as Bernoulli-Euler (BE) theory, which is limited to small deflections and implies no coupling between in-plane displacement ($u_c$ in Figure 2.2) and out-of-plane displacement ($w_c$ in Figure 2.2). The second kinematic relationship is nonlinear and often referred to as Kirchoff theory; this theory is applicable to large deflections involving moderate rotations (to be defined) and leads to in-plane coupling between $u_c$ and $w_c$. As it is difficult to induce the more complex nonlinear theory from BE theory, the full nonlinear theory is presented first and then reduced to the BE theory in a final step that neglects higher-order terms.

Consider the segment of a beam shown in Figure 2.2; following large-deformation conventions, the reference system is described using capital letters. The beam lies along the $X$-axis, with the $Y = 0$ reference point located somewhere in the beam. The choice of the reference axis can be exploited to simplify the analysis: for the moment, suffice it to say that the $Y = 0$ location in the beam does not need to be at the centerline. The position of a given point in the beam in the undeformed configuration is given by $(X, Y)$. The point initially on the reference axis (labeled as point $c$ in Figure 2.2) at position $X$ displaces in the $X$- and $Y$-directions and lies as the deformed location given by $X + u_c(X)$ and $Y + w_c(X) = w_c(X)$ (since the point $c$ lies at $Y = 0$).

A point that is not initially on the reference axis, for example, point $a$ in Figure 2.2, experiences the same displacements as point $c$, *plus* an additional displacement due to the fact that the initially vertical plane containing points $a$ and $c$ rotates by some angle $\theta$ that is dictated by the motion of the neutral axis. Therefore, the deformed position of a point initially at $(X, Y)$ is

$$x = X + u_c(X) - Y \sin \theta \tag{2.12}$$
$$y = w_c(X) + Y \cos \theta \tag{2.13}$$

This implies the displacements of the point lying at the initial position $(X, Y)$ are given by

$$u(X) = u_c(X) - Y \sin \theta \tag{2.14}$$
$$w(X) = w_c(X) - Y (1 - \cos \theta) \tag{2.15}$$

The angle of rotation of the initially vertical plane (in the reference configuration) is related to the angle of rotation of the reference axis (going from the initial to the deformed configuration). Noting that the shape of the reference axis in the deformed coordinate system is given simply by $y(x) = w_c(X)$ (since $Y = 0$ for the reference axis), the rotation is described by

$$\cos \theta = \frac{1}{\sqrt{1 + (w_c'(X))^2}}; \quad \sin \theta = \frac{w_c'(X)}{\sqrt{1 + (w_c'(X))^2}} \tag{2.16}$$

where primes indicate derivatives with respect to $X$. To compute the strains (using an appropriate large deformation definition), the displacement gradients are needed; for example, the strain $\eta_x$ is given by

$$\eta_x = \frac{\partial u}{\partial X} + \frac{1}{2}\left[\left(\frac{\partial u}{\partial X}\right)^2 + \left(\frac{\partial w}{\partial X}\right)^2\right] \tag{2.17}$$

The displacement gradients are given by

$$\frac{\partial u}{\partial X} = u'_c(X) - Yw''_c(X) \left( \frac{1}{\sqrt{1 + (w'_c(X))^2}} - \frac{w'_c(X)}{\sqrt{1 + (w'_c(X))^{3/2}}} \right) \qquad (2.18)$$

$$\frac{\partial w}{\partial X} = w'_c(X) \left( 1 - \frac{Yw''_c(X)}{\sqrt{1 + (w'_c(X))^{3/2}}} \right) \qquad (2.19)$$

The usual simplifying assumption is to assume that $u'_c(X)$, $w'_c(X)$ *and* the maximum value of $Yw''_c(X)$ (which occurs at $Y = h$ where $h$ is a characteristic thickness of the beam) are much smaller than unity. The latter implies the radius of curvature of the reference axis is much larger than the film thickness, $\rho = 1/w''_c(X) \gg h$, that is, the film will never be curled into a sharp arc with radius close to the film thickness. In that case, the strain in the axial direction is given by

$$\eta_x = u'_c(X) - Yw''_c(X) + \frac{1}{2} \left( w'_c(X) \right)^2 \qquad (2.20)$$

It turns out this is the only strain needed for the analysis, because the other strain components will be

$$\eta_y = \frac{\partial w}{\partial Y} + \frac{1}{2} \left[ \left( \frac{\partial u}{\partial Y} \right)^2 + \left( \frac{\partial w}{\partial Y} \right)^2 \right]$$

$$= \cos\theta - 1 + \frac{1}{2} \left[ (1 - \cos\theta)^2 + \sin^2\theta \right] = 0 \qquad (2.21)$$

$$\eta_{xy} = \frac{1}{2} \left[ \frac{\partial u}{\partial Y} + \frac{\partial w}{\partial X} + \frac{\partial u}{\partial X}\frac{\partial u}{\partial Y} + \frac{\partial w}{\partial X}\frac{\partial w}{\partial Y} \right]$$

$$= \frac{1}{2} \left[ \frac{1}{2} \left( w'_c(X) \right)^3 - w'_c(X) \left( u'_c(X) - Yw''_c(X) \right) \right] \qquad (2.22)$$

$$= O[w'_c(X) \cdot \eta_x]$$

That is, the shear strain implied by the assumed displacement distribution scales with the slope times the axial strain and can be neglected.

The beam theory is formulated in terms of the deformation of the reference axis, and for this reason, the notation from here on is simplified in the following manner. Instead of $(X, Y)$, the more conventional notation $(x, y)$ will denote points in the undeformed configuration with $w_c(X)$ replaced by $w(x)$ and $u_c(X)$ replaced by $u(x)$. Further, note from (2.20) that

$$E_x = u' + \frac{1}{2}(w')^2 \qquad (2.23)$$

is the approximation for the Lagrangian strain of the reference axis, and

$$\kappa = w'' \qquad (2.24)$$

is the curvature of the reference axis in the deformed state for this approximation. The Lagrangian strain in this nonlinear beam theory is given by

$$\epsilon_x = E_x - \kappa y = u' + \frac{1}{2}(w')^2 - \kappa y \tag{2.25}$$

It is customary to say that the strain is a combination of the stretching strain $E_x$ and the bending strain $-\kappa y$, as quantified by (2.25). The nonlinearity in this theory is the contribution from the rotation of the reference axis through $(w')^2/2$ to the stretching strain.

It is worth emphasizing the assumptions and implications associated with this beam theory:

- The shear strain is small compared to the stretching and bending strains, such that planes perpendicular to the reference axis in the undeformed state remain perpendicular to the reference axis in the deformed state.
- All strains are small, no larger than 0.1 and usually less than 0.01, for materials relevant to the applications considered in this book. As a consequence, the Lagrangian strain $\epsilon_x$ can be regarded as 'the' strain of the line element in the deformed state that was parallel to the $x$-axis in the undeformed state.
- Both the stretching strain of the axis $E_x$ and the off-axis bending strain contribution are small, of magnitude no greater than the total strain described above.
- The slope $w'$ of the deformed reference axis can be inherently larger than the strain, but $(w')^2$ must be no larger than the magnitude of the strain. A useful rule of thumb is that (2.23) retains reasonable accuracy as long as the magnitude of the slope of the beam is not larger than about 15 to 20 degrees.
- If the quadratic term in the slope in (2.23) is neglected, the expression (2.25) reduces to that of classical BE beam theory. The requirement for the validity of employing linear beam theory is $(w')^2 \ll \|\epsilon_x\|$.

It is critical to note that the kinematics statements above describe *total* strains. The strains that relate to the stresses are the *elastic strains*. The total strain is the superposition of all strains, as in

$$\epsilon_x(x, y) = E_x - \kappa y = \epsilon_x^e(x, y) + \theta(x, y) \tag{2.26}$$

with $\theta(x, y)$ as the distribution of misfit strain as introduced earlier. The elastic strain component $\epsilon_x^e$ is related to the corresponding stress component $\sigma_x$ in a manner that depends on the three-dimensional context, as discussed in Section 2.1. For each of the three cases of primary interest in this book (plane strain, plane stress and equi-biaxial straining), (2.26) can be expressed as

$$\epsilon_x = E_x - \kappa y = \frac{1}{\bar{E}}\sigma_x(x, y) + \bar{c}\theta(x, y) \tag{2.27}$$

where $\bar{E}$ and $\bar{c}$ are listed in Table 2.1.

**Table 2.1** Summary of Effective Moduli and Misfit Strain Coefficients Associated with Common States of Deformation/Stress in Thin Films

| State | Assumption | Effective modulus, $\bar{E}$ | $\theta$ coefficient, $\bar{c}$ |
|---|---|---|---|
| *Plane Stress* | $\sigma_y = 0, \sigma_z = 0$ | $E$ | 1 |
| *Plane Strain* | $\sigma_y = 0, \epsilon_z = 0$ | $\frac{E}{1-v^2}$ | $1 + v$ |
| *Equi-biaxial* | $\sigma_y = 0, \sigma_x = \sigma_z, \epsilon_x = \epsilon_z$ | $\frac{E}{1-v}$ | 1 |

## 2.4    Governing Equations for the Beam Theory via the Principle of Virtual Work

Bilayers and multilayers will be analyzed in Chapters 4 and 5. In this subsection, some of the details of the derivation of the governing equations will be illustrated for a 'beam' comprising a single homogeneous layer of uniform thickness $h$ with Young's modulus $E$ and Poisson's ratio $v$. The neutral axis is the midline of the beam and is taken as the reference $y = 0$. The nonlinear strain displacement relations of (2.23)–(2.25) will be employed. The misfit strain $\theta$ will be taken to be uniform in the derivation. To illustrate the role of the Principle of Virtual Work in generating the equilibrium equations, consider the specific example of a cantilver beam loaded at its right end by a vertical force per unit depth $P$, a horizontal force per unit depth $F$ and, along its length, a vertical force per unit area $p(x)$ as sketched in Figure 2.3.

With reference to Figure 2.3, define the force/depth (i.e., force per unit depth in the $z$-direction) and moment/depth carried by the beam at position $x$ by

$$N = \int_{-h/2}^{h/2} \sigma_x(x, y) dy \quad \text{and} \quad M = -\int_{-h/2}^{h/2} \sigma_x(x, y) y dy \qquad (2.28)$$

The derivation below applies to plane stress, plane strain or equi-biaxial straining. For plane strain and equi-biaxial straining the beam is always what is referred to as a wide plate. For plane stress, the total force and total moment are given by $N \cdot b$ and $M \cdot b$, respectively, where $b$ is the out-of-plane depth of the beam, which would normally be less than $h$. Substitution of (2.27) into the two definitions of (2.28) gives the beam

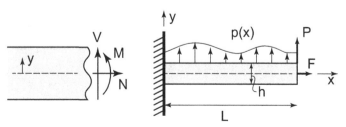

**Figure 2.3** Conventions for a beam with the cantilever beam as a specific example.

constitutive relations (for a uniform misfit strain):

$$N = \bar{E}h\,(E_x - \bar{c}\theta) \quad \text{and} \quad M = \frac{\bar{E}h^3}{12}\kappa = D\kappa \tag{2.29}$$

where $\bar{E}$ and $\bar{c}$ are defined in Table 2.1 and $D$ is defined as the bending modulus.

For virtual displacements $\delta u$ and $\delta w$ with associated virtual strains $\delta E_x = \delta u' + w\delta w'$ and $\delta \kappa = \delta w''$, the internal virtual work is

$$\delta W_{int} = \int_V \sigma_x \delta \epsilon_x dV = \int_0^L (N\delta E_x + M\delta \kappa)\,dx \tag{2.30}$$

For the loads on the cantilver beam shown in Figure 2.3, the external virtual work is

$$\delta W_{ext} = \int_0^L p\delta w dx + P\delta w(L) + F\delta u(L) \tag{2.31}$$

The Principle of Virtual Work requires that $\delta W_{int} = \delta W_{ext}$ for all admissible virtual displacements, which for the clamped cantilver beam requires that $\delta w$ and $\delta w'$ are continuous together with $\delta u(0) = \delta w(0) = \delta w'(0) = 0$. Equating (2.30) and (2.31), integrating by parts and making use of the admissibility conditions, one obtains

$$\int_0^L (-N'\delta u + (M'' - (N'w)' - p)\delta w)dx$$
$$= P\delta w(L) + F\delta u(L) - N\delta u(L) - M\delta w'(L) + ((M' - Nw')\delta w)|_0^L \tag{2.32}$$

Enforcing this condition for all admissible $\delta u$ requires $N' = 0$ on $0 \le x \le L$ and $N(L) = F$; together these imply $N = F$ and $N$ is independent of position ($x$). Enforcing the condition for all admissible $\delta w$ gives the primary equilibrium equation on $0 \le x \le L$,

$$M'' - (Nw')' = p \tag{2.33}$$

together with $M(L) = 0$ and $(M' - Nw')_{x=L} = -P$. In terms of $w$, the ordinary differential equation (2.33) and boundary conditions for the problem in Figure 2.3 are

$$Dw'''' - Fw'' = p$$
$$w(0) = w'(0) = w''(L) = 0 \tag{2.34}$$
$$Dw'''(L) - Fw'(L) = -P$$

Because the right end of the beam is free to expand, the uniform misfit strain $\theta$ does not affect $w(x)$, although it does contribute to $u(x)$. If, however, the right end were also constrained against horizontal displacement, $\theta$ would play a role through its influence on $N$. For the cantilever beam subject to a sufficiently large compressive horizontal load, with $F < 0$, the boundary problem posed by (2.34) governs the buckling response. The vertical shear force/depth $V$ shown in Figure 2.3 does not appear explicitly in this theory because the transverse shear strains do not contribute to the elastic energy in the beam or to the internal virtual work. Direct equilibrium considerations reveal that $V = -M' + Nw'$. With $p$, $P$ and $F$ regarded as prescribed, the potential energy of the cantilever beam system subject to uniform $\theta$ is

$$PE(u, w) = \frac{1}{2} \int_0^L \left( \bar{E}h \left( E_x - \bar{c}\theta \right)^2 + D\kappa^2 - 2pw \right) dx - Pw(L) - Fu(L) \qquad (2.35)$$

## 2.5    Two-Dimensional Nonlinear Plate Theory for a Homogeneous Single Layer

The theory derived in previous sections is the one-dimensional specialization of the widely used von Kármán-Föppl nonlinear plate theory. The two-dimensional theory has been employed to model various thin film problems, especially the unusual patterns seen when buckling delamination occurs for films subject to equi-biaxial stress (as discussed further in Chapter 10). In the remainder of this chapter, the two-dimensional von Kármán-Föppl plate theory will be presented for a homogeneous plate of uniform thickness $h$, Young's modulus $E$ and Poisson's ratio $v$. We will not derive the nonlinear strain-displacement relationships as done earlier for the beam, but the derivation follows similar lines. The primary additional assumption made in deriving the strain displacement relation for a plate is that the rotation about the normal to the plate is on the order of the strain due to the large in-plane stiffness of the plate. This means that the nonlinearity arises in the theory through the squares of the slopes of the plate, analogous to the results for the beam.

We follow the conventional notation for plate theory, using indicial notation for vector and tensor quantities, and making a change in the orientation of the Cartesian coordinates. Now, as depicted in Figure 2.4, the $(x_1, x_2)$ coordinates are the locations of the material points on the plate midplane in the undeformed state. The coordinate $x_3$ is perpendicular to the midplane of the undeformed plate with $x_3 = 0$ on the midplane. The horizontal and vertical displacements of the middle surface are denoted by $u_\alpha(x_1, x_2)$, $\alpha = 1, 2$ and $w(x_1, x_2)$.

The approximation to the (small) Lagrangian strain-displacement relation for in-plane stretch and shear of elements on the mid-plane is

$$E_{\alpha\beta} = \frac{1}{2}(u_{\alpha,\beta} + u_{\beta,\alpha}) + \frac{1}{2}w_{,\alpha}w_{,\beta} \qquad (2.36)$$

where the standard notation for partial differentiation is used, $(\ )_{,\alpha} = \partial(\ )/\partial x_\alpha$. As is also standard, Greek subscripts indices range from 1 to 2, and a repeated Greek indice denotes the sum from 1 to 2. The bending strain-displacement relationship is defined using

$$K_{\alpha\beta} = w_{,\alpha\beta} \qquad (2.37)$$

with the in-plane strain distribution throughout the plate given by

$$\epsilon_{\alpha\beta} = E_{\alpha\beta} - x_3 K_{\alpha\beta} \qquad (2.38)$$

The resultant stresses and moments are defined as

$$N_{\alpha\beta} = \int_{-h/2}^{h/2} \sigma_{\alpha\beta} dx_3 \quad \text{and} \quad M_{\alpha\beta} = -\int_{-h/2}^{h/2} \sigma_{\alpha\beta} x_3 dx_3 \qquad (2.39)$$

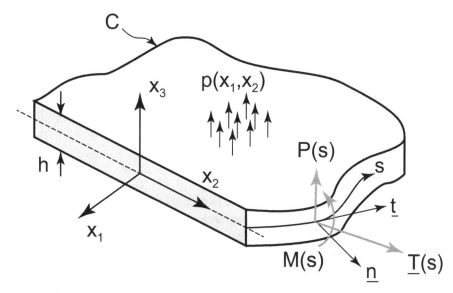

**Figure 2.4** Conventions for a plate assuming plane sections remain plane.

The constitutive relation for the plate relating the in-plane strains to the in-plane stresses is

$$\epsilon_{\alpha\beta} = \frac{1}{E}\left[(1+v)\sigma_{\alpha\beta} - v\sigma_{\gamma\gamma}\delta_{\alpha\beta}\right] + \theta\delta_{\alpha\beta} \tag{2.40}$$

$$\sigma_{\alpha\beta} = \frac{E}{1-v^2}\left[(1-v)\epsilon_{\alpha\beta} + v\epsilon_{\gamma\gamma}\delta_{\alpha\beta} - (1+v)\theta\delta_{\alpha\beta}\right] \tag{2.41}$$

and in this section an arbitrary spatial distribution of $\theta$ is considered. It follows from (2.38)–(2.41) that the constitutive relations for the plate are

$$N_{\alpha\beta} = \frac{Eh}{1-v^2}\left[(1-v)E_{\alpha\beta} + vE_{\gamma\gamma}\delta_{\alpha\beta}\right] - \frac{Eh}{1-v}\bar{\theta}\delta_{\alpha\beta}$$

$$\bar{\theta}(x_1,x_2) = \frac{1}{h}\int_{-h/2}^{h/2}\theta dx_3$$

$$M_{\alpha\beta} = D\left[(1-v)K_{\alpha\beta} + vK_{\gamma\gamma}\delta_{\alpha\beta}\right] - \frac{Eh}{1-v}\bar{\chi}\delta_{\alpha\beta} \tag{2.42}$$

$$\bar{\chi}(x_1,x_2) = \frac{1}{h}\int_{-h/2}^{h/2}\theta x_3 dx_3$$

with $D = Eh^3/\left[12(1-v^2)\right]$.

The internal virtual work in the plate is

$$W_{int} = \int_V \left(\sigma_{\alpha\beta}\delta\epsilon_{\alpha\beta}\right)dV = \int_A \left(N_{\alpha\beta}\delta E_{\alpha\beta} + M_{\alpha\beta}\delta K_{\alpha\beta}\right)dA \tag{2.43}$$

where the first integral is over the volume of the plate $V$ and the second is over the area of the midplane $A$. Denote the vertical force/area acting on the plate surface by

$p(x_1, x_2)$. The edge of the plate in the undeformed state is given by the curve $C$ with outward pointing unit normal $n_\alpha$ and distance along this curve $s$ measured from some reference point on the curve and increasing in the counter clockwise direction. The unit vector $t_\alpha$ is tangent to $C$ pointing in the direction of increasing $s$. Denote the components of the force/length acting on the edge which are work conjugate to $u_\alpha$ and tangent to the $(x_1, x_2)$-plane by $T_\alpha(s)$, denote the vertical force/length work conjugate to $w$ by $P(s)$, and denote the torque/length acting on the edge that is work conjugate to the local rotation about the edge tangent, $w_{,n} = w_{,\alpha} n_\alpha$ by $M_n$. The external virtual work is

$$W_{ext} = \int_A p \delta w \, dA + \int_C (T_\alpha \delta u_\alpha + P \delta w + M_n \delta w_{,n}) \, ds \quad \text{no sum on } n \qquad (2.44)$$

Additional applied loads could be included, such as a force/area tangent to the plane of the undeformed mid-surface, but those in (2.44) are the ones most commonly addressed.

Equating the internal and external virtual work for all virtual displacements gives the following sets of equations:

$$N_{\alpha\beta,\beta} = 0 \quad \text{and} \quad M_{\alpha\beta,\alpha\beta} - N_{\alpha\beta} w_{,\alpha\beta} = p, \qquad (2.45)$$

with the admissible boundary condition pairs (either can be prescribed but not both) given by

$$T_\alpha = N_{\alpha\beta} n_\beta, \quad u_\alpha$$
$$M_n = M_{\alpha\beta} n_\alpha n_\beta, \quad w_{,n} = w_{,\alpha} n_\alpha$$
$$P = -M_{\alpha\beta,\beta} n_\alpha + N_{\alpha\beta} n_\beta w_{,\alpha} - \frac{dM_t}{ds}, \quad w \qquad (2.46)$$

The first equilibrium equation in (2.45) has been used to simplify the second, and $M_t = M_{\alpha\beta} t_\alpha n_\beta$ in (2.46). The reader is referred to texts on nonlinear plate theory for further details.

## 2.6    A Review of Heat Transfer in One Dimension

In this section, we review the basics of heat transfer simply to set the stage for later analyses that consider nonuniform temperature distributions in multilayers. In two dimensions, the governing equation for heat conduction in a solid is given by

$$\rho c_p \frac{\partial T(x, y, t)}{\partial t} = k_y \frac{\partial^2 T(x, y, t)}{\partial y^2} + k_x \frac{\partial T(x, y, t)}{\partial x^2} \qquad (2.47)$$

where $T(x, y, t)$ is the temperature distribution as a function of space and time, $t$, $c_p$ is the specific heat capacity of the material (SI units of J/(kg · K)), $\rho$ is the material density (SI units of kg/m$^3$), and $k_x$ and $k_y$ are the thermal conductivities of the material in each direction (SI units of J/(m · K · s) or W/(m · K)). A common practice is to use thermal diffusivity of the material, defined as $\kappa_T = k/(\rho c_p)$, which implies the governing equation (though not necessarily the boundary conditions) can be expressed in terms of a single constant when the material is isotropic. In this text, we choose to use the specific heat capacity, $c_\rho = \rho c_p$ with SI units of J/(m$^3$K), and the thermal conductivity, $k$.

This text primarily considers heat transfer in one dimension, which implies the thermal boundary conditions (to be described) do not vary significantly in both directions defining the multilayer surface: for example, the $(x, z)$-plane in Figure 2.1. A limited number of solutions for two-dimensional heat transfer are considered in Chapter 13, which considers transient heat transfer surrounding a 'hot spot', that is, a region of localized heating in the $x$-direction.

For one-dimensional heat transfer, the heat flux (in SI units of W/m$^2$) through a surface with a surface normal pointing in the positive $y$-direction is given by :

$$q = -k\frac{\partial T}{\partial y} \tag{2.48}$$

which simply says that heat flows in the opposite direction of the thermal gradient, that is, heat flows from higher temperatures to lower temperatures. For a surface at temperature $T_s$ that is exposed to the ambient temperature $T_\infty$, the convective heat flux is related to the coefficient of heat transfer for the surface, $h$, according to

$$q = h\left(T_s - T_\infty\right) \tag{2.49}$$

where the property $h$ is determined by convective flow conditions near the surface; it has units of W/(m$^2$K).

For simplicity, consider heat conduction in one direction: upon neglect of derivatives with respect to $x$, (2.47) becomes a PDE with respect to time and space (one coordinate). To solve this, one must specify the initial temperature distribution through the domain, and two boundary conditions. One can prescribe either a fixed temperature on both external surfaces, or the heat flux across both surfaces, or a combination of the temperature on one surface and the heat flux on the other.

Heat conduction across an interface requires special consideration, as the interface may introduce additional thermal resistance. For an interface with finite thermal conductance but negligible heat capacity (i.e., the interface cannot store heat due to its infinitesimal thickness), the heat flux across the interface is given by

$$q = -k_{INT}\left(T^+ - T^-\right) \tag{2.50}$$

where $q$ is the heat flux through the interface in the positive $y$-direction, $T^+$ is the temperature at the interface approached from $y^-$ and $T^-$ is the temperature of the interface approached from $y^+$, with $y$ being the position of the interface. The interface property $k_{INT}$ has SI units of W/(m$^2$K), which implies the units of $k_{INT}$ are equal to the units of $k$ divided by a characteristic length scale, which does not need to be defined when $k_{INT}$ is specified directly.

At steady state, the temperature does not change with time; for one-dimensional heat transfer, only the first term on the right of (2.47) remains and is equal to zero. Since the second derivative with respect to space is zero, the temperature in any layer with uniform thermal conductivity has a linear spatial distribution. In a multilayer, the steady-state temperature distribution is therefore a collection of piecewise linear functions:

they are continuous only across interfaces with very large thermal conductance such that $T^- \approx T^+$. Since at steady state the heat flux is constant throughout the multilayer, one can write a linear system of equations for the steady-state temperature distribution that yield the temperature at the top and bottom of a given layer, $(T_i^+, T_i^-)$, where $i$ indicates the layer. This numerical procedure is described in detail in Chapter 14, which also covers the associated software provided with this book.

For transient problems, the temperature distributions within any layer are no longer linear (since the heat flux and temperature gradient evolve with time), and the solution is determined by the solution to the PDE given as (2.47). A straightforward numerical approach is a one-dimensional finite element method, as described in Chapter 15. Software is provided to solve for time-dependent temperature distributions in arbitrary multilayers.

# 3 Linear Elastic Fracture Mechanics

A 'fracture mechanics' approach to predicting failure essentially boils down to a very simple concept: one calculates a parameter that characterizes the distribution of stresses and strains near the crack tip, and assumes that the crack will extend when this parameter reaches a critical value that depends only on the material or interface properties at hand. As one might expect, the calculated parameter will depend on the geometry of the component, the geometry of the crack, any external loads (be they mechanical or imposed temperature fields) and the constitutive law of the material (i.e., its modulus, Poisson's ratio etc.)

The power of the approach is that while the calculated parameter depends on the problem at hand, the critical value is a material or interface property that does not. The critical parameter is the toughness of the material or interface – it is measured, not predicted. Thus, one can measure the critical value of the fracture parameter in an experiment based on a convenient geometry, but assess the crack stability in a totally different geometry, provided one can calculate the intensity parameter for that geometry. The key to the enormous success of fracture mechanics is this tight connection between the experimental measurement of the critical value of the intensity parameter and the evaluation of the intensity parameter for other structural geometries and loadings of interest.

Fracture can occur in three different *modes*, which refer to the relative motion of the crack faces very close to the crack tip as the crack advances. These are shown in Figure 3.1; mode I is often referred to as the opening mode, mode II is typically called the shear or sliding mode, and mode III is also a shearing mode often called the tearing mode. In many problems of interest mode III (the tearing mode) does not come into play, but there are exceptions that will be discussed in later chapters. For homogenous materials with isotropic fracture behavior, the nature of the separation process at the crack tip tends to select a mode I trajectory when the crack advances, and this explains the heavy emphasis placed on mode I behavior and mode I toughness in many structural applications. However, the films, coatings and layered materials of primary interest in this book have interfaces which can greatly alter fracture behavior. If a crack resides within an interface with sufficiently low toughness, it may be trapped and propagate within the interface. The interface may be the path of least resistance.

Under these circumstances, it is typical that combinations of mode I and mode II, the opening and sliding modes, are present. Such mode combinations are called mixed mode. An essential feature of interface fracture mechanics is that the interface toughness

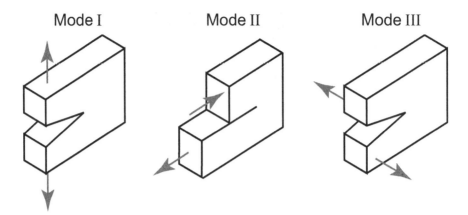

**Figure 3.1** Pure cracking modes in isotropic specimens: opening (mode I), sliding (mode II) and tearing (mode III).

governing crack advance usually depends strongly on the specific mode combination, as is spelled out later in this chapter. There is also the possibility that an interface crack will choose to kink into one of the materials adjoining the interface. This competition between advance in the interface and kinking out of the interface involves the relative toughness of the interface to that of the adjoining materials, in addition to the mode mix itself. The mechanics governing kinking of an interface crack out of an interface is discussed in Chapter 7.

## 3.1     Isotropic Elastic Crack Tip Fields

In an isotropic elastic material, the stress distributions very close to the tip of a crack are given by the well-known asymptotic stress fields. Here, the fields are stated assuming a semi-infinite crack whose tip lies at the origin of a polar coordinate system, with the crack faces lying at $\theta = \pm\pi$ as shown in Figure 3.2:

$$\sigma_{ij}(r,\theta) = \frac{K_I}{\sqrt{2\pi r}}\tilde{\sigma}_{ij}^{I}(\theta) + \frac{K_{II}}{\sqrt{2\pi r}}\tilde{\sigma}_{ij}^{II}(\theta) + T\delta_{1i}\delta_{j1} \tag{3.1}$$

where $K_I$ is the *mode I stress intensity factor*, $K_{II}$ is the *mode II stress intensity factor*, $(r,\theta)$ is the position in polar coordinates and $T$ is a constant with the dimensions of stress that, like the intensity factors, depends on the solution to the specific crack problem. Further details can be found in general fracture references, such as those by Tada, Paris & Irwin (2000), Kanninen & Popelar (1985) and Anderson & Anderson (2005).

    The stress contributions associated with the intensity factors have the characteristic inverse square root singularity, and, sufficiently close to the tip, they dominate the contribution from $T$. Nevertheless, $T$ has been included in (3.1) because it can play a role in some problems, such as kinking (Chapter 7), and in the evaluation of crack tip plasticity which is not addressed in this book. Note the separable form of the terms involving the

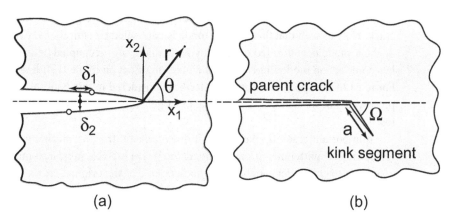

**Figure 3.2** Coordinates and variables used to describe crack tip fields in isotropic materials: (a) a straight crack showing crack opening profile in the deformed state (i.e., the undeformed state is a slit with no opening), (b) a kink segment formed at the tip of a much larger parent crack.

intensity factors (i.e., $\sigma(r, \theta) = f(r)g(\theta)$); the dimensionless functions $\tilde{\sigma}_{ij}(\theta)$ describe the angular dependence.

The stress intensity factors are parameters with units of $N/m^{3/2}$ (or $stress\sqrt{length}$) that are calculated from the elasticity solution for the complete problem and depend on the crack geometry, component geometry, loading, constraints etc. The crack advances when some combination of $K_I$ and $K_{II}$ reaches a critical value, which is a material property describing the material's ability to resist crack growth and is discussed in detail in what follows. Scenarios where one of the stress intensity factors is zero are often referred to as 'pure Mode I' or 'pure Mode II', while scenarios in which both are nonzero are referred to as 'mixed-mode'.

The angular functions are normalized such that $\tilde{\sigma}_{22}^{I}(\theta = 0) = 1$, and $\tilde{\sigma}_{12}^{II}(\theta = 0) = 1$. The elasticity solution for a crack in an isotropic material implies that $K_I$ entirely controls the tensile stress ahead of the crack with $K_{II}$ making no contribution to the direct stress, while $K_{II}$ entirely controls the shear stress ahead of the crack with $K_I$ making no contribution to the shear stress. This decoupling is important: we will see that while this clean decoupling is preserved for a crack on the interface of an important subset of materials, it is not preserved for general elastic mismatch.

The asymptotic elastic solution also yields expressions for the displacements around the crack tip, from which one can calculate the crack face opening displacements. The mode I crack tip opening displacement is often denoted as CTOD. With the displacements given by $u_i(r, \theta)$, the crack opening displacements are $\delta_i(r) = u_i(r, \pi) - u_i(r, -\pi)$, and are given by

$$(\delta_2, \delta_1) = \frac{8 (K_I, K_{II})}{\bar{E}} \sqrt{\frac{r}{2\pi}} \tag{3.2}$$

where $\bar{E} = E/(1 - v^2)$ for plane strain and $\bar{E} = E$ for plane stress. Note that in the polar coordinate system, with the crack faces lying at $\theta = \pm\pi$, the $\delta_2$ displacement is normal to the crack faces and hence represents the opening (or crack face separation), while $\delta_1$

represents the relative tangential sliding displacement of two points on either side of the crack. Note as well that the opening mode is controlled by entirely $K_I$, while the sliding or shear mode is controlled entirely by $K_{II}$. Again, this decoupled behavior (where $K_{II}$ has no effect on mode I opening and $K_I$ has no effect on mode II sliding) is important. For most of the specific bimaterial interface considered in this book, this decoupling is preserved, but we will also discuss the most general class of bimaterials for which it is lost.

An important crack tip parameter in linear elastic fracture mechanics is the energy release rate $G$ with units of energy/area $(\mathrm{J/m^2})$. For a crack subject to mixed mode and extending to create new crack area $A$ in the plane of the original crack, Irwin's relation is

$$G \equiv -\frac{\partial U}{\partial A} = \frac{K_I^2 + K_{II}^2}{\bar{E}} \tag{3.3}$$

where $U$ is the energy of the system (elastic energy in the body plus potential energy of the loads) and $\bar{E} = E$ for plane stress and $\bar{E} = E/(1 - v^2)$ for plane strain. For straight crack fronts in plate problems with an out-of-plane thickness of the solid defined as $b$, $\partial(\ )/\partial A = (1/b)\partial(\ )/\partial a$ where $a$ is the crack length. For mode I cracks, the critical values of the energy release rate and stress intensity factor for the onset of crack advance are denoted by $G_c$ and $K_c$; by (3.3) they are related by $G_c = K_c^2/\bar{E}$. It is common practice to use the term 'toughness' to characterize both $G_c$ and $K_c$ with the context or dimensions making clear which of the two is being referenced.

An important convention in linear elastic fracture mechanics is that the notation, $K_{IC}$ and $G_{IC}$, is usually reserved for plane strain mode I toughness measured under tightly restricted conditions established by the American Society of Testing Materials. These restrictions governing what is referred to as 'a valid $K_{IC}$ test' are laid out in many texts on fracture mechanics (e.g., Knott 1973; Kanninen & Popelar 1985; Anderson & Anderson 2005; Broek 2012). For metals, the restrictions specify the size of the test specimen such that the plastic zone at the onset of crack advance is fully embedded within the crack tip fields (3.1). This ensures that all information concerning the specimen geometry and loads are conveyed to the crack tip by the amplitudes of the stress intensity factors. An additional restriction specifies the specimen thickness needed to attain plane strain conditions along most of the crack front.

## 3.2    Crack Advance in an Isotropic Material by Kinking for a Crack Subject to Mixed Mode Loading

It is common when considering cracking in homogeneous, nominally isotropic materials to focus on mode I, to the point that mode II or mixed-mode fracture behavior fade into the background. The reason is fairly simple: the separation processes at the tip of a crack in a brittle, or relatively brittle, material with nominally isotropic fracture properties usually select a path of crack advance that is mode I. This is in accord with physical intuition because separation is promoted by tensile stresses and mode I behavior generates large tensile stresses directly ahead of the crack tip as it advances. For many brittle, homogenous, nominally isotropic materials it is not possible to measure any toughness other than mode I toughness. For example, if a cracked specimen is subject to mixed-mode loading with a combination of $K_I$ and $K_{II}$, the new crack surface emanating from the preexisting crack tip kinks in the direction that generates mode I conditions, or nearly so. This topic is addressed in this section because it is especially relevant as background for the subject matter in this book. It sets the stage for a study for interface cracks of the competition between continued crack advance in the interface and kinking out of the interface.

Consider the scenario depicted in Figure 3.2(b) where a short putative crack of length $a$ kinks in the direction specified by $\Omega$ emanating from the tip of a parent crack loaded by $(K_I, K_{II}, T)$. The elastic crack problem for the putative kink crack has been solved by several methods (Cherepanov 1962; Bilby, Cardew & Howard 1977; Hayashi & Nemat-Nasser 1981; He & Hutchinson 1989b). For the limit in which $a$ is much less than the length of the preexisting interface crack, and all other relevant in-plane lengths of the geometry, the stress intensity factors associated with the tip of the putative crack, $K_I^t$ and $K_{II}^t$, depend on the load applied to the parent crack through $(K_I, K_{II}, T)$ and on the kink angle $\Omega$. This dependence can be expressed functionally as (for both plane stress and plane strain):

$$K_I^t = c_{11}(\psi, \Omega)K_I + c_{12}(\psi, \Omega)K_{II} + b_1(\Omega)T\sqrt{a}$$
$$K_{II}^t = c_{21}(\psi, \Omega)K_I + c_{22}(\psi, \Omega)K_{II} + b_2(\Omega)T\sqrt{a} \tag{3.4}$$

where the measure of the mode mix acting on the parent cracks is

$$\psi = \tan^{-1}\left(\frac{K_{II}}{K_I}\right) \tag{3.5}$$

such that $\psi = 0$ is mode I with $K_I > 0$, and combinations with $\psi = \pm\pi/2$ are mode II.

The dimensionless functions $c_{\alpha\beta}(\psi, \Omega)$ and $b_\alpha(\Omega)$ will be spelled out in Chapter 7; suffice it to say, they have been computed for a broad range of mode-mix ($\psi$) and kink angle ($\Omega$) and can be determined using software provided with this book. The role of the $T$-stress contributions will be discussed in Section 7.4. For sufficiently short putative kink cracks, that is, $\|\eta\| \ll 1$ with $\eta = T\sqrt{a/(\bar{E}G)}$, the $T$-stress contribution can be neglected. In this case, with $G^t = (K_I^{t2} + K_{II}^{t2})/\bar{E}$ and $\psi^t = \tan^{-1}(K_{II}^t/K_I^t)$ as the energy release rate and mode mix of the fields at the putative crack tip, an equivalent characterization to (3.4) with $T = 0$ is

$$\frac{G^t}{G} = g(\psi, \Omega) \quad \text{and} \quad \psi^t = f(\psi, \Omega) \tag{3.6}$$

Here, $g$ and $f$ are dimensionless functions which can be evaluated in terms of the constants $c_{\alpha\beta}(\psi, \Omega)$. A plot of $G^t/G$ is presented in Figure 3.3. If the parent crack loading is pure mode I ($\psi = 0$), the maximum $G^t$ over all possible kink angles is $G$ at $\Omega = 0$. However, if nonzero mode II loading is present ($\psi \neq 0$), the maximum $G^t$ is greater than $G$ and is attained a nonzero kink angle. For all possible loading combinations on

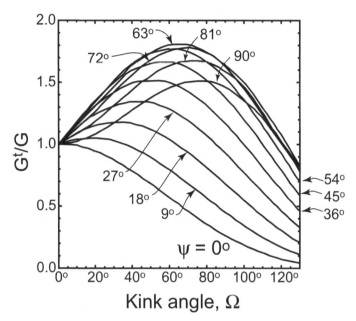

**Figure 3.3** For a crack in an isotropic material, the ratio of the energy release rate of the putative kink crack to the energy release rate of the parent crack, $G^t/G$, as a function of kink angle $\Omega$ for various mixed-mode loadings $\psi$ on the parent crack. Adapted from He & Hutchinson (1989b).

the parent crack, the maximum $G^t$ is $\approx 1.8G$ occurring when $\psi \approx 60°$ at a kink angle of $\Omega \approx 70°$.

Criteria for propagation of a kink crack from a preexisting straight crack in an isotropic homogeneous elastic material with isotropic fracture behavior and subject to mixed-mode loading were considered early in the development of elastic fracture mechanics, before accurate solutions for kinked cracks became available Erdogan & Sih (1963). Once accurate solutions became available, the following two criteria emerged as the primary contenders governing the onset and direction of kinking for a parent crack subject to a mixed-mode loading:

$$\text{Criterion (i):} \quad \Omega \text{ is such that } K_{II}^t = 0 \ \left(\psi^t = 0\right) \text{ with } K_I^t = K_C \tag{3.7}$$

$$\text{Criterion (ii):} \quad \Omega \text{ maximizes } G^t \text{ with } G_C = \frac{K_C^2}{\bar{E}} \tag{3.8}$$

With $\psi > 0$, both criteria imply that the putative crack kinks downward, $\Omega > 0$, as depicted in Figure 3.2(b). If Criterion (ii) led to a significant mode II component of the advancing putative crack, the rationale for retaining $G_C = K_C^2/\bar{E}$ would be questionable. However, as will be discussed below, Criterion (ii) predicts the putative kink crack is very nearly mode I.

Plots of the kink angle, $\hat{\Omega}$, predicted by the two criteria above, are presented in Figure 3.4(a) based on the accurate solutions of He & Hutchinson (1989b) for the full range of mixed-mode loading ($0 < \psi \leq 90°$) of the parent crack. There is almost no difference between the two predictions. Even under pure mode II loading ($\psi = 90°$) the difference

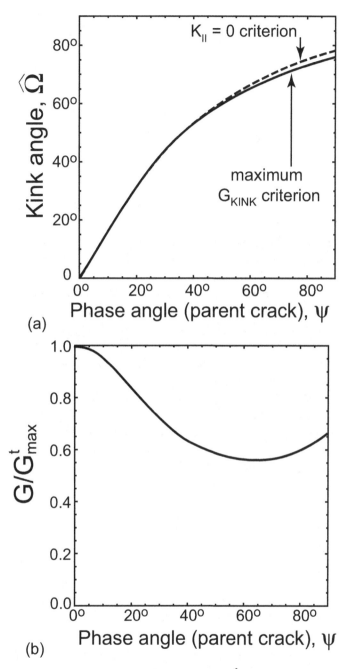

**Figure 3.4** (a) For a crack in an isotropic material, kink angle $\hat{\Omega}$ as a function of mode mix predicted by the two criteria governing kinking, (3.7) and (3.8). (b) $G/G_{max}^{t}$ as a function of mode mix where $G_{max}^{t}$ reflects the maximum driving force with respect to $\Omega$. The value of $G^{t}$ associated with $K_{II}^{t} = 0$ is nearly identical.

between the two predicted kink angles is no more than $2°$. To our knowledge, no experiments have yet been published with accuracy sufficient to discriminate between these two criteria. Moreover, the difference between the two predictions of $G^t$ is even smaller due to the fact that it is being evaluated at, or very near, the smooth maximum with respect to $\Omega$. The ratio, $G/G^t_{max}$, is plotted in Figure 3.4(b).

## 3.3     Bimaterial Interface Crack Tip Fields in Plane Strain and Plane Stress

The stress and strain fields at the tip of a crack on an interface separating two dissimilar isotropic elastic materials have many features in common with the fields discussed in Section 3.1 for a crack in a homogeneous isotropic material. Indeed, for an important class of elastic mismatch between two materials identified later, the relation between the stress fields and the stress intensity factors for the interface crack is identical to that for the homogeneous materials. However, for general elastic mismatch, important differences emerge which must be taken into account in the extension of linear elastic fracture mechanics to interfaces. In particular, the reader is alerted to the fact that the clear cut definition of mode I and mode II fields is lost when there is general elastic mismatch. These issues in the development of interface fracture mechanics will be developed in this section.

Consider a bimaterial (see Figure 3.5) comprising two isotropic elastic materials with Young's modulus and Poisson's ratio $(E_1, v_1)$ for the material above the interface and $(E_2, v_2)$ for that below. Dundurs (1969) has established the following useful general result for all boundary value problems for bimaterials in plane strain or plane stress in which traction boundary conditions are prescribed. Any dimensionless quantity derived in the solution to such a boundary value problem can be expressed in terms of the following two dimensionless mismatch parameters:

$$\alpha_D = \frac{\bar{E}_1 - \bar{E}_2}{\bar{E}_1 + \bar{E}_2}; \qquad \beta_D = \frac{\mu_1 (\kappa_2 - 1) - \mu_2 (\kappa_1 - 1)}{\mu_1 (\kappa_2 + 1) + \mu_2 (\kappa_1 + 1)} \qquad (3.9)$$

where, as defined earlier, $\bar{E} = E/(1 - v^2)$ in plane strain and $\bar{E} = E$ in plane stress. Further, $\mu = E/[2(1 + v)]$ is the shear modulus, and $\kappa = 3 - 4v$ for plane strain and $\kappa = 1 + v$ for plane stress.

These Dundurs' mismatch parameters are important for two reasons. First, they reduce the solution dependence from three parameters to two. Second, and of particular importance for our purposes, for many interface crack problems the dependence on $\beta_D$ is much less important than on $\alpha_D$, effectively reducing the mismatch dependence to just one parameter. Moreover, we will see that the choice $\beta_D = 0$ leads to a clean separation of the crack tip fields into mode I and mode II components.

In plane strain, the range of $\alpha_D$ and $\beta_D$ is restricted to a parallelogram in the $(\alpha_D, \beta_D)$ space, bounded by $\alpha_D = \pm 1$ and $\alpha_D - 4\beta_D = \pm 1$; see Figure 3.6 for various bimaterial combinations with the stiffer material taken to be material no. 1, that is, $\bar{E}_1 > \bar{E}_2$, such that $0 \leq \alpha_D \leq 1$. The range of $\beta_D$ is more restricted in the case of plane stress. In plane strain, $\beta_D = 0$ if both materials are incompressible ($v_1 = v_2 = 1/2$) and $\beta_D = \alpha_D/4$

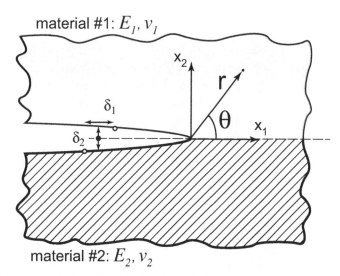

**Figure 3.5** Conventions for a crack on the interface of a bimaterial comprising two isotropic materials.

if $v_1 = v_2 = 1/3$. Note from Figure 3.6 that many material combinations with $\alpha_D \geq 0$ fall in the range $0 \leq \beta_D \leq \alpha_D/4$.

The asymptotic crack tip fields for a crack lying on the interface between two materials for plane strain or plane stress have been given in various forms by several authors (Williams 1959; Cherepanov 1962; England 1965; Rice & Sih 1965; Erdogan 1965;

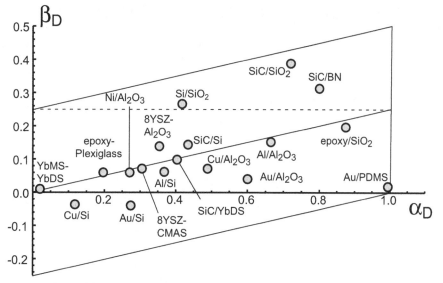

**Figure 3.6** Values of the Dundurs, parameters in plane strain for selected combinations of materials. The parallelogram marks the boundary of possible parameter values for combinations with the stiffer material as material no. 1, i.e., $\bar{E}_1 \geq \bar{E}_2$. Similar plots were first presented in Dundurs (1969) and Suga, Elssner & Schmauder (1988).

Rice 1988). For the most part, the notation and definitions used here follow that of Rice (1988) and a follow-up article by Rice, Suo & Wang (1990) and draw heavily from the article by Hutchinson & Suo (1992). Polar coordinates are defined with the conventions in Figure 3.5 such that the crack face associated with material 1 has $\theta = \pi$ and that with material 2 has $\theta = -\pi$. The expressions for the crack tip fields for in-plane stress components $(\sigma_{11}, \sigma_{12}, \sigma_{22})$ are the same for plane strain and plane stress and most efficiently expressed using complex variables in the form

$$\sigma_{ij} = \frac{Re\left[Kr^{i\epsilon}\right]}{\sqrt{2\pi r}} \tilde{\sigma}_{ij}^{I}(\theta, \epsilon) + \frac{Im\left[Kr^{i\epsilon}\right]}{\sqrt{2\pi r}} \tilde{\sigma}_{ij}^{II}(\theta, \epsilon) \tag{3.10}$$

$$\epsilon = \frac{1}{2\pi} ln\left(\frac{1 - \beta_D}{1 + \beta_D}\right) \tag{3.11}$$

with $i = \sqrt{-1}$ and the complex interface stress intensity factor defined with real and imaginary components $K = K_1 + iK_2$. The real variables, $K_1$ and $K_2$, serve as the stress intensity factors for the bimaterial crack, as will be detailed below. The fields $\tilde{\sigma}_{ij}^{I}$ and $\tilde{\sigma}_{ij}^{II}$ are given in the paper by Rice et al. (1990). The associated asymptotic displacement fields surrounding the crack tip are provided in compact form by Matos, McMeeking, Charlambides & Drory (1989) and are restated in the Appendix for convenience.

For general mismatch, the fields $\left(\tilde{\sigma}_{ij}^{I}, \tilde{\sigma}_{ij}^{II}\right)$ are normalized such that on the interface ahead of the crack tip $(\theta = 0)$, (3.11) becomes

$$\sigma_{22} + i\sigma_{12} = \frac{K}{\sqrt{2\pi r}} r^{i\epsilon}, \quad \text{or}$$

$$\sigma_{22} = \frac{Re\left[Kr^{i\epsilon}\right]}{\sqrt{2\pi r}} \quad \text{and} \quad \sigma_{12} = \frac{Im\left[Kr^{i\epsilon}\right]}{\sqrt{2\pi r}} \tag{3.12}$$

The associated asymptotic displacement fields surrounding the crack tip are provided in Appendix A, in the same form reported by Matos et al. (1989). The associated crack face displacements at a distance $r$ behind the crack tip, $\delta_i = u_i(r, \pi) - u_i(r, -\pi)$, are given by Rice (1988)

$$\delta_2 + i\delta_1 = \frac{8r^{i\epsilon}}{(1 + 2i\epsilon) \cosh(\pi\epsilon)} \frac{K}{E_*} \sqrt{\frac{r}{2\pi}} \tag{3.13}$$

with

$$\frac{1}{E_*} = \frac{1}{2}\left(\frac{1}{\bar{E}_1} + \frac{1}{\bar{E}_2}\right) \tag{3.14}$$

The relation between the energy release rate and the intensity factors is given by (Malyshev & Salganik 1965)

$$G = \frac{1 - \beta_D^2}{E_*}\left(K_1^2 + K_2^2\right) \tag{3.15}$$

For all material combinations having $\beta_D = 0$ such that $\epsilon = 0$ and $r^{i\epsilon} = 1$, (3.10) reduces to the result (3.1) for the singular fields of a crack in a homogeneous material:

$$\sigma_{ij}(r, \theta) = \frac{K_1}{\sqrt{2\pi r}} \tilde{\sigma}_{ij}^I (\theta) + \frac{K_2}{\sqrt{2\pi r}} \tilde{\sigma}_{ij}^{II} (\theta) \tag{3.16}$$

while (3.13) and (3.15) simplify to

$$(\delta_2, \delta_1) = \frac{8 (K_1, K_2)}{E_*} \sqrt{\frac{r}{2\pi}} \quad \text{and} \quad G = \frac{K_1^2 + K_2^2}{E_*} \tag{3.17}$$

Thus, even for bilayers with very large elastic mismatch, for example, $|\alpha_D| \approx 1$, the singular crack tip stress fields are the same as those of homogeneous material if $\beta_D = 0$. Moreover, the stress intensity factors can be interpreted in the same manner as for homogeneous material with the notation $K_1 = K_I$ and $K_2 = K_{II}$, and a clear notion of mode I and mode II carries over to the bimaterial. The expressions for crack opening displacements and the energy release rate differ from their counterparts in the homogeneous material only by $E_*$ replacing $\bar{E}$. For bimaterial combinations with $\beta_D = 0$, the crack tip fields are very similar to those for homogeneous materials, and we will exploit this similarity in developing and applying interface fracture mechanics whenever possible.

## 3.4     Characterization of Interface Toughness for Bimaterial Interfaces with $\beta_D = 0$

In this section and the next it will be assumed that the interface crack is the crack path of least resistance in the sense that, regardless of the combination of stress intensity factors experienced by the interface crack in Figure 3.5, the crack will advance in the interface when the intensity level is sufficiently large. The possibility of kinking out of an interface into one or the other the materials is taken up in Chapter 7. Some bimaterial interfaces adhere at the atomic scale. However, many interfaces have structure at a scale finer than that represented by the bimaterial elasticity model on which our theory will be based. The development below will characterize the effective toughness of the interface, which generally must be measured by experiment in the same way that the toughness of a homogeneous structural material is almost always a measured quantity. Predicting toughness, interface or otherwise, based on mechanics and physics at a finer scale is a challenge not addressed in this book.

The characterization of the interface laid out below is limited to each particular bimaterial and the specific conditions under which the two materials are bonded. For example, the interface might actually be a very thin adhesive layer bonding the two materials. Furthermore, the interface crack might propagate in the interior of the adhesive layer, or it might propagate at the surface where the adhesive bonds to one of the materials. The interface toughness measured in an experiment characterizes the toughness that results from the detailed fine scale processes, but our treatment does not delve into the fine-scale processes themselves. In this sense, interface fracture mechanics is no different from linear elastic fracture mechanics.

Interface cracking experiments on numerous bimaterial interfaces have revealed that cracks do indeed propagate along the interface even when there are significant components of mode II. Significantly, it has been found that many of these systems display higher toughness when a significant mode II component is present. In other words, there is clear evidence from interface cracking experiments that the interface toughness depends on the mode mix. As in (3.5), define the measure of the mode mix by $\psi = \tan^{-1}(K_2/K_1)$. The interface crack propagation criterion for an ideally brittle interface is

$$G = \Gamma_{INT}(\psi) \tag{3.18}$$

with $\Gamma_{INT}(\psi)$ termed the mixed mode interface toughness (in units J/m$^2$).

Ideally brittle in this context implies a fracture process embedded within the zone of dominance of the crack tip fields and with (3.18) applying equally to the onset of crack advance and continuing advance. In the parlance of elastic-plastic fracture mechanics, these two aspects are referred to as small-scale yielding and no increasing resistance to crack growth with crack advance, that is, a flat R-curve. When plasticity occurs at the crack tip in one, or possibly both, of the materials joined at the interface, mode dependence arises due to the strong dependence of the size and shape of the plastic zone on $\psi$, as considered by Tvergaard & Hutchinson (1993). As in conventional fracture mechanics, small-scale yielding requires that the plastic zone is embedded within the crack tip fields. For layered systems, this usually requires that the plastic zone is small compared to the thicknesses of the layers at the interface. Even layered systems which generate no plasticity can display mode dependence. One mechanism is the interaction between the rough interfaces on either side of the crack interface when $\psi \neq 0$, as modeled by Evans & Hutchinson (1989).

An example of mode-dependent toughness is shown in Figure 3.7, for an interface between plexiglass (no. 1) and epoxy (no. 2); the data are those obtained by Wang & Suo (1990) using a Brazil nut specimen. For this material combination, $\alpha_D = -0.15$, $\beta_D = -0.029$ and $\epsilon = 0.009$. As will become evident in the next section, the value of $\beta_D$ for this system is sufficiently small that the approximation $\beta_D = 0$ introduces a negligible error. For example, comparing (3.15) and (3.17), one sees that the error in $G$ from this replacement is less than 0.1%. The toughness data for this interface reveal that the toughness increases as the mode II component increases and the mode II toughness is roughly four times the mode I toughness. Additional studies using the Brazil nut specimen to measure interface toughness have been conducted by Bank-Sills, Travitzky, Ashkenazi & Eliasi (1999).

A phenomenological representation of mixed mode interface toughness that has proved useful for modeling purposes is

$$\Gamma_{INT}(\psi) = \Gamma_I f(\psi) \text{ with } f(\psi) = 1 + \tan^2[(1-\lambda)\psi] \tag{3.19}$$

where $\Gamma_I$ is the mode I toughness and $\lambda$ is a parameter setting the ratio of mode II to mode I toughness:

$$\frac{\Gamma_{II}}{\Gamma_I} = 1 + \tan^2\left[\frac{\pi}{2}(1-\lambda)\right] \tag{3.20}$$

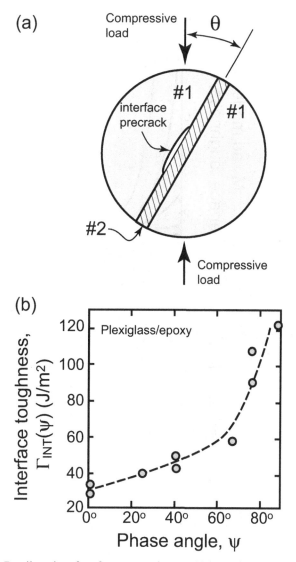

**Figure 3.7** (a) The Brazil nut interface fracture specimen, which consists of a precrack along the interface between a film embedded in a circular specimen; mode mix depends on the orientation between compressive loading plane and the interface plane. (b) Mode dependence of interface toughness for a plexiglas (no. 1)/epoxy (no. 2) interface measured using Brazil nut specimens by Wang & Suo (1990).

The limit $\lambda = 1$ is the mode-independent toughness $\Gamma_{INT}(\psi) = \Gamma_I$ for all $\psi$. Figure 3.8 illustrates this representation. The function $f(\psi)$ is symmetric in $\psi$ implying that $\Gamma_{INT}$ is symmetric with respect to $\pm K_2$. Although a symmetric dependence may be a reasonable approximation for some interfaces, one must be cognizant of the fact that the fracture processes at the interface crack tip may be inherently asymmetric with respect to $\psi$. The data for the plexiglass/epoxy interface suggest this interface has some asymmetry in its interface toughness.

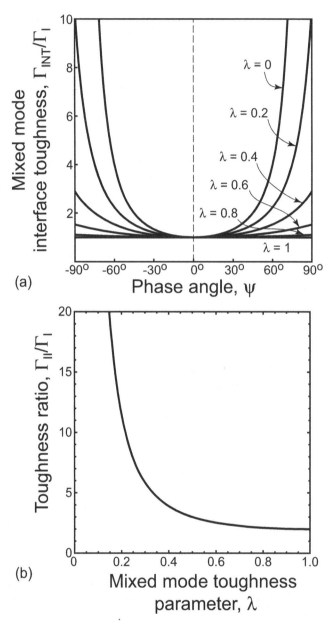

**Figure 3.8** (a) The phenomenological mixed mode interface toughness characterized by (3.19). (b) The ratio of mode II interface toughness to the mode I interface toughness as a function λ, given by (3.20).

Asymmetry can arise from the asymmetry associated with plasticity in one of the adjoining layers, or, in systems free of plasticity, due to the roughness of the local interface cracking process, which is influenced by the sign of $\psi$. Specifically, in line with the crack kinking behavior discussed in Section 3.1 and for interfaces in Chapter 7, positive $\psi$ drives a small kink crack downward and negative $\psi$ drives it upward. Thus, for example, if the material below the interface is very tough and $\psi$ is positive, the interface crack is likely to hug the interface and generate very little roughness. By contrast, if $\psi$ is negative, there may be frequent upward kinking excursions of microcracks which generate a relatively rough fracture surface.

## 3.5    Characterization of Interface Toughness with $\beta_D \neq 0$

The crack tip fields (3.10) for a bimaterial interface for which the second Dundurs' parameter $\beta_D$ is nonzero require additional consideration with regard to the definition of the mode mix. While the normal stress acting on the interface in the singular fields with $\beta_D = 0$ is directly tied to $K_1$ and the shear stress is tied to $K_2$, this is no longer the case when $\beta_D \neq 0$. On the interface ahead of the crack tip, (3.12) gives

$$\sigma_{22} = \frac{1}{\sqrt{2\pi r}} \left( K_1 \cos\left(\epsilon \ln r\right) - K_2 \sin\left(\epsilon \ln r\right)\right)$$
$$\sigma_{12} = \frac{1}{\sqrt{2\pi r}} \left( K_1 \sin\left(\epsilon \ln r\right) + K_2 \cos\left(\epsilon \ln r\right)\right) \tag{3.21}$$

which assumes the decoupled form only when $\epsilon = \beta_D = 0$. We can no longer associate mode I with $K_1$ and mode II with $K_2$. In addition, the ratio of $\sigma_{12}/\sigma_{22}$ is not constant as the crack tip is approached but instead varies slowly in an oscillatory manner as $r \to 0$. In the application to most interfaces, the fact that the singular field has this oscillatory behavior is not important, but the fact that the stress ratio $\sigma_{12}/\sigma_{22}$ depends on the distance ahead of the crack does matter.

To define a measure of the mode mix, use the stress ratio at $r = \ell$ on the interface ahead of the crack tip such that, by (3.12),

$$\psi = \tan^{-1}\left[ \left(\frac{\sigma_{12}}{\sigma_{22}}\right)_{r=\ell}\right] = \tan^{-1}\left[ \frac{Im\left[K\ell^{i\epsilon}\right]}{Re\left[K\ell^{i\epsilon}\right]}\right] \tag{3.22}$$

Thus, this measure depends on the choice of $\ell$. It is readily shown that if $\psi_1$ is associated with $\ell_1$ and $\psi_2$ with $\ell_2$, then the two measures are related by

$$\psi_2 = \psi_1 + \epsilon \ln\left(\frac{\ell_2}{\ell_1}\right) \tag{3.23}$$

As we will illustrate below, one must make a choice of $\ell$ when $\beta_D \neq 0$ to define the mode mix, but it is straightforward to transform to any other choice using (3.23). Note that $\psi$ becomes independent of $\ell$ when $\epsilon = \beta_D = 0$, and we will see that for most interfaces the dependence of $\psi$ on the choice $\ell$ is quite weak even when $\beta_D \neq 0$.

An example which helps to illustrate the mathematical character of bimaterial crack solutions when $\beta_D \neq 0$ is the classical problem shown in Figure 3.9(a) for a crack of length $2a$ on the interface between two semi-infinite half spaces with the stresses in the half spaces far from the crack denoted as $\sigma_{22}^{\infty}$ and $\sigma_{12}^{\infty}$. The solutions for the crack tip stress intensity factors and energy release rate are (England 1965; Rice & Sih 1965;

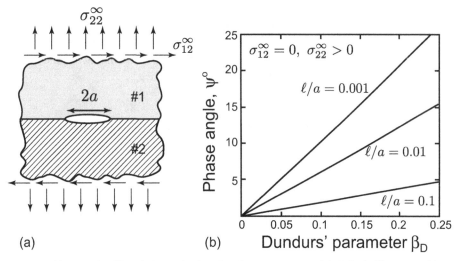

**Figure 3.9** (a) A crack of length $2a$ on the interface between two semi-infinite half spaces subject to a remote tension $\sigma_{22}^\infty$ and remote shear stress $\sigma_{12}^\infty$. (b) Mode mix in degrees as dependent on $\beta_D$ and $\ell/a$ for the case with remote tension.

Erdogan 1965)

$$K = K_1 + iK_2 = \left(\sigma_{22}^\infty + i\sigma_{12}^\infty\right)(1 + 2i\epsilon)(2a)^{-i\epsilon}\sqrt{\pi a}$$
$$G = \frac{\left(1 - \beta_D^2\right)\left(1 + 4\epsilon^2\right)}{E_*}\left[\left(\sigma_{22}^\infty\right)^2 + \left(\sigma_{12}^\infty\right)^2\right] \tag{3.24}$$

For this problem, the stress intensity factors apply to both plane strain and plane stress, and they do not depend on $\alpha_D$, although the energy release rate does.

If present in either half space, a remote component $\sigma_{11}^\infty$ produces no crack tip stress intensity. Conventional stress intensity factors have dimension $N/m^{3/2}$. When $\beta_D \neq 0$, the dimensions of the stress intensity factors are unusual: $\frac{N}{m^{3/2+i\epsilon}}$. The expression (3.24) is dimensionally consistent with the expressions for the stresses in the crack tip field (3.10). For example, combining (3.12) and (3.24), one finds the stresses on the interface ahead of the crack tip to be

$$(\sigma_{22} + i\sigma_{12}) = \frac{Kr^{i\epsilon}}{\sqrt{2\pi r}} = \left(\sigma_{22}^\infty + i\sigma_{12}^\infty\right)(1 + 2i\epsilon)\left(\frac{r}{2a}\right)^{i\epsilon}\sqrt{\frac{a}{2r}} \tag{3.25}$$

For the case $\sigma_{12}^\infty = 0$, the stress intensity factors are

$$K_1 = \sigma_{22}^\infty\sqrt{\pi a}\left[\cos\left(\epsilon\ln(2a)\right) + 2\epsilon\sin\left(\epsilon\ln(2a)\right)\right]$$
$$K_2 = \sigma_{22}^\infty\sqrt{\pi a}\left[2\epsilon\cos\left(\epsilon\ln(2a)\right) - \sin\left(\epsilon\ln(2a)\right)\right] \tag{3.26}$$

and the mode mix (3.22) based on evaluation at $r = \ell$ is

$$\psi = \tan^{-1}\left(\frac{2\epsilon\cos\left(\epsilon\ln\frac{\ell}{2a}\right) + \sin\left(\epsilon\ln\frac{\ell}{2a}\right)}{\cos\left(\epsilon\ln\frac{\ell}{2a}\right) - 2\epsilon\sin\left(\epsilon\ln\frac{\ell}{2a}\right)}\right) \tag{3.27}$$

The asymmetry with respect to the crack plane introduced by the difference between the properties of the two materials induces the mode II component. The extent to which the mode mix depends on $\beta_D$ and the choice of $\ell/a$ is displayed in Figure 3.9. Recall that in plane strain, the largest possible value of $\beta_D$ is 0.25, and for most interfaces $\beta_D$ is considerably smaller. Figure 3.9(b) reveals that the dependence of $\psi$ on $\ell/a$ is relatively weak, and even for bilayers with the largest possible $\beta_D$, the shift from mode I due to elastic mismatch is likely to be less than $20°$ to $30°$, but depending on $\ell/a$. The results in Figure 3.9(b) are typical, and they reflect the fact that, while the effect of $\beta_D$ in the definition of the mode mix should be taken into account if $\beta_D$ is not small, its effect on the mode mix is often small and in many cases much smaller than uncertainties associated with experimental data representing the interface toughness.

Another aspect of the crack solutions for bimaterial interfaces with $\beta_D \neq 0$ is that (3.13) predicts that the crack faces come into contact and overlap at some distance ahead of the crack tip, that is, $\delta_2 < 0$ at a distance even nearer to the tip. This unphysical behavior is due to the fact that the boundary value problem posed for the crack tip fields assumes zero tractions on the crack faces without regard as to whether or not the faces come into contact. Efforts to remedy this deficiency were made not long after the crack tip fields were produced by accounting for contact (Comninou 1977). However, while the crack face contact must be taken into account if it occurs, the focus on near-tip contact has not proved the most effective way to deal with the problem. The solution to the interface crack problem in Figure 3.9 provides some insights. Combining (3.13) and (3.24), one finds

$$\frac{\delta_2}{a} = \frac{8}{E_* \cosh(\pi \epsilon)} \left(\frac{r}{2a}\right)^{1/2} \left[\sigma_{22}^\infty \cos\left(\epsilon \ln\left(\frac{r}{2a}\right)\right) - \sigma_{12}^\infty \sin\left(\epsilon \ln\left(\frac{r}{2a}\right)\right)\right]$$

(3.28)

and first contact occurs ($\delta_2 = 0$) as the tip is approached at the first value of $r$ satisfying

$$\frac{r}{a} = 2 \exp\left[\frac{1}{\epsilon} \tan^{-1}\left(\frac{\sigma_{22}^\infty}{\sigma_{12}^\infty}\right)\right]$$

(3.29)

The contact point is plotted in Figure 3.10 as a function of $\beta_D$ using (3.11). If the remote stress is dominated by tension, that is, $0 \leq \sigma_{12}^\infty/\sigma_{22}^\infty \leq 0.1$, the contact point is so small that it can have no physical consequence. When the remote stress is dominated by shear, that is, $\sigma_{12}^\infty/\sigma_{22}^\infty > 10$, physically relevant contact induced by the elastic mismatch may occur if the magnitude of $\beta_D$ is sufficiently large. Even without elastic mismatch, contact occurs along the entire length of the crack if $\sigma_{22}^\infty = 0$.

Further insights into the role of elastic mismatch generating crack face contact for any bimaterial interface crack problem have been provided in the papers by Rice (1988) and Wang & Suo (1990). The discussion presented here follows the specific development in Hutchinson & Suo (1992), which draws on the earlier papers. By combining the definition of the mode mix $\psi$ tied to $\ell$ in (3.22) with the general expression for the

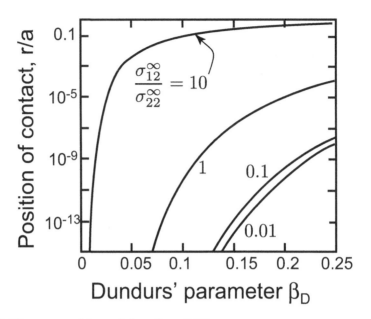

**Figure 3.10** First contact of the crack faces from (3.29).

crack opening displacements in (3.13), one can obtain

$$\delta_2 = \frac{8}{1+4\epsilon^2}\sqrt{\frac{Gr}{2\pi E_*}}\left[\cos\left(\psi + \epsilon\ln\left(\frac{r}{\ell}\right)\right) + 2\epsilon\sin\left(\psi + \epsilon\ln\left(\frac{r}{\ell}\right)\right)\right] \qquad (3.30)$$

Here we identify $\ell$ with the size of the fracture process zone at the crack tip where the deformation departs from linear elastic behavior. The crack tip fields are not expected to describe behavior for $r < \ell$. Let $L$ be in the in-plane length defining the geometry such as a layer thickness or possibly the crack length itself which establishes the domain of validity of the crack tip singular fields. We will assume the crack tip fields retain some accuracy for $r$ as large as $L/10$. The requirement imposed on (3.30) is that the crack is open, $\delta_2 > 0$, within the domain of relevance of the crack tip fields, $\ell < r < L/10$.

If $\beta_D > 0$ such that, by (3.11), $\epsilon < 0$, the requirement that the crack is open on the interval $\ell < r < L/10$ is

$$-\frac{\pi}{2} + 2\epsilon - \epsilon\ln\left(\frac{L}{10\ell}\right) < \psi < \frac{\pi}{2} + 2\epsilon \qquad (3.31)$$

while, if $\beta_D < 0$ and $\epsilon > 0$, the requirement becomes

$$-\frac{\pi}{2} + 2\epsilon < \psi < \frac{\pi}{2} + 2\epsilon - \epsilon\ln\left(\frac{L}{10\ell}\right) \qquad (3.32)$$

In arriving at these conditions it has been assumed that $L > 10\ell$ and the approximation $\tan^{-1}(2\epsilon) \approx 2\epsilon$ has been used, as is consistent with the fact that the largest possible magnitude of $\epsilon$ is $|\epsilon| = 0.175$ as dictated by the extreme limits $\beta_D = \pm 1/2$. As measured by $\psi$, the relevant crack tip region is open over most of the range of mode mix with the limits at near mode II behavior altered by $\epsilon$ and $L/(10\ell)$ as given by (3.31) and (3.32).

Crack tip fields for the bimaterial interface with $\beta_D \neq 0$ are sometimes characterized as having an oscillatory singularity owing to the stresses and displacements that vary according to terms like $\sin(\epsilon \ln(r/a))$. The illustrations above suggest that this designation is misleading. The variation of $\sigma_{12}/\sigma_{22}$ along the interface as the tip is approached is very slow, and the definition of mode mix depends only weakly on the choice of $r = \ell$ at which it is evaluated. Moreover, crack face contact induced primarily by elastic mismatch near the crack tip is usually of concern only when the loading is near mode II, when the possibility contact must be considered even without elastic mismatch. A number of important delamination problems for films and coatings are mode II, or near mode II, and it will necessary to take into account crack face contact. Generally, this is done on a case by case basis independent of the details illustrated above based on the crack tip fields.

If the crack tip is open and crack advance occurs in the interface, the criterion for crack advance for a bilayer with $\beta_D \neq 0$ is (generalizing (3.18))

$$G = \Gamma_{INT}(\psi, \ell) \qquad (3.33)$$

explicitly emphasizing the definition of $\psi$ depends on $\ell$. It follows from the transformation formula (3.23) that this criterion can be transformed from one definition pair $(\psi_1, \ell_1)$ to another $(\psi_2, \ell_2)$ if

$$\Gamma_{INT}(\psi_2, \ell_2) = \Gamma_{INT}\left(\psi_1 + \epsilon \ln\left(\frac{\ell_2}{\ell_1}\right), \ell_1\right) \qquad (3.34)$$

Liechti & Chai (1991) carried out one of the earlist and most complete sets of experiments measuring interface toughness for an epoxy/glass interface. The authors designed and analyzed a test rig depicted in the insert of Figure 3.11 which subjects a bilayer bonded to rigid platens to prescribed combinations of normal and tangential displacements. The authors analyzed the specimen to obtain the interface stress intensity factors as a function of the combination of the two imposed displacements and the elastic mismatch. By varying the combination of the two displacements, Liechti and Chai were able to generate toughness data over nearly the entire range from negative to positive mode II (Liechti & Chai 1991; Liechti & Chai 1992).

With material no. 1 as the epoxy ($E_1 = 2.07$ GPa, $v_1 = 0.37$), material no. 2 as the glass ($E_2 = 68.9$ GPa, $v_2 = 0.20$) and $h = 12.7$ mm, the plane strain elastic mismatch parameters for this system are

$$\alpha_D = -0.935; \quad \beta_D = -0.188; \quad \epsilon = 0.060 \qquad (3.35)$$

To define the mode mix $\psi_1$ used to plot the toughness $\Gamma_{INT}(\psi_1, \ell_1)$ in Figure 3.11, Liechti & Chai chose the thickness of the layers as the reference length, $\ell_1 = h = 12.7$ mm. The authors report that the plastic zone size in the epoxy at the onset of crack growth is on the order of 1 $\mu$m for $\psi_1 = 0$ and approximately 140 $\mu$m for $\psi_1 \approx 90°$. To evaluate the mode mix $\psi_2$, suppose we choose a distance ahead of the crack tip that is representative of the size of the fracture process, $\ell_2 = \ell_1/100 = 127$ $\mu$m, which in

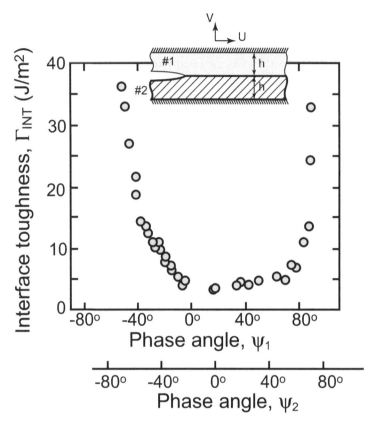

**Figure 3.11** Mixed-mode interface toughness of an epoxy/glass interface adapted from the experimental data of Liechti & Chai (1991). The two measures of the mode mix, $\psi_1$ and $\psi_2$, are described in the text. The insert is a schematic of the sample geometry and loading used to carry out the test.

this case is approximately the plastic zone size of the epoxy for pure mode II loading. Then, by (3.23), the shift in the mode mix, $\epsilon \ln(\ell_B/\ell_A)$, is $\psi_2 = \psi_1 - 15.8°$. Figure 3.11 illustrates the evaluation of the interface toughness (3.34) is a simple translation of the $\psi$-axis.

The choice $\ell_2 = 127\ \mu$m based on the scale of the fracture process generates a dependence of $\Gamma_{INT}$ on $\psi$ that is more symmetric than the choice based on $\ell_1 = h$, but there is some inherent asymmetry regardless of choice. In spite of the asymmetry, the toughness of the epoxy/glass interface is moderately well described by the phenomenological relation (3.19) with $\Gamma_I = 4$ J/m$^2$, $\lambda \approx 0.2$ and $\psi \to \psi_2$. The toughness ratio, mode II to mode I, for this interface is $\Gamma_{II}/\Gamma_I \approx 10$.

In summary, the example for the epoxy/glass interface illustrates how the choice of reference length, $\ell$, affects the definition of the interface toughness dependence for cases in which $\beta_D \neq 0$. Any value of $\ell$ can be used to define the mode mix, even one such as $\ell = h$ in this example, which is almost certainly larger than the domain of accuracy of the crack tip fields. However, the value of $\ell$ (and $\beta_D$) must be recorded along with the toughness $\Gamma_{INT}(\psi)$. This information allows the toughness to be converted for any other choice of $\ell$ by (3.34).

## 3.6    Interface Toughness Measurements

Interface toughness is measured in the same way as bulk toughness, at least conceptually. First, for a given specimen geometry with a predefined crack, one utilizes elasticity solutions to describe the relationship relationship between applied loads (or displacements) and the crack tip parameters $G$ and $\psi$ (or $K_1$ and $K_2$). Second, one measures the critical values of the applied loads (or displacements) that advance the crack. The elasticity solution is then used to translate these to critical values of the crack tip parameters, which define the interface toughness. This was the procedure used to generate the toughness results shown in Figure 3.7, the testing of Brazil nut specimens by Wang & Suo (1990) – and Figure 3.11, the testing of bimaterial strips by Liechti & Chai (1991). Many analogous efforts have been made to measure delamination toughness in laminted composites, for example, those by Banks-Sills, Freed, Eliasi & Fourman (2006) and Sørensen, Jørgensen, Jacobsen & Østergaard (2006). Other test configurations are possible (with many relevant solutions distributed throughout this book), but the overarching process is always the same.

While conceptually straightforward, there are several significant challenges inherent to testing interfaces. This section is *not* meant to provide a thorough review of ways to address these challenges, which might require an entire book in and of itself. The purpose of this section is simply to highlight critical testing considerations and a few methods that are favored for their simplicity. This section focuses on thin films, coatings and multilayers, though many of the comments that follow apply equally to bulk materials joined through diffusion bonding or other processes that introduce adhesion layers with very small thickness. A more complete review of key references with regard to interface toughness measurements has been provided by Chen & Bull (2010).

First and foremost, there is the challenge of creating the interface itself: the process that bonds the two materials together must be the same process used to create the components placed in service. Typically, thin films are fabricated using techniques such as chemical vapor deposition (CVD), arc plasma spray (APS), physical vapor deposition (PVD) and so forth (For an excellent review, see the book by Freund & Suresh (2003).) Alternatively, relevant interfaces may be created via chemical reactions at a surface, such as occurs in thermally grown oxides (TGOs). These processes more often than not introduce practical constraints that limit the thickness of the film/coating to less than a millimeter. Further, the details of these processes (such as deposition rate and temperature) can play a critical role in the types of bonds that are formed on the interface, not to mention the resulting microstructure in the layers, which can profoundly affect the layer's mechanical properties.

Naturally, one strategy to circumvent the challenge of testing very thin coatings or films is to bond fixtures to the multilayer that facilitate the application of loads or the measurement of displacements. Practically speaking, however, this can be quite a challenge for several reasons. Obviously, the strategy requires that the new interfaces introduced by the attachments be significantly stronger than the interface of interest. The bonding strategy must not alter the physical properties of the thin film, which can be a problem if attaching the grips involves thermal treatments or chemical agents. Moreover, the addition of fixtures often introduces compliance into the specimen, which can significantly alter the relationship between applied loads and crack tip loading parameters. In the experiments of Liechti and Chai shown in Figure 3.11, considerable effort was made to avoid this and ensure that the displacements applied to the fixtures accurately reflected those on the upper and lower surfaces of the specimen. If this is not possible, the elasticity solution relating applied loads, geometry and crack tip parameters must be adjusted, typically through detailed finite element analysis.

This complication is even worse if the fixtures are attached with epoxy or other adhesives that often experience plastic deformation, which further alters the relationship between remote loading and crack tip conditions. As in the Liechti and Chai work, such challenges can be (and have been) addressed through careful, detailed analysis and development of testing fixtures and sample mounting procedures. Unfortunately, the effort made to characterize one type of interface often cannot be translated to another system, due to changes in compliances, chemical compatibility, the required interface strengths and so forth.

In a nutshell, absent a large development effort to target a specific system, one often is forced to test specimen geometries that consist of flexible multilayers, which limits the range of loading that can be translated directly to the interface. Figure 3.12 shows schematic illustrations of two test configurations that are widely used due to the fact they utilize common coupon specimens and require relatively simple instrumentation: the bend test and the blister test.

The development of the bend test shown in Figure 3.12(a) for interface fracture toughness measurements is largely attributed to Tony Evans and co-workers (Cao & Evans 1989; Charalambides, Lund, Evans & McMeeking 1989; Charalambides, Cao, Lund & Evans 1990; Evans, Rühle, Dalgleish & Charalambides 1990). The testing by Dauskardt, Lane, Ma & Krishna (1998) a decade or so later is also notable for its impact in extending the method. Numerous refinements of the method with applications to various material systems can be found easily in the literature. For bend testing, four-point loading is generally preferred over three-point loading, as the former produces a constant moment between the inner rollers, which makes subsequent analysis very straightforward.

Frequently, bend specimens utilize additional layers to increase the energy release rate, which otherwise scales with film thickness; for very thin films, the substrate may crack or plastically deform prior to the interface energy release rate reaching its critical value. Adding additional 'stiffener' layers increases the energy release rate for fixed displacement, implying the crack will advance prior to unwanted substrate deformation (such as yielding or cracking). The use of stiffener layers is prone to the same issues discussed above with regard to attaching fixtures: they must exhibit stronger bonding than the interface of interest, exhibit well-known mechanical properties and so forth. If stiffeners are added to multilayers with preexisting residual stress or stress-free

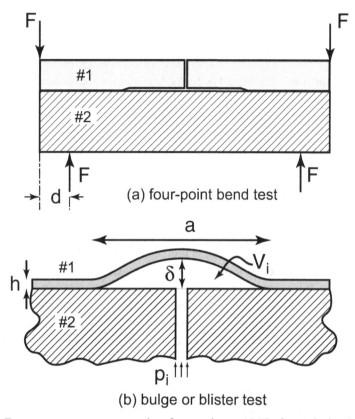

**Figure 3.12** Two common tests to measure interface toughness. (a) The four-point bend test is very common but requires a method to cause a penetrating film crack to deflect into the interface. (b) For the blister test, the challenge is to introduce a hole that just reaches the interface of interest.

curvature, it is critically important to know whether the process of attaching the stiffener alters the state of the stress in the original multilayer. For example, if a stiffener is added by diffusion bonding, this typically requires pressure that flattens the multilayer from an inititially curled state. In this case, one must be sure to keep track of the stress-free reference state for each layer, as will be discussed in detail in Chapters 4 and 5.

In the blister test, the surface of a film that is created by a delamination precrack is loaded via pressure or injected volume through an access hole in the substrate, as shown in Figure 3.12(a). The interface toughness can be inferred from mechanics solutions that relate the film displacement, the pressure (or the injected volume) and the energy release rate. The mechanics of interface debonding for this type of testing was first described in significant detail by Jensen (1991) and Jensen & Thouless (1993). This work was extended by Sofla, Seker, Landers & Begley (2010), who provide a comprehensive set of solutions for the circular blister test that spans the full range of plate and membrane film behaviors.

In blister testing, the size of the delamination crack does not need to be directly observed, as it can be inferred from the pressure/displacement relationship or volume/

displacement relationship. Under controlled volume injection, the propagation of the interface crack is stable, providing multiple opportunities to quantify the interface toughness for a single interface. The paper by Hohlfelder, Luo, Vlassak, Chidsey & Nix (1996) provides a clear illustration of blister testing of interfaces, wherein access to the interface is achieved by backside etching of the substrate. For many systems, this is not feasible; the difficulty in accessing the interface to pressurize the blister is likely a reason the four-point bend test is far more common.

A downside of the configurations shown in Figure 3.12 is that they enable toughness measurements only for a fairly limited range of mode mix. The mode mix for both configurations can be modified by adjusting layer thickness, but only by a rather modest amount. For example, the mode mix for a compliant film used in a blister test varies between $-70° < \psi < -40°$ when the film thickness to span ratio varies $0.001 < h/a < 0.1$. Similar moderate ranges can be probed in bend specimens by adjusting the thickness of individual layers.

This represents a common trade-off amongst all interface testing specimens: those that are easy to fabricate, easy to load and easy to interpret offer a narrow rage of mode mix. This point is driven home by the Brazil nut specimen shown in Figure 3.7 Wang & Suo (1990); it provides access to a broad range of mode mix but is more difficult to fabricate, particularly with regard to introducing the initial interface flaw. Similarly, the bimaterial strip utilized by Leichti and Chai (Figure 3.11) is easy to fabricate (for some systems), but difficult to load in a manner that makes the analysis straightforward (Liechti & Chai 1991, 1992).

Regardless of the specimen geometry, a central challenge is the need to introduce a predefined interface crack into the specimen. This is less of a problem for relatively weak interfaces, since a crack that propagates through a layer will often naturally deflect into the interface. This implies that the bend specimens shown in Figure 3.12(b) are relatively easy to prepare for weak interfaces: a scratch on the top surface can be propagated to the interface via bending and will naturally deflect into the interface. (This process is discussed in detail in Chapter 8.) The blister test shown in Figure 3.12(a) provides a direct means to load the interface and introduce the interface precrack, provided one can drill or etch a hole in the substrate to reach the interface in a manner that does not damage the film.

For tough interfaces, introducing the precrack is often not simple. A common technique involves depositing an extremely thin sacrificial layer over a controlled pattern on the surface of the layer below the interface of interest, prior to deposition of the upper layer(s). If the sacrificial layer leads to weak bonding with one of the adjacent layers (e.g., carbon powder), it may serve as a de facto precrack by debonding during the early stages of loading, hopefully injecting a crack along the interface of interest. For some systems, the sacrificial layer needs to be removed by heating or chemical etching. This obviously introduces the concern that the process to prepare fracture specimens alters the nature of the system, for example, through annealing of residual stresses during heating, or through chemical contamination of the interface.

An interesting and potentially promising technique to introduce well defined interface cracks has recently emerged, although it is limited to optically transparent films, which notably include many TGOs. For some systems, femto-second laser pulses will pass through the film and be absorbed by the substrate: for low laser fluences, the net effect is that the interface debonds with essentially no changes to either the film or the substrate. This creates significant opportunities for subsequent interface testing. In more limited circumstances, higher laser fluences can be used to both debond the interface *and* excite the film in a manner than drives debonding and can be related to the interface toughness, as described in Jorgensen, Pollock & Begley (2015).

Finally, a frustratingly persistent challenge to interface toughness measurements is keeping the crack on the interface, that is, avoiding kinking. The problem is particularly acute when attempting to measure mixed mode toughness with significant mode II components (i.e., high phase angles), since the elevated interface toughness for this mode may be higher than that of the bulk toughness of the adjacent layers. It should be emphasized that in many applications (such as those dominated by thermal misfit strains), the interface will be loaded close to pure mode II. Simply put, mode II is often the most relevant to an application but can be very difficult to measure via externally imposed loading while avoiding kinking. Chapter 7 addresses the details of when kinking can be expected to occur from an interface crack, and as such provides the conceptual framework to assess the likelihood of kinking during interface toughness measurements.

## 3.7    Interface Toughness Measurements for Thermal Barrier Coatings

We conclude this chapter with some substantial recent efforts to measure the delamination toughness of thermal barrier coatings (TBCs), which highlight some of the issues raised in the previous section. TBCs are an essential feature of blades and hot section components of aircraft and power generating turbines, which will be discussed further in Section 5.6 and in Chapters 12 and 13. In brief, TBC multilayers comprise an outer ceramic thermal barrier overlaying a ceramic oxidation protection layer which grows from a metallic layer bonded to the superalloy blade or component. Internal cooling of the blade, or backside cooling of a plate or shell component, results in a significant temperature drop across the TBC, enabling higher gas temperatures.

TBCs are prone to delamination and spalling due to the temperature variations and associated thermal mismatch stresses experienced in each cycle of heating and cooling occuring in the turbine. Depending on the specific TBC, delamination usually occurs at or near the interface between the outer thermal barrier and the thermally grown oxide, or at the interface between the thermally grown oxide and the metallic bonding layer. Lifetime and reliability advancement of TBCs hinge on improvements in delamination toughness, placing a high priority on the measurement of interface toughness for these systems. The highest stresses driving delamination usually occur during cool-down when the higher thermal contraction of the metal blade relative to the ceramic coating gives rise to compression in the coating. A consequence is that mode II, or near-mode II, edge delamination becomes important. For reasons such as those alluded to in Section 3.6, measurement of TBC delamination toughness has been notoriously difficult and slow to be realized, especially over the relevant range of mode mix.

An overview of some of the tests used to measure TBC delamination under different modes of loading was given by Hutchinson & Hutchinson (2011). Contributions include the employment of a stiffener bonded to the coating to enhance the delamination driving force (Hofinger, Occhsner, Bahr & Swain 1998), indentation-induced delamination (Begley, Mumm, Evans & Hutchinson 2000; Vasinonta & Beuth 2001), efforts to develop a bar shear-off test (Guo, Mumm, Karlsson & Kagawa 2005), stiffener enhanced bend specimens which were precursors to the specimens discussed below (Théry, Poulain, Dupeux & Braccini 2009) and in situ tests involving highly refined micromachining (Eberl, Wang, Gianola, Nguyen, He, Evans & Hemker 2011).

Recent progress shortly preceding the writing of this book is the comprehensive experimental effort conducted at ONERA by Vaunois (2014) and Vaunois, Poulain, Kanoute & Chaboche (2016). Their study measured the mode dependence of interface delamination for a specific TBC system as a function of thermal cycling and addressed the problem of making use of such data to predict coating lifetimes. Here, as a complement to the other interface toughness data presented in this chapter, we present a selection of the data from the above authors, especially to highlight the mode dependence and the degradation of toughness with thermal cycling. The reader is referred to the original work for full details and additional data.

The TBC system considered by Vaunois et al. (2016) is an electron beam, physical vapor – deposited yttria-stablized zirconia (YSZ) coating on a Ni-Pt-Al bond coat that was deposited on a superalloy substrate. Three types of specimens were developed (see top of Figure 3.13) for achieving a range of mode mix: a four-point bend specimen (4PB), an inverted four-point bend specimen (I4PB) and a four-point bend end-notched flexure specimen (4ENF). Each test made use of a steel plate stiffener glued to the coating to significantly enhance the energy release rate acting on the coating/bond coat interface. The authors carried out extensive finite element analyses of the specimens to evaluate both the energy release rate and the mode mix.

For the 4PB specimen, $\psi = 21°$; for the I4PB, $\psi = -72°$, while for the 4ENF specimen $\psi = -79°$. The delamination tests were conducted at room temperature. Specimens were tested to obtain the room temperature delamination toughness prior to thermal cycling and following $N$ thermal cycles, denoted by $G_c(N)$. Each thermal cycle for the data shown in Figure 3.13 consisted of an 8 minute heat-up to 1100°C, followed by a 60 minute dwell at 1100°C, with a 4 minute forced-air cool-down. Following Vaunois et al. (2016), the toughness data in Figure 3.13 have been normalized by the toughness measured with the 4PB specimen prior to any thermal cycling, denoted by $\Gamma_C(0)_{4PB}$.

As noted by Vaunois et al. (2016), the degradation of the delamination toughness with thermal cycling is exponential. Here, we make use of the phenomenological mixed mode toughness dependence (3.19), incorporating an exponential dependence of thermal cycles $N$ to fit the solid line for the 4PB (with $\psi = 21°$) in Figure 3.13:

$$\Gamma_C(\psi, N) = \Gamma_{IC} f(\psi) 10^{-0.002N} \text{ with } f(\psi) = 1 + \tan^2\left[(1 - \lambda)\psi\right] \qquad (3.36)$$

such that for any other mode mix,

$$\frac{\Gamma_C(\psi, N)}{\Gamma_C(0)_{4PB}} = \frac{f(\psi)}{f(21°)} 10^{-0.002N} \qquad (3.37)$$

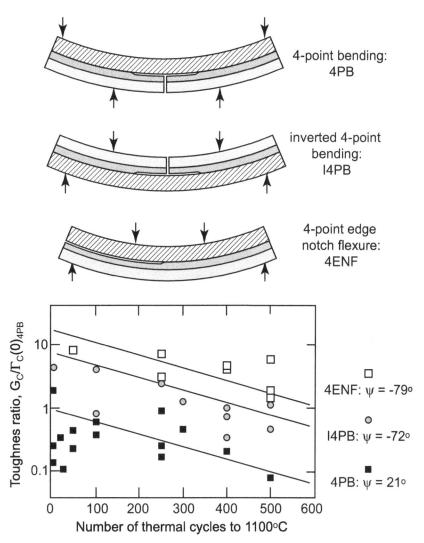

**Figure 3.13** The degradation of delamination toughness of thermally cycled thermal barrier coatings measured for three mode mixes by Vaunois (2014) and Vaunois et al. (2016) illustrating the strong influence of mode mix. Each experimental point is the interface toughness measured at room temperature after the specimen has been exposed to thermal cycling (see text). Included in the data are some specimens for which partial debonding of the stiffner/coating interface contributed to the scatter. The solid trend line for *4PB* specimen has been fit to the thermal cycling data following Vaunois et al. (2016). The trend lines for the data for the other two test configurations are predictions based on the phenomenological mode mix toughness form in (3.37) with $\lambda = 0.02$.

The solid lines for the two other mode mixes in Figure 3.13(a), $\psi = -72°$ and $\psi = -79°$, have been generated with the parameter choice $\lambda = 0.02$. Similar trend lines were generated by Vaunois et al. (2016) with a different phenomenological mixed mode toughness function involving two parameters. In the notation of Figure 3.13, the room temperature delamination criterion for a given mode mix after $N$ thermal cycles is

$G_C = \Gamma_C(\psi, N)$. It is worth noting that the mode mix dependence $f(\psi)$ employed here is symmetric in $\psi$ such that the different sign of the 4PB mode mix from the other two specimens does not come into play. Finally, the reader is referred to Vaunois et al. (2016) for additional data associated with cycling to 1070°C and 1150°C, which reveals temperature dependence of the exponential degradation of the interface toughness.

# 4    Steady-State Delamination
of Bilayers

A significant fraction of the insight and understanding needed to understand mixed-mode fracture can be obtained by considering a simple elastic bilayer consisting of two thin layers bonded together, with a semi-infinite crack lying on the interface. The concepts and results generated for the bilayer in this chapter are extended to multilayers consisting of many layers in Chapter 5. Additional aspects of interface fracture in bilayers are discussed throughout subsequent chapters, but notably Chapter 7 (kinking out of the interface), Chapter 9 (finite-sized edge flaws) and Chapter 12 (temperature gradients).

## 4.1    A Basic Solution

A basic solution for a crack on the interface of a bilayer presented in this section illustrates a number of aspects of the mechanics of interface cracks. The solution for the stress intensity factors and energy release to the problem depicted in Figure 4.1 will be presented and applied in this section. The solution was obtained within the context of linear elasticity for an infinitely long bilayer with a semi-infinite crack emerging from the left by Suo & Hutchinson (1990). In this section, we focus on the key results of the bilayer system and their implications: though not shown here, the derivation of the results in this section is identical to that presented in Chapter 5 for a multilayer with many layers.

For the equilibrated load system shown, the stress intensity factors and energy release rate are independent of the position of the crack tip–what is often referred to as steady-state cracking. Within the context of plane stress or plane strain, the energy release rate can be computed exactly using simple energy arguments accounting for the energy difference of the bilayer system far ahead and behind the crack tip. (Here, we focus on the solution for the bilayer and its implications, leaving the underlying derivation for the more general multilayer treatment in Chapter 5.) Determination of the mode mix requires a sophisticated elasticity analysis, which was carried out in the reference cited above and tabulated.

Software accompanying this text generates the full solution for any material combination, using these tabulated values. Alternatively, one can compute the needed results via the finite element code provided with this book, as described in Chapter 16. While the choice between the two computational approaches is somewhat arbitrary, the tabulated

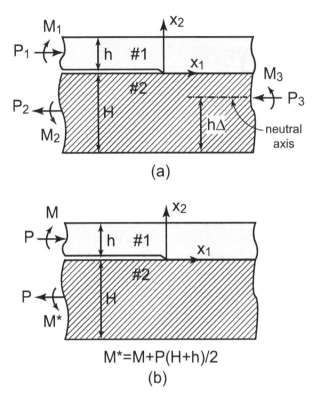

**Figure 4.1** (a) The bilayer under an equilibrated set of loads (forces and moments per unit depth perpendicular to the plane): a semi-infinite crack lies on the interface between the two isotropic elastic materials. (b) Equivalent loading giving identical stress intensity factors; see text for definitions of $P$ and $M$.

results for bilayers are highly accurate and avoids some of the numerical challenges associated with analyzing long, slender layers (as described in Chapter 17).

Overall equilibrium of the load system shown in Figure 4.1 requires that forces/depth and moments/depth satisfy

$$P_3 = P_1 - P_2$$
$$M_3 = M_1 - M_2 + P_1 h \left( \frac{1}{2} + \frac{1}{\eta} - \Delta \right) + P_2 h \left( \Delta - \frac{1}{2\eta} \right) \tag{4.1}$$

with

$$\eta = \frac{h}{H}; \quad \Delta = \frac{1 + 2\Sigma\eta + \Sigma\eta^2}{2\eta(1 + \Sigma\eta)}; \quad \Sigma = \frac{\bar{E}_1}{\bar{E}_2} = \frac{1 + \alpha_D}{1 - \alpha_D} \tag{4.2}$$

Here $\bar{E} = E$ for plane stress, $\bar{E} = E/(1 - v^2)$ for plane strain and $\alpha_D$ is the first Dundurs' mismatch parameter defined by (3.9) as appropriate for plane stress or plane strain.

The energy release rate of the interface crack is

$$G = \frac{1}{2\bar{E}_1} \left( \frac{P_1^2}{h} + 12 \frac{M_1^2}{h^3} \right) + \frac{1}{2\bar{E}_2} \left( \frac{P_2^2}{H} + 12 \frac{M_2^2}{H^3} - \frac{P_3^2}{Ah} - \frac{M_3^2}{Ih^3} \right) \tag{4.3}$$

where

$$I = \Sigma \left[ \left( \Delta - \frac{1}{\eta} \right)^2 - \left( \Delta - \frac{1}{\eta} \right) + \frac{1}{3} \right] + \frac{\Delta}{\eta} \left( \Delta - \frac{1}{\eta} \right) + \frac{1}{3\eta^2}$$

(4.4)

$$A = \frac{1}{\eta} + \Sigma$$

Now, as shown by Suo & Hutchinson (1990), the stress intensity factors, and thus also the energy release rate, for the load system shown in Figure 4.1(b) are identical to those shown in Figure 4.1(a) where the force/depth $P$ and moment/depth $M$ are given by

$$P = P_1 - C_1 P_3 - C_2 \frac{M_3}{h}; \quad M = M_1 - C_3 M_3$$

(4.5)

with

$$C_1 = \frac{\Sigma}{A}; \quad C_2 = \frac{\Sigma}{I} \left( \frac{1}{\eta} + \frac{1}{2} - \Delta \right); \quad C_3 = \frac{\Sigma}{12I}$$

(4.6)

In these variables, the stress intensity factors and energy release rate are given by

$$K_1 + i K_2 = h^{-i\epsilon} \left( \frac{1 - \alpha_D}{1 - \beta_D^2} \right)^{1/2} \left( \frac{P}{\sqrt{2hU}} - i\epsilon^{i\gamma} \frac{M}{\sqrt{2h^3 V}} \right) e^{i\omega}$$

$$G = \frac{1 - \beta_D^2}{E_*} \left( K_1^2 + K_2^2 \right) = \frac{1 - \alpha_D}{E_*} \left( \frac{P^2}{2hU} + \sin\gamma \frac{PM}{h^2 \sqrt{UV}} + \frac{M^2}{2h^3 V} \right)$$

(4.7)

with

$$\frac{1}{U} = 1 + \Sigma \eta \left( 4 + 6\eta + 3\eta^2 \right)$$

$$\frac{1}{V} = 12 \left( 1 + \Sigma \eta^3 \right)$$

(4.8)

$$\frac{\sin\gamma}{\sqrt{UV}} = 6\Sigma \eta^2 \left( 1 + \eta \right)$$

Here, $\beta_D$ is the second Dundurs' mismatch parameter, $E_*^{-1} = \left( \bar{E}_1^{-1} + \bar{E}_1^{-1} \right)/2$ is an interface modulus, and $\epsilon = \ln \left[ (1 - \beta_D) / (1 + \beta_D) \right] / (2\pi)$.

The stress intensity factors are fully determined apart from the phase factor $\omega(\alpha_D, \beta_D, \eta)$, which was computed and tabulated by Suo & Hutchinson (1990). The stress intensity factors depend on elastic mismatch, $\alpha_D$ and $\beta_D$, through $\epsilon$ and $\omega$, while $G$ depends only on $\alpha_D$. The curves in Figure 4.2 show the dependence of the phase factor $\omega$ on a wide range of elastic mismatch. All cases are covered by thickness ratios $\eta$ falling between $\eta = 0$ (a thin film on a thick substrate) and $\eta = 1$ (layers of equal thickness) by interchanging the roles of the layers if the top layer is thicker than the bottom layer.

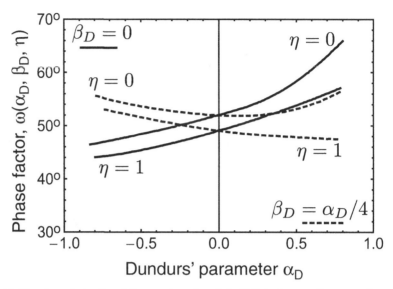

**Figure 4.2** The phase factor for a bilayer, $\omega(\alpha_D, \beta_D, \eta)$, in (4.7) in degrees. For $\alpha_D = \beta_D = 0$ and $\eta = 1$, the exact value is $\omega = \sin^{-1}\sqrt{4/7} = 49.107°$. For $\alpha_D = \beta_D = 0$ and $\eta \to 0$, $\omega \to 52.07°$.

Expression (4.7) illustrates the dependence of the stress intensity factors on the unusual dimensional term $h^{-i\epsilon}$, as discussed in Section 3.2. For bilayers with $\beta_D = \epsilon = 0$, (4.7) yields

$$K_1 = \sqrt{1 - \alpha_D}\left(\frac{P}{\sqrt{2hU}}\cos\omega + \frac{M}{\sqrt{2h^3V}}\sin(\gamma + \omega)\right)$$

$$K_2 = \sqrt{1 - \alpha_D}\left(\frac{P}{\sqrt{2hU}}\sin\omega - \frac{M}{\sqrt{2h^3V}}\cos(\gamma + \omega)\right)$$

(4.9)

## 4.2     Extended Application of Bilayer Results for Plane Strain Delamination

Several examples below illustrate the use of the above results to determine the stress intensity factors and energy release rate for some basic problems. Additional examples are given in Hutchinson & Suo (1992). However, before considering specific examples, we take a moment to emphasize important issues related to the extended applicability of these results.

The distinction between plane stress and plane strain in the solution has already been noted and is accounted for in the definitions of $\bar{E}$ and the Dundurs' parameters. But the applicability is broader than just plane stress or plane strain, as will be illustrated by the example of a thin film on a deep substrate, as shown in Figure 4.3. With proper interpretation, the plane strain result also applies to the case where the residual stress state in the film prior to delamination is any biaxial stress state and not just a plane strain state, as long as the cracking process is plane strain. This example will also illustrate how one makes use of the bilayer solution.

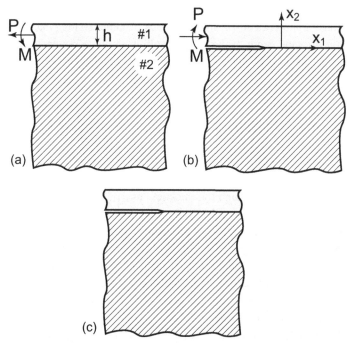

**Figure 4.3** (a) Resultant $P$ and $M$ associated with the stress distribution in the film. (b) Loads, $P$ and $M$, producing the stress intensity factors and energy release rate of the interface crack. (c) The problem of an interface crack advancing from a free edge under steady-state conditions. The stress distribution far ahead of the crack is characterized by the resultants in (a).

Suppose the residual stress state in the film is uniform with components $\sigma_{11} \neq 0$, $\sigma_{22} = 0$ and $\sigma_{33} \neq 0$. This initial state would be equilibrated by $P = \sigma_{11}h$ and $M = 0$ in Figure 4.3(a). The delamination crack on the interface in Figure 4.3(c) results from the superposition of the loads in Figures 4.3(a) and 4.3(b). Thus, assuming this is plane strain delamination with no change in strain in the $x_3$-direction during cracking, the stress intensity factors and energy release rate are provided by the solution to the problem with the loads in Figure 4.3(b). (The loads in the problem of Figure 4.3(a) give rise to no stress intensities since they exactly balance the residual stress. One could cut along the interface and the film would not move.)

*The important conclusion of this argument is that the stress intensity factors and the energy release rate depend on $\sigma_{11}$ but not on $\sigma_{33}$, assuming the delamination process is plane strain.* In particular, the plane strain bilayer results apply when the stresses in the uncracked bilayer are generated under conditions of biaxial straining but the changes during delamination occur subject to plane strain. More insight into the question of whether the cracking process should be regarded as plane strain will be provided when we discuss fully three-dimensional cracking problems in Chapter 11. However, it is not too early to anticipate that plane strain cracking is often a relevant mode of cracking due to the constraint imposed on the body by displacements parallel to the crack front. For the thin film in Figure 4.3(c), the separated portion of the film is constrained to maintain the same out-of-plane strain $\epsilon_{33}$ after cracking that it experiences before cracking. This constraint breaks down when the crack becomes so long that is comparable to the actual out-of-plane depth of the object itself. Then three-dimensional effects come into play, as discussed in Chapter 11. Discussion of the plane strain film problem is continued in Section 4.3.

## 4.3      General Results for a Thin Film on an Infinitely Thick Substrate

Consider the thin film on a semi-infinite substrate in Figure 4.3(a) where, prior to interface cracking, the residual stress in the bonded film far from the edge is independent of $x_1$ but varies through the thickness as $\sigma_{11} = \sigma(x_2)$ (as noted above the out-of-plane component, $\sigma_{33}$, has no bearing on the results that follow for delamination subject to plane strain). An edge-crack propagates along the interface such that no forces act on the separated segment of the film, as depicted in Figure 4.3(c). As previously discussed, the solution to the problem in Figure 4.3(c) is the superposition of the two problems in Figures 4.3(a) and 4.3(b). The stress intensity factors and energy release rate are given by the problem in Figure 4.3(b) whose loads are equal and opposite to those in the bonded film. Specifically, with reference to the sign conventions indicated in Figure 4.3(b),

$$P = \int_0^h \sigma(x_2)dx_2 \quad \text{and} \quad M = \int_0^h \sigma(x_2)\left(x_2 - \frac{h}{2}\right)dx_2 \qquad (4.10)$$

For an infinitely deep substrate, the parameters defining the bilayer solution are $H \to \infty, \eta \to 0, A \to \infty, I \to \infty, \gamma = 0$, such that (4.3) and (4.7) yield

$$G = \frac{1}{2\bar{E}_1}\left(\frac{P^2}{h} + 12\frac{M^2}{h^3}\right)$$

$$K = h^{-i\epsilon}\left(\frac{1 - \alpha_D}{1 - \beta_D^2}\right)^{1/2}\left(\frac{P}{\sqrt{2h}} - i\frac{\sqrt{6}M}{h^{3/2}}\right)e^{i\omega} \qquad (4.11)$$

If $\beta_D = \epsilon = 0$, the mode mix is given by

$$\tan\psi = \frac{Ph\tan\omega - 2\sqrt{3}M}{Ph + 2\sqrt{3}M\tan\omega} \qquad (4.12)$$

The bilayer solution is derived under the tacit assumption that the crack faces are not predicted to interpenetrate (except possibly in an exceedingly small zone near the crack tip when $\epsilon \neq 0$). For systems with $\beta_D = \epsilon = 0$, the condition that the crack faces do not penetrate is $K_1 \geq 0$. By (4.12), the combination of film stresses ensuring no crack face penetration is

$$Ph + 2\sqrt{3}M\tan\omega \geq 0 \qquad (4.13)$$

Pure mode II pertains if the equality holds in the above expressions.

A uniform film stress, $\sigma$, gives $P = \sigma h$ and $M = 0$, such that

$$G = \frac{\sigma^2 h}{2\bar{E}_1} \qquad (4.14)$$

This is the widely used result for the energy release rate for thin films subject to uniform residual stress. If $\beta_D = 0$, the condition (4.13) for no interpenetration is $\sigma > 0$, and from (4.12), $\psi = \omega$ where $\omega(\alpha_D, \beta_D, \eta = 0)$ is plotted in Figure 4.2. The mode mix ranges from about $\psi = 45°$ to $\psi = 65°$, depending on $\alpha_D$, and is $\psi = 52.07°$ with no mismatch.

Thus, interface delamination of a film with a uniform tensile stress on a thick substrate is mixed mode with somewhat more mode II than mode I, depending on the mismatch. If the film is subjected to uniform compression, (4.11) gives $K_1 < 0$, implying the physically unacceptable result that the crack faces interpenetrate. This case is not covered by the general bilayer solution presented in this section. The problem must be addressed on its own merits accounting for crack face contact. A complexity of behaviors can arise. If friction between the faces is negligible, the crack tip will usually be subject to near-mode II conditions ($K_1 \sim 0$, $K_2 < 0$), with $G$ given by (4.14) to a good approximation (Stringfellow & Freund 1993; Balint & Hutchinson 2001). Appreciable friction will reduce $G$ below (4.14). For systems where interface bonding is due to reversible adhesion, such as PDMS on glass, re-adhesion can occur behind the advancing crack tip on mode II delamination, significantly complicating the estimation of $G$ (Collino, Philips, Rossol, McMeeking & Begley 2014; Begley, Collino, Israelachvili & McMeeking 2014).

If there is a gradient of stress through the film such that $P = 0$ and $M > 0$ then, by (4.13), the crack is open and by (4.12) $\psi = \omega - \pi/2$. Thus a film with tension at the surface varying linearly to equal and opposite compression at the interface ($M > 0$) delaminates under mixed-mode conditions that are somewhat more mode I than mode II, with $-45° < \psi < -25°$, with $K_2 < 0$. A film with $P = 0$ and $M < 0$ again falls outside the scope of the general bilayer solution because $K_1 < 0$ is implied. Problems of this type have been considered by Hutchinson & Hutchinson (2011).

By (4.12) and (4.13), a pure mode I interface crack is generated by a nonuniform film stress distribution having $M = (Ph/\sqrt{6})\tan\omega$ with $P > 0$. The average stress in the film is tensile and the gradient is such that the tensile stress is larger at the surface than at the interface.

The mode mix for cases for which $\beta_D$ cannot be neglected can be dealt with in a manner similar to that described for the example in the next section. By (4.11), the mode mix is

$$\tan\psi = \frac{Im\left(K\ell^{i\epsilon}\right)}{Re\left(K\ell^{i\epsilon}\right)} = \frac{Ph\tan\left(\omega + \epsilon\ln\frac{\ell}{h}\right) - 2\sqrt{3}M}{Ph + 2\sqrt{3}M\tan\left(\omega + \epsilon\ln\frac{\ell}{h}\right)} \qquad (4.15)$$

## 4.4 Equal Thickness Bilayer with Elastic Mismatch Subject to Moments

In the absence of elastic mismatch ($\alpha_D = \beta_D = 0$), the configuration and loading shown in Figure 4.4 is symmetric with respect to the crack and the interface. In the notation

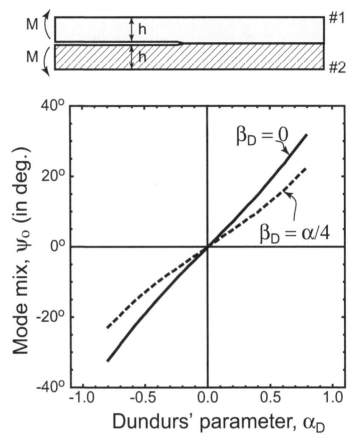

Figure 4.4 A bilayer with symmetric geometry and loading but nonsymmetric elastic properties.

of Figure 4.1a and (4.3), the nonzero loads are $M_1 = M_2 = M$. Assuming $M > 0$, the crack experiences mode I conditions. Using the above results one finds

$$K_1 = 2\sqrt{3}Mh^{-3/2} \text{ and } K_2 = 0, \text{ or } G = \frac{12M^2}{\bar{E}h^3} \text{ and } \psi = 0 \qquad (4.16)$$

If the layers in Figure 4.4 have different elastic properties, (4.7) yields

$$G = \frac{12M^2}{E_*h^3} \text{ and } K = h^{-i\epsilon}e^{i(\gamma+\omega-\pi/2)}\left(\frac{12}{1-\beta_D^2}\right)^{1/2}Mh^{-3/2} \qquad (4.17)$$

with $E_*^{-1} = \left(\bar{E}_1^{-1} + \bar{E}_2^{-1}\right)/2$ and $\sin\gamma = \sqrt{3}\left(1+\alpha_D\right)/\sqrt{7+6\alpha_D}$. The mode mix, as defined in (3.22) and based on distance ahead of the tip $\ell$, is

$$\tan\psi = \frac{Im\left[K\ell^{i\epsilon}\right]}{Re\left[K\ell^{i\epsilon}\right]} = \tan\left(\gamma + \omega - \frac{\pi}{2} + \epsilon\ln\left(\frac{\ell}{h}\right)\right) \qquad (4.18)$$

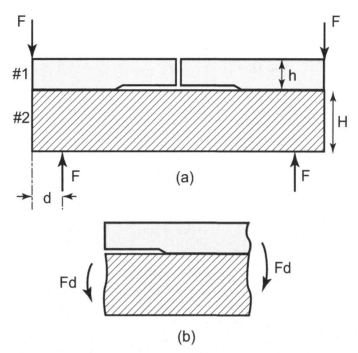

**Figure 4.5** (a) A four-point bend specimen. (b) The half-segment of the specimen within the loading points characterizing the interface cracking problem.

or

$$\psi = \gamma + \omega - \frac{\pi}{2} + \epsilon \ln\left(\frac{\ell}{h}\right) \equiv \psi_o + \epsilon \ln\left(\frac{\ell}{h}\right) \qquad (4.19)$$

For $\alpha_D = 0$ and $\beta_D = \epsilon = 0$, (4.19) reduces to $\psi = 0$ in accord with (4.16). Elastic mismatch renders the system nonsymmetric, even though the geometry and loading are symmetric. The dependence of $\psi_o$ (i.e., on $\psi$ for $\ell = h$) on elastic mismatch is illustrated in Figure 4.4. For mismatch satisfying $\|\alpha_D\| < 0.5$, the asymmetry is relatively small, that is $|\psi_o| < 10°$. However, for larger elastic mismatch in the range $0.75 < |\alpha_D| < 1$, the mode mix can be as large as $|\psi_o| \approx 40°$. The length $\ell$ at which $\psi$ is evaluated produces an additional shift in mode mix by $\epsilon \ln(\ell/h)$ as specified by (4.19).

## 4.5    Four-Point Bend Delamination Specimen

Another important application of the bilayer solution is the four-point bend specimen in Figure 4.5, which is widely used for measuring interface toughness (e.g., Cao & Evans 1989; Charalambides et al. 1989; Charalambides et al. 1990; Dauskardt et al. 1998). The bilayer system in Figure 4.1(a) applies to the specimen between the four loading points with $M_3 = -Fd$, $M_2 = Fd$ and with the other load quantities set to zero. By (4.3),

$$G = \frac{1}{2\bar{E}_2 H^3}\left(12 - \frac{1}{I\eta^3}\right)(Fd)^2 \qquad (4.20)$$

where $\eta$ is defined in (4.2) and $I$ using (4.2)–(4.4). Using (4.7) one can show

$$
\tan \psi = \frac{Im\left[K\ell^{i\epsilon}\right]}{Re\left[K\ell^{i\epsilon}\right]} = \frac{C_2 \sin\left(\omega + \epsilon \ln\left(\frac{\ell}{h}\right)\right) - \sqrt{UV}C_3 \cos\left(\gamma + \omega + \epsilon \ln\left(\frac{\ell}{h}\right)\right)}{C_2 \cos\left(\omega + \epsilon \ln\left(\frac{\ell}{h}\right)\right) + \sqrt{UV}C_3 \sin\left(\gamma + \omega + \epsilon \ln\left(\frac{\ell}{h}\right)\right)}
$$

$$(4.21)$$

where the constants $C_2$ and $C_3$ are provided in (4.6), while $U$, $V$ and $\gamma$ are defined in (4.8). For general elastic mismatch and thickness ratio, the energy release rate and mode mix can be evaluated numerically from formulas (4.20) and (4.21).

As the simplest example, consider a specimen with no elastic mismatch ($\alpha_D = \beta_D = 0$) and a thickness ratio of unity ($\eta = 1$). From (4.20) and (4.21), one finds

$$
G = \frac{21}{4} \frac{F^2 d^2}{\bar{E} h^3} \quad \text{and} \quad \tan \psi = \frac{\sqrt{21} \sin \omega - \cos(\gamma + \omega)}{\sqrt{21} \cos \omega + \sin(\gamma + \omega)}
\tag{4.22}
$$

with $\omega = \sin^{-1}\left(\sqrt{4/7}\right) = 49.11°$ and $\gamma = \sin^{-1}\left(\sqrt{3/7}\right) = 40.89°$. The mode mix is $\psi = \gamma = 40.89°$. This assumes $F > 0$ such that $K_1 > 0$ and the crack faces do not make contact. The software accompanying this text includes a module for general four-point bend specimens (with arbitrary thickness ratios and elastic mismatch).

For the inverted specimen with $F < 0$, the crack faces make contact away from the crack tip, propping open the crack tip. The basic bilayer solution in this section does not apply to that scenario. The mode mix with $F < 0$ has a larger relative component of mode II (with $K_2 < 0$) than for the case with $F > 0$ Hutchinson & Hutchinson (2011).

## 4.6    Bilayer with Thermal Expansion Mismatch Subject to a Uniform Temperature

As a final illustration of the application of the basic solution, consider a bilayer with no external loads or residual stress at temperature $T_o$ that is subject to a uniform temperature change $\Delta T$. The bilayer is assumed to have thermal expansion mismatch, $\alpha_1 \neq \alpha_2$, as well as elastic mismatch. Stress intensity factors and energy release rate are sought for steady-state edge delamination along the interface of the prestressed bilayer shown in Figure 4.6(c). There are no external loads applied to the bilayer, thus the resultant force and moment acting on the prestressed bilayer are zero.

A two-step solution procedure is required, similar to that employed earlier for the thin film with a general stress distribution. First, the stresses and resultant loads $P$ and $M$ in the intact bilayer in Figure 4.6(a) are determined. Second, equal and opposite loads to those shown in Figure 4.6(a) are applied to the cracked bilayer in Figure 4.6(b). The desired solution to the problem in Figure 4.6(c) is the superposition of the problems in Figures 4.6(a) and 4.6(b); the stress intensity factors are obtained from the problem in Figure 4.6(b).

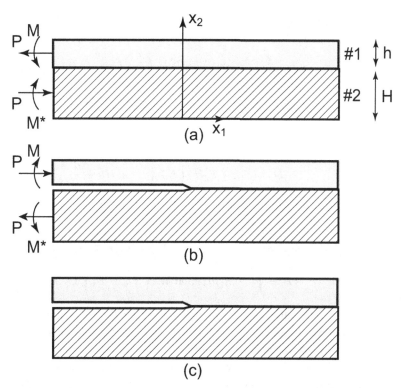

**Figure 4.6** A semi-infinite interface crack that lies on the interface face between two isotropic elastic materials subject to stressing, but no thermal loads can be represented as the superposition of two cases: (a) The equivalent forces and moments associated with the intact prestress bilayer, and (b) equal and opposite forces and moments applied to the cracked layer. The superposition of (a) and (b) produces (c), the thermal stress problem with no external loads. The stress intensity factors are obtained by solving the problem in *(b)*; note that $M^* = M + P(h + H)/2$.

The *thermal* strains are $\epsilon_{11}^T = \epsilon_{33}^T = \alpha_1 \Delta T$ in layer 1 and $\epsilon_{11}^T = \epsilon_{33}^T = \alpha_2 \Delta T$ in layer 2. The stresses in the intact bilayer depend on the out-of-plane constraint. To illustrate this constraint and how it affects the stress intensity factors and energy release rate of the interface crack, three out-of-plane conditions are considered in computing the stress: plane stress ($\sigma_{33} = 0$), plane strain ($\epsilon_{33} = 0$) and equi-biaxial stresses ($\sigma_{11} = \sigma_{33}$). Equi-biaxial stressing applies to a three-dimensional bilayer that is free to expand and bend about both the $x_1$- and $x_3$-axes; the *total* strains $\epsilon_{11} = \epsilon_{33}$ vary linearly in the $x_2$-direction. The stress component of interest in all three cases can be expressed as

$$\sigma_{11} = \bar{E}_1 \left( \epsilon_o + \kappa x_2 - \bar{c}_1 \alpha_1 \Delta T \right) \text{ in no. 1}$$
$$\sigma_{11} = \bar{E}_2 \left( \epsilon_o + \kappa x_2 - \bar{c}_2 \alpha_2 \Delta T \right) \text{ in no. 2} \tag{4.23}$$

where $\bar{E}$ and $\bar{c}$ are defined in Table 2.1 for plane stress, plane strain or equi-biaxial straining.

Overall equilibrium, $\int \sigma_{11}dx_2 = 0$ and $\int \sigma_{11}x_2dx_2 = 0$ (i.e., the resultant force and moment on the cross section are zero), gives $\epsilon_o$ and $\kappa$:

$$\epsilon_o = \bar{c}_2 u_2 \Delta T + \frac{\bar{E}_1 h \Delta T \left(\bar{c}_1 \alpha_1 - \bar{c}_2 \alpha_2\right)}{D}\left(\bar{E}_1 h\left(H^2 + Hh + \frac{h^2}{3}\right) + \frac{1}{3}\bar{E}_2 H^3\right)$$

$$\kappa = -\frac{\bar{E}_1 h \Delta T}{D}\left(\bar{c}_1 \alpha_1 - \bar{c}_2 \alpha_2\right)\left(\bar{E}_1 h\left(H + \frac{h}{2}\right) + \frac{1}{2}\bar{E}_2 H^2\right)$$

(4.24)

with

$$D = (\bar{E}_1 h + \bar{E}_2 H)\left(\bar{E}_1 h\left(H^2 + Hh + \frac{h^2}{3}\right) + \bar{E}_2\frac{H^3}{3}\right)$$
$$- \left(\bar{E}_1 h\left(H + \frac{h}{2}\right) + \frac{1}{2}\bar{E}_2 H^2\right)^2$$

The loads in Figure 4.6 (see also Figure 4.1(b)) can be computed as

$$P = \int_H^{H+h} \sigma_{11}dx_2 \text{ and } M = \int_H^{H+h} \sigma_{11}\left(y - H - \frac{h}{2}\right)dx_2$$

(4.25)

The computation above is straightforward, but the expressions that emerge are lengthy and unrevealing for the general bilayer. For general bilayers, numerical results are generated by the accompanying software. An explicit solution will be presented for two special cases: a bilayer with thermal expansion mismatch but no elastic mismatch ($\alpha_D = \beta_D = 0$) and equal layer thickness ($\eta = 1$), and a thin layer on an infinitely thick substrate with both thermal and elastic mismatch.

For the *equal thickness bilayer with no elastic mismatch* ($\bar{c}_1 = \bar{c}_2 \equiv \bar{c}$), one finds

$$\epsilon_o = \bar{c}\alpha_2\Delta T + \frac{\bar{c}}{2}\left(\alpha_1 - \alpha_2\right)\Delta T$$

$$\kappa = \frac{3\bar{c}\left(\alpha_1 - \alpha_2\right)\Delta T}{4h}$$

$$P = -\frac{\bar{E}h\bar{c}\left(\alpha_1 - \alpha_2\right)\Delta T}{8}$$

(4.26)

$$M = \frac{\bar{E}h^2\bar{c}\left(\alpha_1 - \alpha_2\right)\Delta T}{16}$$

$$M_* = -M = \frac{Ph}{2}$$

The stress intensity factors for all three out-of-plane conditions from (4.7) are exactly

$$K_1 = 0 \text{ and } K_2 = -\frac{\bar{E}\sqrt{h}\bar{c}\left(\alpha_1 - \alpha_2\right)\Delta T}{4}$$

(4.27)

This problem is pure mode II. The energy release rate, (3.3), assuming plane strain crack advanced governed by $\bar{E} = E/(1 - v^2)$, for *both* plane strain and equi-biaxial precracking conditions, is

$$G = \frac{(1 + v)Eh\left[(\alpha_1 - \alpha_2)\Delta T\right]^2}{16(1 - v)}$$

(4.28)

For plane stress prestressing and plane stress crack advance, the result is

$$G = \frac{Eh\left[(\alpha_1 - \alpha_2)\,\Delta T\right]^2}{16} \qquad (4.29)$$

Plane strain crack advance for a bilayer with prestresses computed under equi-biaxial constrain is valid as long as the interface crack is short compared to the out-of-plane width of the bilayer. (Recall that the steady-state approximation also requires the crack length be large in comparison to the film thickness $h$.) When the crack length begins to become comparable to the out-of-plane width, three-dimensional considerations become important, and the assumption of the plane strain crack advance must be modified as will be illustrated in Chapter 11.

For a *layer on an infinitely thick substrate with thermal and elastic mismatch*, one finds $\kappa = M = 0$ and

$$\epsilon_o = \bar{c}_2 \alpha_2 \Delta T; \quad P = -\bar{E}_1 h\,(\bar{c}_1 \alpha_1 - \bar{c}_2 \alpha_2)\,\Delta T \equiv \sigma h \qquad (4.30)$$

The thin layer is under uniform stress $\sigma_{11} = \sigma$ in the intact state. This is precisely one of the examples considered earlier in Section 4.3 for the thin film on a deep substrate–the case of a uniform film stress whose energy release rate and mode mix are discussed in connection with (4.14). Again, note for this example: the stress intensity factors and energy release rate under plane strain crack advance are identical at the same prestress $\sigma_{11} = \sigma$ whether it is evaluated under plane strain or equi-biaxial stressing.

## 4.7 Soft Material Substrates: A Stiff Film on an Infinitely Deep, Highly Compliant Substrate

While the general form of the basic bilayer solution presented in Section 4.1 is valid for arbitrarily large elastic mismatch between the layers, the results for determining the mode mix, c.f. (4.7) or (4.9), require knowledge of $\omega$, which was computed by Suo & Hutchinson (1990) only for elastic mismatches in the range of $|\alpha_D| \leq 0.8$. In particular, the numerical results in Section 4.1 do not extend to metal or ceramic films on substrates of soft materials such as elastomers and polymers for which the first Dundurs' parameter is nearly unity, that is, $0.99 < \alpha_D < 1$. It is evident from Figure 4.2 that $\omega$ is likely to be a strong function of $\alpha_D$ in the range $\alpha_D > 0.8$. The purpose of this section is to present results for $\omega$ applicable to thin stiff films on deep soft substrates, where $\bar{E}_1 / \bar{E}_2$ can be as large as $10^3$ or more. The other mismatch limit, for a soft material on a stiff substrate ($\alpha_D \simeq -1$), does not present similar challenges because the result for $\omega$ in Figure 4.2 can be approximated with some confidence by extrapolating to $\alpha_D = -1$.

Suo & Hutchinson (1990) generated the numerical results for $\omega$ before soft material bilayers and multilayers had come into the prominence they now enjoy, and thus were not motivated to present results for very large values of $\bar{E}_1 / \bar{E}_2$. Nevertheless, it is also true that the method these authors used to generate the solution to the bilayer elasticity problem encounters numerical difficulties for mismatches with $\bar{E}_1 / \bar{E}_2 \gtrsim 5$. Fortunately, as will be demonstrated here, for $\bar{E}_1 / \bar{E}_2 \gtrsim 5$, the film can be modeled accurately by

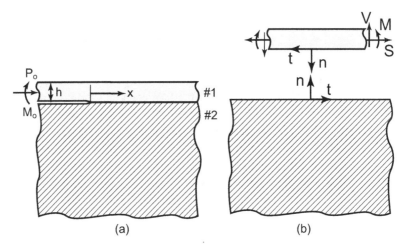

**Figure 4.7** (a) The beam/substrate model of the bilayer for a beam bonded to a semi-infinite substrate. The problem considered is plane strain, and the sign conventions for the stresslike variables are given in (b).

beam theory (or plate theory), and the problem of a beam on a compliant elastic substrate is more readily analyzed for large values of $\bar{E}_1/\bar{E}_2$. The extension to compliant substrates has also been addressed by Freund & Suresh (2003) although not to the large values of $\bar{E}_1/\bar{E}_2$ considered here.

The bilayer problem analyzed in this section is shown in Figure 4.7(a). It is the limit for an infinitely deep substrate of the problem considered in Section 4.1. Plane strain conditions are assumed, and as just noted the film is modeled using beam theory. The problem of a beam on an elastic substrate has been addressed by Shield & Kim (1992), and aspects of the problem are discussed in the book by Freund & Suresh (2003). The formulation and numerical analysis outlined below follows that of Shield & Kim (1992) in most respects, but the emphasis here is on obtaining results applicable to soft material substrates with very large elastic mismatches. For this purpose, a well-selected nondimensionalization of the governing equations provides insights and functional scaling of $\omega$ for large values of $\bar{E}_1/\bar{E}_2$. It will also be seen that the results of the beam model closely approximate those from the full elasticity solutions in the range of intermediate elastic mismatches, and it will be argued the beam model becomes increasingly accurate the larger is $\bar{E}_1/\bar{E}_2$. Details central to these assertions are supplied below because they are not available in the literature.

The plane strain problem for the beam on the infinitely deep substrate is posed in Figure 4.7(a). Following Shield & Kim (1992), standard beam theory is employed with moment/depth $M$, shear force/depth $V$ and resultant in-plane force/depth $S$, with the sign conventions indicated in Figure 4.7(b). The shear and normal tractions on the substrate interface at the bottom of the beam are denoted by $\sigma_{12} = t(x)$ and $\sigma_{22} = n(x)$; these tractions have a square root singularity at the tip of the crack at $x = 0$. For an incompressible substrate with $\nu_2 = 1/2$, the singularity is the conventional square root singularity, while otherwise it has an oscillatory behavior akin to that for the bi-material interface with $\beta_D \neq 0$. The numerical results in this section will be limited to incompressible substrates, which usually serve as a good approximation for many soft materials such as elastomers and gels. However, some attention will be paid to the accuracy of this limit for substrates that are not strictly incompressible.

For the infinitely deep substrate, the energy release rate in (4.11) continues to hold for the beam/substrate model,

$$G = \frac{1}{2\bar{E}_1} \left( \frac{P_o^2}{h} + \frac{12M_o^2}{3} \right) \tag{4.31}$$

For the incompressible substrate, $v_2 = 1/2$, and the stress intensity factors are related to $G$ by

$$G = \frac{1}{2\bar{E}_1} \left( K_I^2 + K_{II}^2 \right) \tag{4.32}$$

This relationship is equivalent to taking the interface modulus defined for the bilayer interface in (3.15) as $E_* = 2\bar{E}_2$, rather than (3.14). The relation (4.32) expresses the work released by the crack tip fields in a small increment of crack growth. The singular crack tip tractions do no work through the associated displacements on the beam side of the interface. Only $\bar{E}_2$ contributes to the interface modulus. The second formula in (4.11) is thereby replaced by

$$K_1 + iK_2 = \sqrt{\frac{\bar{E}_2}{\bar{E}_1}} \left( \frac{P_o}{\sqrt{2h}} - i\frac{\sqrt{6}M_o}{h^{3/2}} \right) e^{i\omega} \tag{4.33}$$

with the mode mix given by

$$\tan\psi = \frac{P_o h \tan\omega - 2\sqrt{3}M_o}{P_o h + 2\sqrt{3}M_o \tan\omega} \tag{4.34}$$

Thus, as seen for the general bilayer, the only missing information in determining the mode mix is knowledge of $\omega$.

With $v_2 = 1/2$, $\omega$ depends only on $\bar{E}_2/\bar{E}_1$ or on $\alpha_D = (\bar{E}_1 - \bar{E}_2)/(\bar{E}_1 + \bar{E}_2)$. If $\omega$ is known for one load combination $(P_o, M_o)$, it is known for all. In specifying the pair of coupled integral equations below for determining the interface tractions and $\omega$, the load combination $P_o > 0$ and $M_o = 0$ will be used. This particular loading is of special interest because it is applicable to the delamination of a uniformly tensioned film. For this loading, by (4.34), $\omega$ is the mode mix, that is, $\psi = \omega$.

The tractions on the interface, $0 < x < \infty$, must satisfy the following pair of coupled integral equations (with $v_2$ not limited to be $1/2$):

$$\int_0^X T(\tilde{X})d\tilde{X} + 6\bar{M}(X) - \frac{2}{\pi}\int_0^\infty \frac{T(\tilde{X})}{\tilde{X} - X}d\tilde{X} + \left( \frac{1 - 2v_2}{1 - v_2} \right) N(X) = 1 \tag{4.35}$$

$$12\int_X^\infty \bar{M}(\tilde{X})d\tilde{X} + \frac{2}{\pi}\left( \frac{\bar{E}_2}{\bar{E}_1} \right)\int_0^\infty \frac{N(\tilde{X})}{\tilde{X} - X}d\tilde{X} - \left( \frac{1 - 2v_2}{1 - v_2} \right)\left( \frac{\bar{E}_2}{\bar{E}_1} \right) T(X) = 0 \tag{4.36}$$

The following dimensionless quantities have been introduced:

$$X = \left(\frac{\bar{E}_2}{\bar{E}_1}\right)\frac{x}{h}; \quad [T(X), N(x))] = \left(\frac{\bar{E}_1}{\bar{E}_2}\right)\left(\frac{h}{P_o}\right)[t(x), n(x)] \qquad (4.37)$$

The moment distribution in the beam is given in terms of $T$ and $N$ by

$$\bar{M}(x) \equiv \frac{M(X)}{P_o h} = \left(\frac{\bar{E}_1}{\bar{E}_2}\right)\int_0^X N(\tilde{X})\left(\tilde{X} - X\right)d\tilde{X} + \frac{1}{2}\int_0^X T(\tilde{X})d\tilde{X} \qquad (4.38)$$

Supplementing the two integral equations are the conditions that no vertical force is applied to the beam and the moment vanishes as $x \to \infty$:

$$\int_0^\infty N(X)dX = 0 \quad \text{and} \quad \lim_{X \to \infty} \bar{M}(X) = 0 \qquad (4.39)$$

After mapping the semi-infinite interval onto the interval $-1 \le u \le 1$ with the transformation $X = (1 + u)/(1 - u)$, the equations (4.30)–(4.32) are reduced to a system of linear algebraic equations using a finite series representation of $(T, N)$:

$$(T(u), N(u)) = \frac{(1 - u)^k}{\sqrt{1 - u^2}}\sum_{j=0}^m (T_j, N_j)P_j(u) \qquad (4.40)$$

with $P_j(u)$ as the Chebyshev polynomial of degree $j$, and with $k \ge 4$ ensuring that the integrals converge for each contribution from the series. This numerical method is otherwise similar to that used by Shield & Kim (1992) but adapted here for the semi-infinite problem.

The numerical results for $\omega$ for the incompressible substrate are presented in Figure 4.8 for the range $1 \le \bar{E}_1/\bar{E}_2 \le 1000$, where they are compared with the full bilayer results from Figure 4.2 in the lower portion of the range $1 \le \bar{E}_1/\bar{E}_2 \le 5$. It is evident that already by $\bar{E}_1/\bar{E}_2 = 5$, the beam substrate model is quite accurate, and it will be argued at the end of this section that the accuracy of the model increases as $\bar{E}_1/\bar{E}_2$ increases. It is also evident from Figure 4.8 that $\omega \to \pi/2 \,(90°)$ as $\bar{E}_1/\bar{E}_2 \to \infty$. It does so with a scaling $(\bar{E}_1/\bar{E}_2)^{1/3}$, as will be discussed below. Recalling that $\psi = \omega$ for the loading $P_o > 0$ and $M_o = 0$, it follows that the delamination problem of a stiff film on a highly compliant incompressible substrate with this loading is dominantly mode II with a small positive mode I component. Conversely, by (4.34), for the loading $P_o = 0$ and $M_o > 0$, the problem is dominant mode I with a small negative component of mode II.

Distributions of the normalized shear and normal tractions for the loading $P_o > 0$ and $M_o = 0$ are plotted in Figure 4.9 using the mapping variable $u$. Recall that $x/h = (\bar{E}_1/\bar{E}_2)X$ and $X = (1 + u)/(1 - u)$. The distribution of the shear traction changes modestly when expressed in terms of $u$ or $X$ as $\bar{E}_1/\bar{E}_2$ increase from 1 to 100, and then it becomes essentially independent of $\bar{E}_1/\bar{E}_2$ for even larger values. This is the limit of the membrane model discussed in the next paragraph. However, the normal traction distribution collapses on $u = -1$ $(X = 0)$ as $\bar{E}_1/\bar{E}_2$ increases. The choice $x/h = (\bar{E}_1/\bar{E}_2)X$ is consistent with scaling for stretching of the beam due to shear transfer from the substrate. Had one chosen to scale consistent with beam bending due to normal tractions,

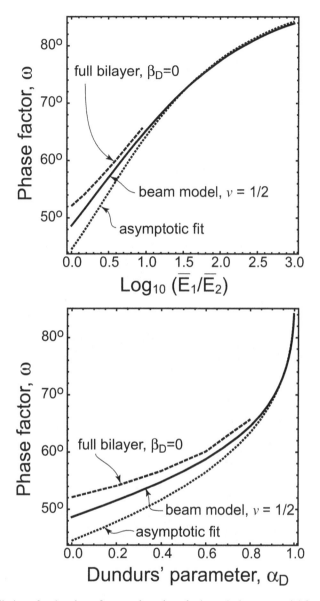

**Figure 4.8** Predictions for the phase factor $\omega$ based on the beam/substrate model for the case of an infinite deep incompressible substrate. Included for comparison are the results from Figure 4.2 based on the elasticity solution to the full bilayer and the fit based on the asymptotic formula (4.43).

one would instead have made the choice $x/h = \left(\bar{E}_1/\bar{E}_2\right)^{1/3} Z$. When $\bar{E}_1/\bar{E}_2$ is large, the portion of the beam over which shear tractions dominates is much larger than that in which bending is important. The difference in scalings is reflected in the shift of the normal distribution towards $u = -1$ in Figure 4.9. The scaling difference leads to the dependence of $\omega$ on $\left(\bar{E}_1/\bar{E}_2\right)^{1/3}$ for large $\bar{E}_1/\bar{E}_2$.

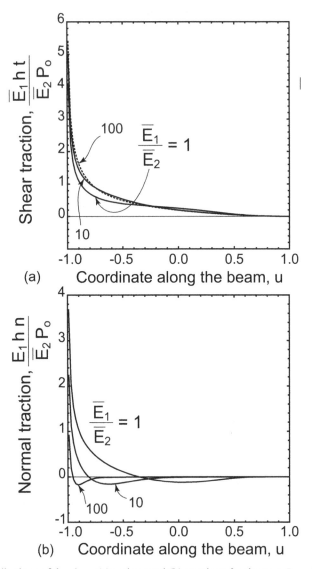

**Figure 4.9** Distributions of the shear (a) and normal (b) tractions for the case $P_o > 0$ and $M_o = 0$ with $v_2 = 1/2$. The coordinate $x$ maps to $u$ by $x/h = (\bar{E}_1/\bar{E}_2)X$ with $X = (1+u)/(1-u)$.

Specifically, focusing on the dependence for large $\bar{E}_1/\bar{E}_2$ and employing the two scaling dependencies just noted, one has

$$K_I = \lim_{x \to 0}\left(\sqrt{2\pi x}\right)n(x) \propto \left(\sqrt{\left(\frac{\bar{E}_1}{\bar{E}_2}\right)^{1/3}}Z\right)\left(\frac{\bar{E}_2}{\bar{E}_1}\right)N(Z) \propto \left(\frac{E_2}{\bar{E}_1}\right)^{5/6} \quad (4.41)$$

$$K_{II} = \lim_{x \to 0}\left(\sqrt{2\pi x}\right)t(x) \propto \left(\sqrt{\left(\frac{\bar{E}_1}{\bar{E}_2}\right)}X\right)\left(\frac{\bar{E}_2}{\bar{E}_1}\right)T(X) \propto \left(\frac{E_2}{\bar{E}_1}\right)^{1/2} \quad (4.42)$$

Thus, as $\bar{E}_1/\bar{E}_2$ becomes large, $K_I/K_{II} \propto (\bar{E}_2/\bar{E}_1)^{1/3}$. A simple asymptotic approximation to $\omega$ for large $\bar{E}_1/\bar{E}_2$ based on this scaling is

$$\omega \simeq \frac{\pi}{2} - 0.793\,(1 - \alpha_D)^{1/3} = \frac{\pi}{2} - 0.999\left(\frac{\bar{E}_2/\bar{E}_1}{1 + \bar{E}_2/\bar{E}_1}\right)^{1/3} \tag{4.43}$$

where the coefficient 0.793 was chosen to fit the numerical results at $\alpha_D = 0.95$. The prediction from this formula is included in Figure 4.8, where it is seen to give a good approximation if $\bar{E}_1/\bar{E}_2 > 10$.

The scaling noted above implies that the gradients of the tractions acting on the beam decrease as $\bar{E}_1/\bar{E}_2$ become large. These are precisely the conditions required for the increasing accuracy of beam theory as a model for the upper film layer, and this is the basis for asserting that the beam/substrate model becomes an increasingly accurate model of the elasticity bilayer problem as $\bar{E}_1/\bar{E}_2$ becomes large. The comparison of the two sets of results in Figure 4.8 illustrates this trend. For the specific case with $P_o > 0$, $M_o = 0$ and $v_2 = 1/2$, it is insightful to demonstrate that, in the limit as $\bar{E}_1/\bar{E}_2$ becomes large, the integral equation (4.35) reduces to the following integral equation involving only the shear traction:

$$\int_0^X T(\tilde{X})d\tilde{X} - \frac{2}{\pi}\int_0^\infty \frac{T(\tilde{X})}{\tilde{X} - X}d\tilde{X} = 1 \tag{4.44}$$

This is the so-called membrane model of the stress distribution in a film derived by accounting only for the streting strain in the film and shear traction at the interface, combined with the assumption that the interface lies along the mid-plane of the film, or equivalently, that the film is a membrane of zero thickness. To see that this equation is implied by (4.35) and (4.36) for large $\bar{E}_1/\bar{E}_2$, neglect the two terms multiplied by $\bar{E}_2/\bar{E}_1$ in (4.36) to obtain $\int_X^\infty \bar{M}(\tilde{X})d\tilde{X} = 0$. This is valid outside the bending boundary layer region with its left end at $X = 0$ and its length on the order of $(\bar{E}_2/\bar{E}_1)^{2/3}$ in the $X$ variable (on the order of $(\bar{E}_1/\bar{E}_2)^{1/3}h$ in the $x$ variable). This result, in turn, implies that $\bar{M}(X) = 0$ outside the bending boundary layer such that (4.36) follows immediately from (4.35). The membrane model applies only outside the boundary layer region. However, the detailed analysis of the beam model shows that the quantities of interest, including the stress intensity factors and the energy release rate, approach the predictions of the membrane model as $\bar{E}_1/\bar{E}_2$ becomes large. Further discussion of the membrane model and its relation to the beam model is given by Freund & Suresh (2003) and Shield & Kim (1992).

Several issues related to the solution in this section remain open, including the role of substrate compressibility, the applicability of the results for the semi-infinite substrate to bilayer systems with finite depth and width, and the role of substrate nonlinearity, which is more likely to be an issue for highly compliant substrates than for stiffer substrates.

Many soft materials are nearly incompressible with Poisson's ratios slightly below 1/2. For the elastic bilayer under plane strain, the second Dundurs' parameter (3.10)

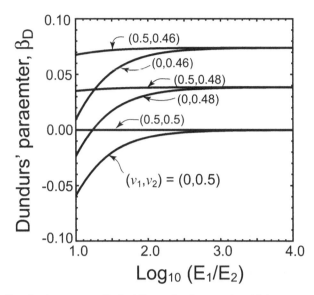

**Figure 4.10** The Dundurs' parameter $\beta_D$ for bilayers in plane strain with incompressible and nearly incompressible highly compliant substrates. $(E_1, v_1)$ pertains to the upper layer and $(E_2, v_2)$ to the substrate.

can be written as

$$\beta_D = \frac{1}{2}\left(\frac{1 - 2v_2 - \frac{E_2(1+v_1)(1-2v_1)}{E_1(1+v_2)}}{1 - v_2 + \frac{E_2(1-v_1^2)}{E_1(1+v_2)}}\right) \tag{4.45}$$

For large $E_1/E_2$, $\beta_D \to (1 - 2v_2)/[2(1 - v_2)]$, which is proportional to the coefficient of the terms omitted in the beam model in (4.35) and (4.36) when $v_2 = 1/2$. A plot of $\beta_D$ as a function of $E_1/E_2$ for $v_2 = 1/2$, 0.48 and 0.46 with $v_1 = 0$ or $1/2$ is presented in Figure 4.10. For these incompressible and nearly incompressible substrates, $\beta_D$ is relatively small, for example, much smaller than $\alpha_D/4$. This suggests that the approximation of substrate incompressibility may be a good one for soft materials with the benefit that the consideration of more complicated crack tip fields associated with nonzero $(1 - 2v_2)/[2(1 - v_2)]$ can be avoided, as has been discussed previously for problems with nonzero but small values of $\beta_D$. However, such decisions should be made for the specific problem at hand based on the level of accuracy required.

Examination of the tractions in Figure 4.9 for the problem with $P_o > 0$ and $M_o = 0$ suggests that the shear traction drops to almost zero for $u > 1/2$, corresponding to $x/h > 3(\bar{E}_1/\bar{E}_2)$. For approximate applicability of the semi-infinite results in this section to a film of finite width $L$ one therefore anticipates that the width should should satisfy $L/h > 3(\bar{E}_1/\bar{E}_2)$. For soft substrate systems $\bar{E}_1/\bar{E}_2 > 10^3$ is not uncommon, and thus the film must be very wide relative to its thickness to be considered infinitely wide. If not, the finite width of the film must be modeled. Similarly, highly compliant substrates place constraints on the depth of the substrate that can be regarded as infinitely deep. With $H$ as a the substrate depth, the general bilayer solution in Section 4.1 indicates the approximation of an infinitely deep substrate becomes valid when $(\bar{E}_1/\bar{E}_2)(h/H) \ll 1$. Thus, the substrate depth must satisfy a condition such as $H/h > 10(\bar{E}_1/\bar{E}_2)$ for the results of this section to be applicable.

To our knowledge, there have been no nonlinear studies along the lines of the present problem published for film/substrate systems where the substrate is modeled with a nonlinear constitutive law applicable to elastomeric materials, such as a neo-Hookean model, within a finite strain framework. The question arises as to the range of validity of the linear results in this section for stiff films on soft material substrates with large values of $\bar{E}_1/\bar{E}_2$. The following estimate of the shear strain in the substrate in the vicinity of the crack tip provides insight to this question. Consider again the loading with $\sigma_o = P_o/h$ and $M_o = 0$ in the range of large $\bar{E}_1/\bar{E}_2$, such that the crack is nearly mode II. Then, by (4.41) and (4.42), $K_{II} = \sqrt{\bar{E}_2/\bar{E}_1} \cdot \sigma_o\sqrt{h}$ and the shear stress at the interface near the tip is

$$\sigma_{12} = \frac{K_{II}}{\sqrt{2\pi x}} = \sigma_o\sqrt{\frac{1}{2\pi}\frac{\bar{E}_2}{\bar{E}_1}\frac{h}{x}} \tag{4.46}$$

It follows that the shear strain in the substrate at the interface near the tip is (with $\nu_2 = 1/2$)

$$\gamma_{12} = \frac{4\sigma_{12}}{\bar{E}_2} = \frac{2\sqrt{2}}{\sqrt{\pi}}\frac{\sigma_o}{\bar{E}_1}\sqrt{\frac{\bar{E}_1}{\bar{E}_2}\frac{h}{x}} \tag{4.47}$$

We evaluate the strain in the substrate at $x = h$, which is physically relevant for delamination, and furthermore, the beam/substrate model is not accurate for smaller $x$. One has

$$\gamma_{12} = 1.60\frac{\sigma_o}{\bar{E}_1}\sqrt{\frac{\bar{E}_1}{\bar{E}_2}} \quad \text{at } x = h \tag{4.48}$$

For a metal or ceramic film on an elastomeric substrate, the largest value of $\sigma_o/\bar{E}_1$ might be 0.01, while a more typical value would be 0.001. With $\sigma_o/\bar{E}_1 = 0.001$ and $\bar{E}_1/\bar{E}_2 = 10^3$, the shear strain (4.39) is $\gamma_{12} = 0.05$; with $\sigma_o/\bar{E}_1 = 0.01$ and $\bar{E}_1/\bar{E}_2 = 10^3$, it is $\gamma_{12} = 0.5$. The latter case probably exceeds the range of applicability of the linear theory, but the former case is almost certainly well within the range of linear applicability. Experience with other applications employing nonlinear elastomeric constitutive laws such as the neo-Hookean model reveals that the range in which linear theory is reasonably accurate usually extends to strains well above 0.1.

# 5   Steady-State Delamination in Multilayers

In this chapter, the framework to analyze a multilayer stack of blanket films is presented. Emphasis is placed on stacks with piecewise-linear distributions of misfit, or 'eigenstrain', strain variations within each layer allowing for the possibility of discontinuities from layer to layer. In addition to being applicable to thermal problems with steady-state thermal distributions through the multilayer, the formulation encompasses layers with residual processing strains which vary from layer to layer and possibly within layers. The principle focus of this chapter is on computing the steady-state energy release rate for a semi-infinite crack. The results of the analysis are algebraic but nevertheless will usually require some computation. As the framework is applicable to any number of layers, it provides the basis to derive the results presented in Chapter 4 for bilayers.

In Section 5.1, basic results for the stresses and strains in a multilayer subject to overall stretching and bending, together with internal misfit strains, are derived. Section 5.2 makes use of the basic results to derive general results for delamination energy release rates in the absence of any misfit strains. Section 5.3 gives an alternative derivation which can be used to predict the energy release rate under general conditions, including misfit strains, and for the special case where the stresses prior to cracking arise under equi-biaxial conditions but the delamination occurs under plane strain conditions. Section 5.4 discusses the computation of the mode mix for the general multilayer. The derivation in Section 5.3 also leads to a bilayer approximation in Section 5.5, which can be used to estimate both the energy release rate and mode mix. Illustrative examples of multilayer analysis are presented in Section 5.6.

Software that implements the specific framework described in this chapter is described in Chapter 14, which provides additional examples beyond those in this chapter. Computation of the mode mix for general multilayer problems requires finite element analysis, which is described in Chapter 16. However, several illustrative examples of mode mix in multilayers are provided here in this chapter, providing some insight regarding the role of layer properties.

## 5.1   Stresses and Strains in a Multilayer Subject to Overall Stretch and Bending with Internal Misfit Strains

The multilayer is a stack of $N$ layers bonded together with each layer having uniform thickness $h_i$ (with $i = 1, N$), using the numbering convention illustrated in Figure 5.1(a).

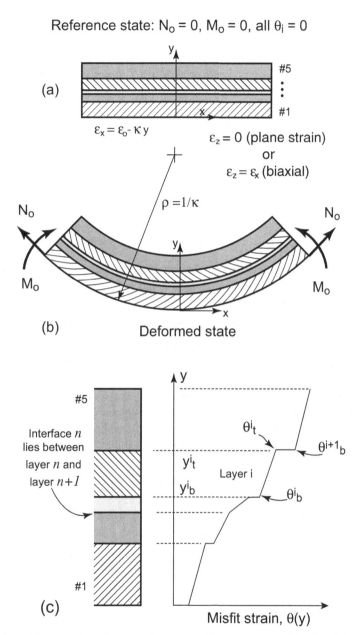

**Figure 5.1** A multilayer stack in the stress-free state, without any external loads and thermal strains equal to zero. (b) The deformed state of the multilayer stack, which experiences stretching and bending as a result of thermal mismatch and external loads, (c) The definition of misfit strains in the layers; the misfit strain in each layer is a piecewise-linear function, with discrete jumps at interfaces. The layer numbering starting at the bottom layer as no. 1 follows that used in the accompanying code. It differs from the numbering for the bilayer in Section 4, which identifies the top layer as no. 1.

The elastic properties of each layer are taken to be isotropic, and the modulus and Poisson's ratio of the $i^{th}$ layer are given by $E_i$ and $v_i$. The total thickness of a finite thickness multilayer is assumed to be small compared to its length, such that it can be considered as a multilayered beam or wide plate. As depicted in Figure 5.1(b), the formulation accommodates a $z$-independent end force/depth $N_o$ and moment/depth $M_o$, as well as misfit strains (or eigenstrains) in each layer $\theta_i(y)$, which may have a piecewise-linear variation in the $y$-direction. The formulation covers the following cases:

- plane stress (the depth in the $z$-direction is small compared to the total thickness)
- plane strain (the depth in the $z$-direction is large compared to the thickness, and the multilayer is constrained against overall straining in the $z$-direction and against bending about the $x$-axis)
- equi-biaxial strain (the depth in the $z$-direction is large compared to the thickness, and the multilayer is unconstrained in the $z$-direction and is free to bend about the $x$-axis)

The following derivation invokes the small deformation Bernoulli-Euler assumption for a beam or wide plate under a combination of bending and stretching described in Chapter 2; plane sections remain plane, and the square of slopes are small compared to total strain. The total strain through the multilayer varies according to

$$\epsilon_x = \epsilon_o - \kappa \cdot y \tag{5.1}$$

where $y$ is zero at the bottom of the stack as shown in Figure 5.1, the curvature $\kappa$ is defined as positive when the multilayer curves upwards, and $\epsilon_o$ is the total strain at $y = 0$. Within the context of the three case noted above, (5.1) produces exact elasticity solutions for the class of problems considered here, and this is the basis for the fact that all of the energy release rates derived in this chapter are exact elasticity results. In all three cases being considered, the shear stresses and the normal stresses $\sigma_y$ are zero. The out-of-plane stress and strain components, $\sigma_z$ and $\epsilon_z$, depend on the particular case and will be specified below.

With $\sigma_x(y)$ as the stress distribution through the multilayer, equilibrium requires

$$N_o = \int_{bottom}^{top} \sigma_x(y)dy; \quad M_o = -\int_{bottom}^{top} \sigma_x(y)ydy \tag{5.2}$$

where $N_o$ and $M_o$ are the applied force/depth and moment/depth shown in Figure 5.1(b). Note that $M_o$ is defined about the bottom of the multilayer ($y = 0$) and not about the neutral axis of the stack.

The final element of the formulation is the relationship between the stress and the strain, which depends on whether one chooses plane stress, plane strain or equi-biaxial strain as a modeling assumption. Table 2.1 defines two coefficients for each layer for each of these three cases which will be used in what follows. As previously noted, the misfit strain within each layer is taken to be piecewise linear and for the $i^{th}$ layer is defined using the notation in Figure 5.1 as

$$\theta_z = \theta_x \equiv \theta_b^i + \frac{y - y_b^i}{h_i}\left(\theta_t^i - \theta_b^i\right) \tag{5.3}$$

where $y_b^i$ is the position of the bottom of the $i^{th}$ layer and $y_t^i$ is the top of the $i^{th}$ layer; $\theta_b^i$ is the misfit strain the bottom of the $i^{th}$ layer and $\theta_t^i$ is the misfit strain at the top of the $i^{th}$ layer. A misfit strain component $\theta_y(y)$ has no effect on any of the results being considered.

The total strain in the layer is the sum of the elastic strain and the misfit strain:

$$\epsilon_x = \epsilon_o - \kappa \cdot y = \epsilon_x^e + \theta \tag{5.4}$$

The coefficients listed in Table 2.1 for the three cases considered here (plane stress, plane strain and equi-biaxial strain) are defined such that in each layer the three-dimensional constitutive relation gives

$$\epsilon_x = \frac{\sigma_x}{\bar{E}} + \bar{c}\theta; \qquad \sigma_x = \bar{E}\left(\epsilon_o - \kappa \cdot y - \bar{c}\theta\right) \tag{5.5}$$

Later it will be important to distinguish between two types of possible delaminations for the case of equi-biaxial straining: equi-biaxial and plane strain. Additional notation will be introduced to deal with the fact that one condition can pertain to prestraining and another to delamination.

Combining (5.2), (5.3) and (5.5) gives two linear equations governing $\epsilon_o$ and $\kappa$:

$$\begin{aligned} a_{11}\epsilon_o + a_{12}\kappa &= b_1 + N_o \\ a_{21}\epsilon_o + a_{22}\kappa &= b_2 + M_o \end{aligned} \tag{5.6}$$

where the coefficients are defined by

$$a_{11} = \sum_{i=1}^{N} \bar{E}_i h_i$$

$$a_{12} = a_{21} = -\sum_{i=1}^{N} \bar{E}_i \left( \frac{\left(y_t^i\right)^2 - \left(y_b^i\right)^2}{2} \right)$$

$$a_{22} = \sum_{i=1}^{N} \bar{E}_i \left( \frac{\left(y_t^i\right)^3 - \left(y_b^i\right)^3}{3} \right) \tag{5.7}$$

$$b_1 = \sum_{i=1}^{N} \frac{\bar{c}_i \bar{E}_i h_i}{2} \left( \theta_b^i + \theta_t^i \right)$$

$$b_2 = -\sum_{i=1}^{N} \bar{c}_i \bar{E}_i h_i \left( \frac{h_i}{3} \left( \theta_t^i - \theta_b^i \right) + \frac{1}{2} \left( \theta_t^i y_b^i + \theta_b^i y_t^i \right) \right)$$

and

$$y_b^1 = 0; \quad y_t^i = y_b^i + h_i \text{ for } 1 \le i \le N; \quad y_b^{i+1} = y_t^i \text{ for } 1 \le i \le N-1 \tag{5.8}$$

If the applied loads $N_o$ and $M_o$ vanish, the overall strain and curvature of the multilayer is due to the misfit strains through $b_1$ and $b_2$. For equi-biaxial straining, the applied loads $N_o$ and $M_o$ act on the ends of the multilayer in both the $x$-direction and the $z$-direction.

Once the elongation and curvature of the stack are computed from (5.6)–(5.8), the stress in each layer can be computed from (5.5).

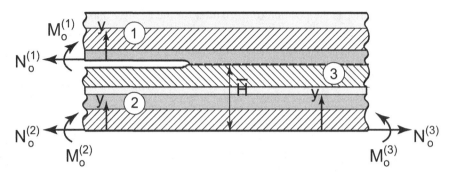

**Figure 5.2** General problem for steady-state delamination of a multilayer showing the submultilayers, no. 1 and no. 2, and the full multilayer, no. 3.

It will also be useful to obtain an expression for the location of the neutral bending axis as identified by the distance $y = \Delta$ from the bottom of the multilayer. Bending and stretching decouple for a force/moment pair applied about the neutral axis. With $\epsilon_N$ as the strain of the neutral axis, the position of the neutral axis and the transformation of (5.6) are

$$\Delta = -\frac{a_{12}}{a_{11}}; \quad a_{11}\epsilon_N = b_1 + N_o; \quad \left(a_{22} - \frac{a_{12}^2}{a_{11}}\right)\kappa = b_2 + M_o + N_o\Delta \qquad (5.9)$$

While all of the coefficients listed above are relatively simple algebraic expressions, it seldom makes sense to attempt to reduce them further by algebraic manipulation for stacks of three layers or more. They are readily evaluated numerically.

## 5.2     Energy Release Rates in the Absence of Any Misfit Strains: All $\theta^i = 0$

We now carry out basic energy accounting to derive expressions for the steady-state energy release rate of a crack propagating along any one of the interfaces of the multilayer or within the interior of any layer parallel to the interfaces. The derivation is exact, and it is also the basis for results reported for the bilayer in Chapter 4. Figure 5.2 depicts an infinitely long multilayer with a semi-infinite delamination crack propagating along one of the interfaces between the layers. The approach and the software also accommodate the computation of the energy release rate and mode mix of a crack propagating within any one of the layers. One simply replaces that layer by two layers having the same properties and the same combined thickness with the new interface chosen to coincide with the desired crack plane.

The discussion that follows deals with the generic situation depicted in Figure 5.2. The forces/depth and moments/depth, $N_o^{(i)}$ and $M_o^{(i)}$ (for $i = 1, 3$), are regard as being prescribed, and the misfit strains $\theta^i(y)$ in 5.3 are assumed to be zero. Overall equilibrium requires

$$N_o^{(1)} + N_o^{(2)} = N_3^{(3)} \text{ and } M_o^{(1)} + M_o^{(2)} - N^{(1)}\bar{H} = M_o^{(3)} \qquad (5.10)$$

where $\bar{H}$ is the thickness of the lower separated portion of the multilayer.

The following energy accounting for steady-state advance of the crack in Figure 5.2 is exact. Denote the elastic energy per unit length of the multilayer per depth in each of the three sections far ahead and far behind the crack tip by $U^{(i)}$, as labeled where

$$U^{(i)} = \frac{1}{2} \int_{bot}^{top} \sigma_{ij}^{(i)} \epsilon_{ij}^{e(i)} dy \text{ for } i = 1, 3 \tag{5.11}$$

Each of the three sections is a multilayer itself, or possibly a single layer, and the integration is from the bottom to the top of the submultilayer making up the respective section in Figure 5.2. The formulas in Section 5.1 generate separate results $(\epsilon_o^{(i)}, \kappa^{(i)}, U^{(i)})$ for each of the three sections. When a crack advances a unit length, the elastic energy change per unit depth in the $z$-direction of the system is $\Delta U = U^{(1)} + U^{(2)} - U^{(3)}$ because the advance of the crack is equivalent to trading a unit slice far ahead of the tip with a unit slice far behind the tip.

For a system such as being considered here with prescribed loads and zero misfit strains, the change in potential energy of the system – the change in strain energy plus the change in potential energy of the loads – is the negative of the change in strain energy. Thus, the energy release rate for the delamination crack is given by

$$G = -\frac{\partial \text{Potential Energy}}{\partial a} = U^{(1)} + U^{(2)} - U^{(3)} \tag{5.12}$$

For the three cases, (5.11) is

$$U^{(i)} = \frac{j}{2} \int_{bot}^{top} \bar{E} \left( \epsilon_o^{(i)} - \kappa^{(i)} y \right)^2 dy \tag{5.13}$$

where the integration is through the thickness of the respective submultilayer with $y$ measured from its bottom. The coefficient $\bar{E}$ is given in Table 2.1, and $j = 1$ for plane stress and plane strain and $j = 2$ for equi-biaxial strain.

For the equi-biaxial case, the prescription above provides the energy release rate for prestresses generated under equi-biaxial conditions with the delaminated segments far behind the crack front also subject to equi-biaxial conditions. The more important case where the stresses in the uncracked layer are generated by equi-biaxial straining but the cracking process occurs under a constraint of plane strain requires special attention. Equation (5.12) applies in this case, but (5.13) does not. It will be expedient to devote an entire section (Section 5.3) to an alternative derivation which applies to this special case and to the development of a bilayer approximation to the present multilayer problem.

The energy release rate provided by the above analysis is exact due to the fact that the Euler-Bernoulli theory generates elasticity solutions for the multilayer far ahead and far behind the crack tip for the loads considered. While the solution is closed form and algebraic, it is generally necessary to generate the solutions numerically, and the accompanying software provides this option. The problem for the mode mix at the crack tip requires the solution of a more complicated elasticity problem. For the multilayer delamination crack, the software accompanying the text employs a finite element method to compute both the energy release rate and the mode mix $\psi$. If the crack lies in the interior of one of the layers, one of the possibilities noted earlier, then the stress intensity factors are conventional and $\psi = \tan^{-1}(K_{II}/K_I)$. If the crack lies on the interface between two dissimilar materials, then the crack tip singularity is that associated with the specific bimaterial interface, as detailed in Chapter 3. Use of the software to generate results will be illustrated later in this chapter.

## 5.3  An Alternative Derivation for Energy Release Rates Including Misfit Strains: Application to Equi-biaxial Stressing Followed by Plane Strain Cracking

The derivation in the previous section does not account for misfit strains, and it does not apply to a special case of interest: a multilayer stressed under equi-biaxial conditions which then delaminates under plane strain conditions. In the latter case, stressing prior to delamination is controlled by the biaxial moduli, $\bar{E} = E/(1 - v)$ (see Table 2.1), while delamination is governed by the plane strain moduli, $\bar{E} = E/(1 - v^2)$. The alternative derivation for the energy release rate given below is general, applying to the cases already considered including misfit strains, and it applies as well to the special case. This alternative approach also leads directly to a bilayer approximation for multilayer delamination given in Section 5.5 which can be used to estimate the mode mix.

The multilayer delamination problem to be solved is shown at the bottom of Figure 5.3 with the misfit strain distribution $\theta(y)$ in (5.3) and the prescribed force/moment pairs shown, $(N_o^{(i)}, M_o^{(i)})$ for $i = 1, 3$, where overall equilibrium requires the pairs satisfy (5.10). The uncracked multilayer in the top part of Figure 5.3 is subject to the same misfit distribution $\theta(y)$ plus the prescribed moment pair $N_o^{(3)}$ and $M_o^{(3)}$. The solution to the problem in the bottom part of the figure for the stresses and strains is the superposition of the solutions to the problems in the top and middle part of the figure. Note that the middle problem specifically addresses the changes in the stresses and strains due to delamination, and it is this problem that gives rise to the stress intensity factors. The misfit strains do not enter into the middle problem except through the applied loads, which are

$$\Delta N_o^{(1)} = N_o^{(1)} - N_{upper}^{(3)}; \quad \Delta N_o^{(2)} = N_o^{(2)} - N_{lower}^{(3)}$$
$$\Delta M_o^{(1)} = M_o^{(1)} - M_{upper}^{(3)}; \quad \Delta M_o^{(2)} = M_o^{(2)} - M_{lower}^{(3)} \tag{5.14}$$

Here, $N_o^{(3)}$ and $M_o^{(3)}$ have been split into an equivalent load system with upper and lower contributions defined by

$$N_{lower}^{(3)} = \int_{bottom}^{\bar{H}} \sigma_x^{(3)} dy; \quad N_{upper}^{(3)} = \int_{\bar{H}}^{top} \sigma_x^{(3)} dy$$
$$M_{lower}^{(3)} = -\int_{bottom}^{\bar{H}} \sigma_x^{(3)} y dy; \quad M_{upper}^{(3)} = -\int_{\bar{H}}^{top} \sigma_x^{(3)} (y - \bar{H}) dy \tag{5.15}$$

The load system for the middle problem also satisfies overall equilibrium ensuring that

$$\Delta N_o^{(1)} + \Delta N_o^{(2)} = 0; \quad \Delta M_o^{(1)} + \Delta M_o^{(2)} - \Delta N_o^{(1)}\bar{H} = 0 \tag{5.16}$$

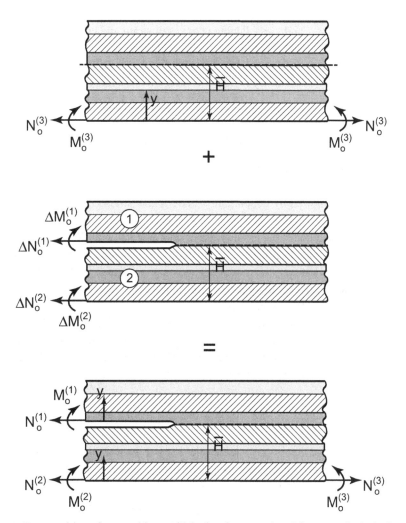

**Figure 5.3** Superposition of two problems which give the general multilayer results including the effect of misfit strains. The stress intensity factors and, therefore, the energy release rate are generated by the middle problem.

Because the stress intensity factors are associated with the middle problem, attention can be directed exclusively to this problem for the solution for the energy release rate and the mode mix. Note that the middle problem is a special case to the solution given in the previous section in connection with Figure 5.2 determined with $\theta(y) = 0$ and the load system $(\Delta N_o^{(1)}, \Delta M_o^{(1)})$, $(\Delta N_o^{(2)}, \Delta M_o^{(2)})$ and $(\Delta N_o^{(3)}, \Delta M_o^{(3)}) = 0$. It follows that the energies per length per depth associated with the middle problem are given by $U^{(3)} = 0$ and

$$
\begin{aligned}
U^{(1)} &= \frac{j}{2} \int_{\bar{H}}^{top} \bar{E} \left( \Delta \epsilon_o^{(1)} - \Delta \kappa^{(1)} \left( y - \bar{H} \right) \right)^2 dy \\
U^{(2)} &= \frac{j}{2} \int_{bottom}^{\bar{H}} \bar{E} \left( \Delta \epsilon_o^{(2)} - \Delta \kappa^{(2)} y \right)^2 dy
\end{aligned}
\tag{5.17}
$$

where $\Delta \epsilon_o^{(i)}$ and $\Delta \kappa^{(i)}$ denote changes in strain and curvatuve for each of the two sub-multilayers in the delaminated region defined and computed in the manner given in the previous section. As before $j = 1$ for plane stress and plane strain, while $j = 2$ for equi-biaxial strain. By (5.12), the energy release rate is

$$G = U^{(1)} + U^{(2)} \tag{5.18}$$

For the three basic cases (plane stress, plane strain and equi-biaxial strain), this approach generates the same values of $G$ as the approach given in the previous section, but now including the misfit strains if they are present. While the approach in the previous section was not applicable to the special class of problems with equi-biaxial strain states followed by delamination under plane strain conditions, the present approach is applicable. For this special case, the load quantities $N_{lower}^{(3)}$, $N_{upper}^{(3)}$, $M_{lower}^{(3)}$ and $M_{upper}^{(3)}$ are computed under the assumption of equi-biaxial conditions using the definitions $\bar{E}$ and $\bar{c}$ from Table 2.1; these terms include the dependence on the misfit strain $\theta(y)$ if it is present. However, the modulus governing the change in stresses associated with delamination in the middle problem in Figure 5.3 is governed by the plane strain modulus such that in (5.17) one must take $\bar{E} = E/(1 - v^2)$ and $j = 1$.

As an aside, it is worthwhile reemphasizing a point made in Section 4.3. The approach detailed in this section based on the stress changes and their energies associated with delamination, that is, the middle problem of Figure 5.3, provides insight as to why the classical formula for plane strain delamination of a film of thickness $h$ on a deep substrate

$$G = \frac{(1 - v^2)\sigma^2 h}{2E} \tag{5.19}$$

is valid for conditions whether the stress is generated under equi-biaxial conditions or under plane strain conditions. In both cases, the *change* in stress in the $x$-direction and the stress intensity factors are governed by the same problem.

## 5.4    Computation of Mode Mix for Steady-State Delamination in Multilayers

The general framework needed to compute the mode mix for any *bilayer* was given in Section 4.1, with simplified results for extremely thick substrates described in Section 4.2. While the results presented for those cases are comprehensive, it should always be kept in mind that numerical solutions to the associated elasticity problem are required to compute mode mix.

Since the general multilayer problem involves too many variables to tabulate results, the software accompanying this book is designed to enable computation of mode mix in as straightforward a manner as possible. A complete description of the underlying numerical method is provided in Chapter 16. However, it is worth commenting on a few of the considerations that come into play when computing mode mix.

First, as finite elements serve as the underlying computational framework, the computation of mode mix requires specification of specimen width and crack length, which are not required to compute the steady-state energy release rate. For very long cracks relative to the largest thickness of any sublayer, and very wide specimens relative to that dimension, the results asymptote to become independent of the chosen dimensions. However, the absolute sizes required to reach this limit may depend on the elastic properties of the layers: generally speaking, when the crack length and remaining ligament width are more than 10 times the thickness of the largest sublayer, the results typically change less than a few percent as the dimensions are increased.

Second, in strict terms, the results depend only on the Dundurs' parameters for problems with traction boundary conditions; for problems with internal misfit strains, the results have a very weak dependence on the Poisson's ratios of the layers. That is, two problems with the same Dundurs' parameters but different Poisson's ratios may yield slightly different results. This is not a significant effect in nearly all circumstances, with computed mode-mix varying only by a few degrees as a result of changes in Poisson's ratio. However, it is important to keep in mind if one seeks to make detailed comparisons between results generated with the present software and those reported elsewhere.

Third, the software described in Chapter 16 explicitly computes the stress intensity factors $K_1$ and $K_2$ for plane strain conditions. One should carefully consider the relevant theory for comparison, taking into account the discussion in Section 5.3. Computation of the phase angle describing mode mix, $\psi$, requires application of the framework described in Section 3.5. That is, one utilizes (3.22) with the complex stress intensity factor $K = K_1 + iK_2$ and a user-defined lengthscale $\ell$.

## 5.5    A Bilayer Approximation and Approximate Estimates of the Mode Mix

The fact that the stress intensity factors and energy release rate of the delamination crack are given by the solution to the middle problem of Figure 5.3 opens the way to the development of a bilayer approximation that can exploit the formulas in Chapter 4 as a way to estimate the mode mix. It should be emphasized that the software accompanying the text computes the mode mix for plane stress or plane strain along the lines discussed in Section 5.4 for the actual multilayer and does not make use of the approximation developed below. However, the approximation developed below may be useful on its own merits for certain modeling purposes, and the discussion will provide insights as to when the mode mix is significantly dependent on the properties of the layers.

The middle problem of Figure 5.3 involves submultilayers, while that for the bilayer in Figure 4.2(b) involves individual isotropic layers; otherwise the two problems are similar. The first step in developing the bilayer approximation is to replace the multilayers in Figure 5.3 by single isotropic layers with equivalent stretching and bending stiffness. The elastic properties of a multilayer are generally orthotropic, not isotropic, so this step necessarily introduces an approximation. The approximation does not affect the overall bending and stretching stiffness in the $x$-direction because these will be precisely reproduced. It does, however, approximate the elastic stiffness in the other two directions as well as the effective in-plane shear stiffness, all of which have some influence on the mode mix. For this reason, there is no point in attempting to assign a definitive value to the second Dundurs' parameter for the bilayer replacement, and for simplicity the choice $\beta_D = 0$ will be made here.

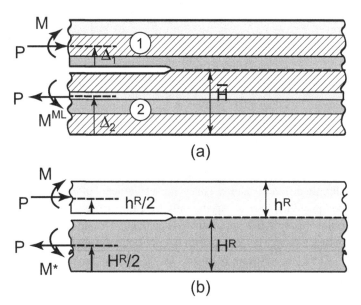

**Figure 5.4** (a) Loads for the multilayer delamination problem; (b) loads for the delamination problem for the the replacement bilayer.

The approximation requires the identification of equivalent isotropic layers for the two submultilayers above and below the delamination in the middle of Figure 5.3. Taking note of the notation employed in the bilayer solution in Figure 4.1(b) and in Figure 5.4(b), let $\bar{E}_1^R$ and $h^R$ be the modulus and thickness of the isotropic replacement layer above the delamination, and let $\bar{E}_2^R$ and $H^R$ be the corresponding values for the lower replacement layer. Here $\bar{E}$ denotes the plane stress or plane strain modulus governing the change in stress occurring during delamination. Next, transform the load pairs in the middle of Figure 5.3 to the system shown in Figure 5.4(a). Here, $\Delta_1$ and $\Delta_2$ define the neutral bending axis of each of the two submultilayers as specified in (5.9), and the shifted equilibrium load system is

$$P = -\Delta N_o^{(1)}; \quad M = \Delta M_o^{(1)} + \Delta N_o^{(1)} \Delta_1; \quad M^{ML} = -\Delta M_o^{(2)} - \Delta N_o^{(2)} \Delta_2 \quad (5.20)$$

Note that in this chaper, $\Delta_1$ and $\Delta_2$ have dimensions of length, as shown in Figure 5.3, while for the bilayer problem in Chapter 4, $\Delta$ is dimensionless with $h\Delta$ giving the location of the neutral axis in Figure 4.1. The requirement that a replacement layer reproduces the stretching and bending stiffness of a multilayer in (5.9) is

$$\bar{E}h = a_{11}; \quad \frac{\bar{E}h^3}{12} = a_{22} - \frac{a_{12}^2}{a_{11}}$$

$$\rightarrow h = \sqrt{\frac{12\left(a_{22} - \frac{a_{12}^2}{a_{11}}\right)}{a_{11}}}; \quad \bar{E} = \sqrt{\frac{a_{11}^3}{12\left(a_{22} - \frac{a_{12}^2}{a_{11}}\right)}} \qquad (5.21)$$

These formulas generate $\bar{E}_1^R$, $h^R$, $\bar{E}_2^R$ and $H^R$ associated with the replacement bilayer in Figure 5.4(b).

The final step in developing the bilayer approximation is to identify the load pairs shown in Figure 5.4(b). The load pair acting on the upper replacement layer is taken to be $(P, M)$, with values defined in (5.20). The load pair acting on the lower replacement layer cannot be taken as $(P, M^{ML})$ because the bilayer would not satisfy overall moment equilibrium. With the pair $(P, M_*)$ shifted to the neutral axis of the lower replacement layer, overall moment equilibrium is satisfied with

$$M_* = M + P\frac{\left(h^R + H^R\right)}{2} \tag{5.22}$$

In summary, the problem for the bilayer replacement defined in Figure 5.4(b) is precisely the bilayer problem analyzed in Figure 4.1(b) in Chapter 4 with $\alpha_D = (\bar{E}_1^R - \bar{E}_2^R)/(\bar{E}_1^R + \bar{E}_2^R)$ and $\beta_D = 0$. The results for the energy release rate and the mode mix presented in that chapter provide an approximation for the corresponding multilayer problem. Because $M_*$ for the bilayer is different from $M^{ML}$ for the multilayer, the energy release rate of the bilayer replacement is usually an approximation. However, for the case in which the lower section of the multilayer is infinitely deep ($\bar{H} \rightarrow \infty$), the bilayer approximation for G is exact. Thus, the energy release rate for the bilayer replacement is exact for delamination of multilayer thin films and coatings on thick substrates.

## 5.6    Examples of Delamination in Multilayers

It should be evident from the general solution for the bilayer problem in Chapter 4 that the task of generating and presenting a complete set of results for the general *trilayer* delamination problem is quite formidable, given the large number of dimensionless parameters characterizing the problem. Nevertheless, some special results for trilayers have been published, focusing, in part, on delamination of a face sheet in a three-layer sandwich plate (Kardomateas et al. 2013).

In this section, we use the above formulations and software accompanying this book to illustrate fundamental behaviors in two different multilayer applications: the 'superlayer' test used to probe the interface toughness of thin films, and a thermal barrier coating system. Following these examples, the asymptotic behavior of a very thin layer sandwiched between two outer layers is used to shed light on the extent to which a thin intermediate layer influences the mode mix.

### The 'Superlayer' Test: Debonding in a Trilayer

Many applications utilize very thin films as adhesion layers: for example, titanium or chromium films in microelectronic multilayers. It can be very challenging to measure the interface toughness between these films and the substrate (typically silicon or silicon oxide), because the inherent driving force for delamination of the film by itself scales with its thickness and can be quite small. It is nonetheless important to quantify the interface toughness between the adhesion layer and substrate, as this interface can control stability in the actual device.

**Table 5.1** Properties Used in the Super-Layer Example

| Layer | Thickness | $E$ | $\upsilon$ | $\alpha$ (ppm/$^\circ$C) | $\theta$ |
|---|---|---|---|---|---|
| Silicon | 500 μm | 170 GPa | 0.20 | 3.1 | 0 |
| Titanium | 50–200 nm | 110 GPa | 0.30 | 9.8 | $7.84 \times 10^{-4}$ |
| Chromium | 1–20 μm | 279 GPa | 0.21 | 10.5 | $\frac{(1-v)\sigma_o}{E}$ |

The 'superlayer' test is designed to extract the interface toughness between very thin films and substrates by artificially increasing the driving force through the deposition of an additional layer (the superlayer), which has large deposition stresses (Bagchi & Evans 1996). In a sense, the superlayer serves as an idealized proxy for the device attached to the adhesion layer. The stresses induced by the superlayer are then exploited to drive delamination at the desired interface. In order to be effective, the superlayer has to be nonreactive with the adhesion layer and have well-known and controllable deposition stresses; chromium and nickel are two commonly used materials. The stress in the superlayer is typically controlled by the partial pressure of argon during deposition; a calibration curve relating stress to argon pressure is generated using wafer curvature measurements (see, e.g., Zheng & Sitaraman 2007).

Figure 5.5 illustrates various results pertaining to delamination in a typical 'superlayer' system: a silicon wafer coated with a thin titanium adhesion layer, which is then coated with a chromium superlayer. Table 5.1 summarizes the properties used in the analysis. In this example, a temperature change is defined for the titanium layer that generates a biaxial stress of 150 MPa, which is representative of typical stresses present after deposition. The change in temperature of the chromium superlayer is defined such that it generates a equi-biaxial misfit stress given by $\sigma_o = E\theta/(1 - v)$.

Figure 5.5(a) shows the steady-state energy release rate for delamination between the silicon substrate and titanium layer, as a function of superlayer thickness. Multiple curves are shown for different adhesion layer thickness: if the superlayer on top is of comparable thickness, its high state of stress dominates the response. The energy release rate scales linearly with the thickness of the superlayer (Figure 5.5(a)) and with the square of the stress in the superlayer (Figure 5.5(b)). The phase angle describing the mode mix is shown in Figure 5.5(c) and 5.5(d), as computed from numerical analyses described in Chapter 16. Figure 5.5(c) illustrates that the relative thickness of the layers and their state of stress strongly influence the mode mix. This is a consequence of the edge load and moment resultants in the released bilayer that is formed by the delamination (i.e., the Ti-Cr bilayer).

Relevant interface toughnesses for such systems are typically in the range of 1–10 J/m$^2$; Figure 5.5 clearly illustrates that superlayer thickness must be in the range of 1–10 μm, with deposition stresses in the range of 0.3–1.0 GPa. The results illustrate that the residual stress and thickness of the titanium adhesion layer have no bearing on the driving force in the regime of interest; this is simply a consequence of the fact that the driving force is dominated by the strain energy in the superlayer associated with the deposition stress. However, the state of stress in the bilayer formed by the delamination

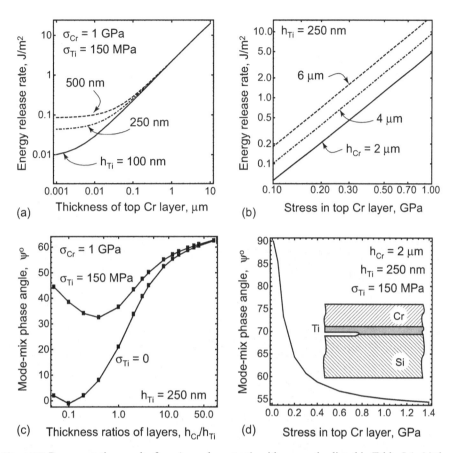

**Figure 5.5** Representative results for a 'superlayer test', with properties listed in Table 5.1: (a) the steady-state energy release rate as a function of superlayer thickness for several adhesion layer thickness values, (b) the energy release rate as a function of stress in the superlayer for several superlayer thickness values, (c) the phase angle as a function of superlayer thickness, for two different states of stress in the adhesion layer, (d) the phase angle as a function of stress in the superlayer for fixed film thicknesses.

does affect the mode mix, as illustrated in Figures 5.5(c–d). Though not shown in Figure 5.5, the energy release rate for delamination of the superlayer/titanium interface (above the one of interest) is essentially equivalent to that at the target interface; hence, the toughness of the superlayer/target film interface must be larger than that of target film/substrate interface in order for the test to be effective.

The results shown in Figure 5.5 were generated using the *LayerSlayer* code that accompanies this text. The energy release rate can be computed either with the semi-infinite film code associated with this chapter (and described in Chapter 14) or with the finite element code described in Chapter 16. Naturally, the finite element computation can be influenced by edge effects, and one must take care to ensure sufficiently large dimensions that capture the asymptotic limit. Here, phase angles are computed assuming $\epsilon(\beta_D) \sim 0$ as a matter of convenience, which eliminates the need to

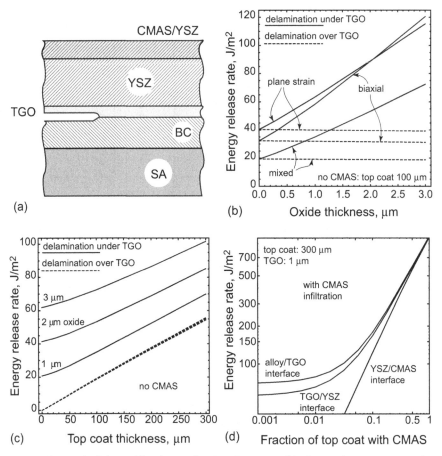

**Figure 5.6**  (a) A typical thermal barrier coating (TBC) system. (b) The steady-state energy release rate as a function of oxide thickness, computed under various assumptions. (c) The steady-state energy release rate as a function of top coat thickness, for several oxide thickness values, assuming equi-biaxial straining followed by plane strain delamination. (d) The increase in ERR associated with penetration of the top coat by molten mineral contaminants (often referred to as CMAS).

choose an explicit length scale. The final example of this section will illustrate that the effect of this simplification can be expected to be small.

## Delamination in a Thermal Barrier Coating System

Thermal barrier coating systems used in gas turbines reduce the operating temperatures of superalloy substrates and protect them from oxidation. As shown in Figure 5.6(a), a typical system consists of superalloy substrate, a metallic bond coat, followed by a porous ceramic oxide top coat. At elevated temperatures, the bond coat oxidizes at the interface with the top coat and forms aluminum oxide, which serves as an effective chemical barrier that protects the superalloy. Further, at elevated temperatures, mineral contaminants ingested into the turbine can melt and penetrate the porous top coat: these contaminants are variations of calcium-magnesium-alumino-silicates or CMASs.

**Table 5.2** Typical Properties for a Thermal Barrier Coating System. The Computations in Figure 5.6 Use Values in the Middle of the Range Unless Otherwise Stated

| Layer | Thickness | $E$ | $v$ | $\alpha$ (ppm/$^\circ$C) | $T_{ref}$ |
|---|---|---|---|---|---|
| Nickel superalloy | 2–5 mm | 180–250 GPa | 0.30 | 13–18 | 1100 |
| Metallic bond coat | 50–150 μm | 180 GPa | 0.30 | 13–18 | 1100 |
| Alumina (TGO) | 0.1–10 μm | 380 GPa | 0.20 | 8.1 | 1100 |
| YSZ | 50–500 μm | 15–50 GPa | 0.20 | 13.5 | 1100 |
| CMAS | 50–500 μm | 50–200 GPa | 0.20 | 10.6 | 1100 |

In this section, the driving forces for delamination are considered assuming uniform temperature changes, which is meaningful for isothermal furnace tests that are often used to test system reliability. The case of through-thickness temperature gradients, which is more realistic for in-service conditions, is considered in Chapter 12. A common assumption when analyzing multilayers used at very high temperature is that stress relaxation occurs at high temperature; this implies that the stress-free reference temperature is the service (or furnace) temperature, and that cracking is driven by compressive stresses that arise during cooling. Typical properties for a thermal barrier coating system are shown in Table 5.2.

Figure 5.6(b) shows the energy release rate for debonding at the bond coat-TGO interface and the TGO-top coat interfaces, as a function of oxide thickness. Several computations are shown: 'plane strain' assumes plane strain conditions exist both prior to and after debonding, and similarly for 'biaxial'. The curves labeled 'mixed' assume biaxial stresses prior to delamination, but that debonding occurs in plane strain. For debonding above the TGO, the strain energy stored in the TGO is not released, and the oxide thickness has little effect. However, for debonding below the TGO, delamination releases considerable energy stored in the oxide, and driving force increases linearly with oxide thickness. Figure 5.6(c) shows that the driving force also increases linearly with top coat thickness. This scaling is identical to that implied by the bilayer results given in Chapter 4.

In order to determine whether delamination occurs above or below the TGO, one must consider the mode mix and the associated toughnesses of those interfaces. On the one hand, while the driving force is highest if the TGO is released, the bond coat/TGO interface tends to have high interface toughness due to plastic deformation in the soft bond coat. While the driving force for delamination at the TGO/top coat interface is smaller, the toughness of this ceramic-ceramic interface tends to be low. The role of the mode mix for this TBC system is discussed in Chapter 16, as an illustration of the numerical computations described in that chapter.

The increase in operating temperatures afforded by advances in TBCs has led to a new problem: dust ingested by the turbine experiences high enough temperatures to melt and infiltrate the porous ceramic top coat. A typical dust composition is calcium-magnesium-alumino-silicate (CMAS), which melts at $\sim 1000^\circ$C. The melted CMAS infiltrates porous yttria-stabilized zirconia top coats and can form either a glassy phase or crystalline phase depending on the details of its composition and infiltration

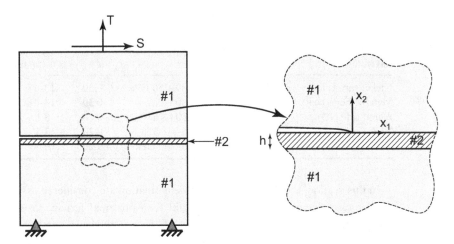

**Figure 5.7** A thin layer (no. 2) sandwiched between two thick layers having the same elastic properties (no. 1), as considered by Suo & Hutchinson (1989). The analysis is limited to either plane stress or plane strain.

process. The CMAS-infiltrated YSZ layer solidifies upon cooling, dramatically increasing its modulus. Figure 5.6(d) shows the impact of this process on the energy release rate for steady-state delamination. In the calculation, the thickness of the top coat is held fixed, but a fraction of it is replaced with the stiff infiltrated layer properties. Results for delamination at the TGO/YSZ interface are shown, along with the results for delamination at the YSZ/infiltration interface. In essence, the compliance of the pristine porous YSZ top coat is lost, and the energy release rate spikes dramatically.

**A Thin Layer Sandwiched between Two Thick Layers**
Here, basic results for a special trilayer are presented which shed light on the extent to which a thin layer sandwiched between two thick layers influences the mode mix. The geometric configuration is shown in Figure 5.7, and the results are taken from Suo & Hutchinson (1989).

In the absence of the thin layer, denote the conventional mode I and mode II stress intensity factors at the crack tip of the finite body on the left in Figure 5.7 by $K_I$ and $K_{II}$ where this body has uniform isotropic elastic properties denoted by material no. 1. The asymptotic problem depicted on the right in Figure 5.7 has been analyzed in which the thickness $h$ of the layer having isotropic properties of material no. 2 is very small compared to all other in-plane lengths of the trilayer. In this asymptotic problem, the outer field of the problem on the right is the crack tip fields associated with $K_I$ and $K_{II}$. The fields at the actual crack tip in the figure on the right in Figure 5.7 are the tip fields of a bimaterial interface crack detailed in Section 3.2 and characterized by $K_1$ and $K_2$.

The energy release rate of the interface crack is the same as that of the crack without the thin layer being present:

$$G = \frac{\left(K_I^2 + K_{II}^2\right)}{\bar{E}_1} \tag{5.23}$$

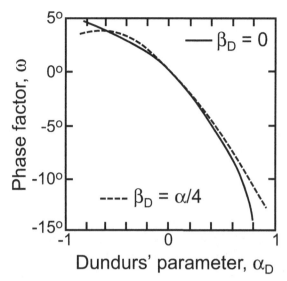

**Figure 5.8** Phase factor $\omega(\alpha_D, \beta_D)$ in degrees for a thin layer sandwiched between two thick layers of the same material.

as one would expect on physical grounds. The relation between the two sets of stress intensity factors is expressed in the complex form

$$K_1 + iK_2 = \sqrt{\frac{1 - \alpha_D}{1 - \beta_D^2}} (K_I + iK_{II}) h^{-i\epsilon} e^{i\omega} \tag{5.24}$$

where $i = \sqrt{-1}$ and $\epsilon = \ln\left((1 - \beta_D) / (1 + \beta_D)\right)/(2\pi)$. The results apply to either plane stress or plane strain with the two Dundurs' parameters $\alpha_D$ and $\beta_D$ defined accordingly as in Chapter 3. The phase factor $\omega(\alpha_D, \beta_D)$ depends on the Dundurs' parameters and is tabulated in the paper by Suo & Hutchinson (1989); it is plotted in Figure 5.8 for $\beta_D = 0$ and $\beta_D = \alpha_D/4$.

With $\psi_o = \tan^{-1}(K_{II}/K_I)$ as the mode mix of the cracked body in the absence of the thin layer and with $\psi$ as the mode mix of the interface crack tip fields defined by (3.22), one finds using (5.24)

$$\psi = \psi_o + \omega(\alpha_D, \beta_D) + \epsilon \ln\left(\frac{\ell}{h}\right) \tag{5.25}$$

where $\ell$ is the distance ahead of the interface crack tip where the stress mix is evaluated for cases with $\beta_D \neq 0$. From Figure 5.8, one can see that the shift in the mode mix due to $\omega(\alpha_D, \beta_D)$ is generally fairly small, amounting to not more than 5° unless the thin layer is highly compliant (compared to the two outer layers), in which case the shift can be as much as 15° or more. This conclusion is the same for any overall applied mode mix $\psi_o$.

One important conclusion to be drawn from this example, as has been seen in some of the earlier examples, is that generally the mode mix is not strongly dependent on the elastic mismatch of the material layers (immediately adjoining the crack). Exceptions to this rule occur when one of the materials at the interface is highly compliant relative to the others. The example also sheds some light on the bilayer steady-state approximation of the previous section as applied to the trilayer in Figure 5.7. For the trilayer with the very thin layer sandwiched between two thick layers, the bilayer approximation is equivalent to simply ignoring the presence of the thin layer and replacing the trilayer with the two thick layers. In other words, the bilayer approximation for this case simply returns $G$ in (5.23) and the conventional mode mix $\psi_o$. As just discussed, this is a good approximation to the trilayer problem as long as the modulus of the thin layer is not too small relative to that of the thick layers. On the other hand, the bilayer approximation for the mode mix becomes suspect when a layer at or near the interface has a modulus much lower than the other layers.

The trilayer in Figure 5.7 takes the outer layers to be infinitely thick, and thus it is natural to ask how thick the outer layers must be such that the results (5.23)–(5.25) are applicable. If there is any doubt on the part of the analyst he or she should make use of the computational software to obtain further insight. As long as the thin layer is not extremely compliant (i.e., $\alpha_D$ not nearly unity), one expects that the results in (5.23)–(5.25) should be applicable if the layer thickness is less than 1% of the thickness of the outer layers.

One important conclusion to be drawn from this example, as has been seen in some of the earlier examples, is that generally the mode mix is not strongly dependent on the elastic Dundurs' parameter $\beta_D$. Exceptions to this rule occur when one of the materials at the interface is highly compliant relative to the others. In the case of the first example of the superlayer test, the elastic mismatch is not extreme, and one can anticipate that taking $\beta_D = 0$ when computing the mode mix will not lead to a significant error. For the second example, the YSZ ceramic coating is quite compliant compared to the thin TGO layer lying beneath it, and thus for delamination at the TGO/YSZ interface there may be a non-negligible influence of the mode mix on $\beta_D$. The software can compute this dependence, as is discussed in more detail in Chapter 14.

# 6  Steady-State Channeling and Tunneling Cracks

Thin films bonded to thick substrates that sustain tensile stress can undergo a number of cracking patterns, some of which are shown in Figure 6.1. The rich variety of crack patterns makes life interesting for the crack analysts but, more importantly, must be taken into account in the design of thin film and coating systems against cracking. As Evans et al. (1988) first emphasized, the concept of steady-state crack propagation plays a central role in design against cracking in these systems by providing a simple but robust approach. There are close parallels with thin film delamination discussed earlier in Chapter 4.

Figure 6.2 illustrates some of the phenomena associated with the propagation of a single crack which will be addressed in this chapter. With $\sigma$ as the tensile stress in the film, $h$ as the film thickness and $\bar{E} = E/(1 - v^2)$ as the plane strain modulus of the film, the steady-state energy release rate for each of the types of cracking in Figure 6.2 can be expressed in the form (Evans et al. 1988; Hutchinson & Suo 1992)

$$G = Z\frac{\sigma^2 h}{\bar{E}} \tag{6.1}$$

where $Z$ is a number depending on the film-substrate elastic mismatch and any effects such as delamination or plasticity accompanying the film through-crack. If cracking is to be avoided, the relevant toughness (to be defined later), $\Gamma$, must satisfy

$$\Gamma > Z\frac{\sigma^2 h}{\bar{E}} \tag{6.2}$$

Condition (6.2) reveals two of the most important drivers of film cracking: tensile stress, which increases susceptibility in proportion to its square, and film thickness. This same dependency has already been encountered and discussed in connection with thin film delamination in Chapter 4. In this chapter, the focus is on through-cracks in thin films which channel across the film. A short section will also be presented on a through-crack in the interior of a multilayer which tunnels across the layer. Both channeling and tunneling cracks can induce accompanying interface delamination or plasticity in the substrate as they propagate. They can also penetrate the substrate or adjoining layers. All these possibilities give rise to a rich array of cracking phenomena. While only a few of these phenomena will be explored in any depth in this chapter, the software accompanying the text enables many of them to be quantitatively analyzed.

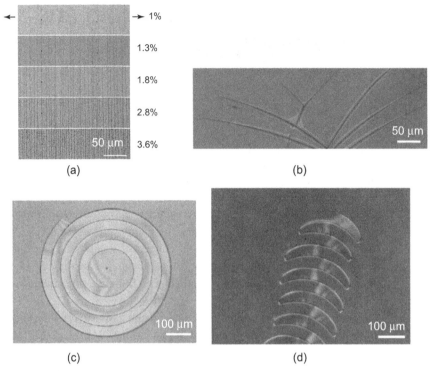

**Figure 6.1** (a) 50-nanometer-thick ZnO monolayer film strips deposited by magnetron sputtering on a PTFE substrate which have been monotonically stretched horizontally to the five strain levels noted. Through-cracks with no accompanying delamination propagate vertically across the entire strip with density increasing with increasing strain. (b) 1.8 μm spin-coated silicate layer on a silicon substrate. The sol-gel condensation produces equi-biaxial tension in the film. The through-cracks propagate accompanied by double-sided interface delamination revealing limited delamination in steady state. (c) 1.1 μm spin-coated silicate layer on a silicon substrate. The sol-gel condensation produces equi-biaxial tension in the film; in this example, a through-crack accompanied by one-sided interface delamination propagates and generates an Archimedes spiral. (d) 1.2 μm spin-coated silicate on a slicon substrate. The sol-gel condensation process produces equi-biaxial tension in the film; in this example, a through-crack accompanied by one-sided interface delamination propagates and generates a crescent-shaped crack. The four photographs have been supplied by Jöel Marthelot (2014); photographs (c) and (d) are similar to those in Marthelot (2014).

Consider the film in Figure 6.3(a) that is bonded to an infinitely deep substrate where the stress state in the film is uniform with $\sigma_{11} = \sigma$ and $\sigma_{13} = 0$. The component $\sigma_{33}$ is arbitrary since it will not influence the results of interest. By symmetry, the crack front of a through-crack propagating in the negative $x_3$-direction is subject to mode I conditions with an energy release rate averaged through the thickness, $G$, that depends on crack length in a manner indicated in Figure 6.3(b). The essential feature is that $G$ increases monotonically approaching from below the steady-state release rate for a long crack, $G_{SS}$.

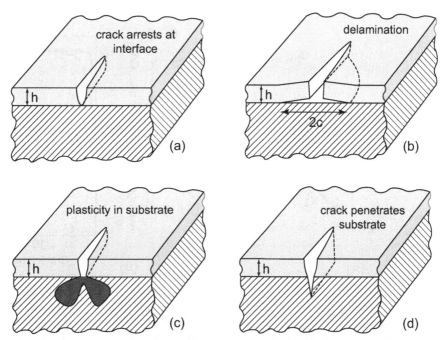

**Figure 6.2** Several film cracking scenarios. (a) A crack arresting at the film/substrate interface and propagating across the film. (b) Delamination of the interface accompanying the propagating film crack. (c) Plasticity induced in the substrate by the propagating film crack. (d) A propagating film crack that penetrates through the interface into the substrate.

The term 'edge crack' refers to cases where one end of the crack lies on the free face of the specimen, as shown in Figure 6.3(a). The term 'interior crack' refers to a flaw whose entire length lies far from an edge of the specimen, in which case Figure 6.3(a) depicts one half of the crack. For systems where the Dundurs' parameter $\alpha_D$ is the range $(-1, -1/2)$, three-dimensional calculations have shown that the steady-state energy release rate is approached for surprisingly short cracks with lengths not much larger than several times $h$ at most. In other words, once the crack length is on the order of $h$, it is already as potent as a very long crack and the steady-state limit applies. Note that interior cracks reach their steady-state limit much faster than those that intersect a free edge.

For relatively brittle films which crack without accompanying effects as in Figures 6.2(b–d), the condition $\Gamma_{IC} > G_{SS}$ excludes the propagation of all such through-cracks, where $\Gamma_{IC}$ is the mode 1 toughness of the film. Conversely, if $\Gamma_{IC} < G_{SS}$ it is highly likely that flaws in the film on the order of its thickness, either along the edge or in the interior, will initiate and propagate across the entire film. It is in this sense that this simple condition serves as a robust design criterion against through-cracking. Of course, if it can be shown there are no flaws as large as $h$, then one may be willing to risk a system having $\Gamma_{IC} < G_{ss}$, but then it would be essential that all flaws remain small if wide spread through-cracking is to be avoided. Thin films have a huge expanse

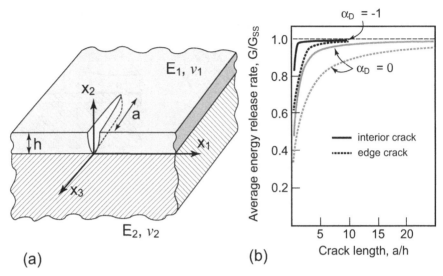

(a)                                    (b)

**Figure 6.3** A crack in the film arrested at the film/substrate interface propagating across the film. (a) Geometry and notation. (b) The average energy release rate along the crack front for edge and interior flaws, computed via three-dimensional finite element analysis (Ambrico & Begley 2002); the energy release rate approaches the steady-state value for a semi-infinite crack after extensions of only on the order of the film thickness, particularly for compliant films on stiff substrates.

compared to their thickness, and for this reason it is not typically overly conservative to assume that critical flaws exist, such that the requirement $\Gamma_{IC} > G_{SS}$ is realistic.

## 6.1    An Isolated Through-Crack at Steady State

Beuth (1992) provided a comprehensive analysis of an isolated crack in a single film on top of a semi-infinite substrate, addressing scenarios in which the crack length in the $x_3$-direction is much larger than its depth in the $x_2$-direction (cf. Figure 6.3(a)). The work identifies the conditions under which a partial through-crack penetrates to the interface, as well the energy release rates associated with channeling of partial through-cracks. Here, we summarize Beuth's results for steady-state propagation of a semi-infinite, *complete* through-crack in an elastic system, which represents the maximum energy release rate for channeling across the film. Brief commentary is provided at the end of this section regarding the role of plasticity in adjacent layers.

As previously illustrated for steady-state delamination in bilayers and multilayers, the energy released by the propagation of the crack by a unit distance is the difference in the energy per unit length far ahead of the crack and that far behind the crack tip (with both problems characterized by plane strain conditions). The steady-state energy release rate obtained by Beuth by computing this difference of energies provides the value of the energy release rate averaged across the crack front. In this chapter, the notation $G$ will be used to denote this average, and the subscripts denoting steady state will be

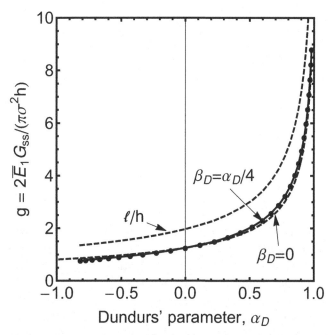

**Figure 6.4** Beuth's prefactor determining the steady-state energy release rate of a channeling crack through a film on a substrate, as a function of the elastic mismatch; the results were computed with the software in Chapter 16 and are identical to Beuth's original results (Beuth 1992). The length introduced in the model of film cracking in Section 6.2 is also shown.

suppressed unless there is a reason to draw a distinction with conditions which are not steady state. It should be noted that for a brittle film material with a local propagation condition, $G = \Gamma_{IC}$, the crack front through the thickness will generally be curved. The condition $G = \Gamma_{IC}$ will also be satisfied as an average through the thickness.

The steady-state energy release rate for the problem shown in Figure 6.3(a) has the form Beuth (1992):

$$G = \frac{\pi}{2} \frac{\sigma^2 h}{\bar{E}_1} g(\alpha_D, \beta_D) \tag{6.3}$$

with $\bar{E}_1 = E_1/(1 - v_1^2)$, where Beuth's values for $g$ are plotted for plane strain in Figure 6.4 for two combinations of Dundurs' parameters. The first parameter, $\alpha_D$, has the dominant effect revealing that the energy release rate is strongly enhanced if the film is very stiff compared to the substrate. The results shown in Figure 6.4 were generated using the accompanying finite element software (see Chapter 16) and are identical to those originally presented by Beuth (1992).

The limit of extreme mismatch between a stiff film and compliant substrate when $\alpha_D \sim 1$ corresponds to metals or ceramics on soft polymers, such as rubber. Such systems are becoming increasingly common with the development of flexible electronics for sensing and displays. Figure 6.5 reveals that the energy release rate for channel cracking in such systems scales with $(1 - \alpha_D)^{-1}$, implying dramatic increases in the

**Table 6.1** Film Roperties Used in the Channel Cracking Study with Extreme Elastic Mismatch; The Substrate has Thickness 63.5 μm, Elastic Modulus 1.5 *GPa* and Poisson's Ratio 0.49 for All Cases

| Case | Thickness $h_f$ | Modulus $E_f$ | Poisson's ratio $v_f$ | $1 - \alpha_D$ | $\beta_D$ | $\frac{\bar{E}_f h_f}{\bar{E}_s h_s}$ |
|---|---|---|---|---|---|---|
| A | 12.5 nm | 110 GPa | 0.35 | $3.2 \times 10^{-5}$ | 0.020 | 12.5 |
| B | 250 nm | 110 GPa | 0.35 | $3.2 \times 10^{-5}$ | 0.020 | 250 |
| C | 50 nm | 91 GPa | 0.39 | $3.8 \times 10^{-5}$ | 0.020 | 42.8 |
| D | 3 μm | 4 GPa | 0.4 | $8.3 \times 10^{-4}$ | 0.020 | 114 |
| E | 1 μm | 4 GPa | 0.4 | $8.3 \times 10^{-4}$ | 0.020 | 38 |
| F | 1 μm | 0.33 GPa | 0.4 | 0.01 | 0.019 | 3.1 |

energy release rate for $\alpha_D \to 1$ (Begley & Bart-Smith 2005); a list of properties for systems corresponding to the data points is given as Table 6.1. An important consideration for such systems is that such extreme modulus mismatch implies that substrate thickness or crack spacing can play an important role, even when their values are orders of magnitude larger than the film thickness.

For example, for a metal on silicone rubber, with $E_1 \sim 10^5 E_2$, the film and substrate have comparable stretching stiffness even when the substrate is $10^5$ times thicker than the film. This implies that the limit of a 'semi-infinite' substrate can be impractical

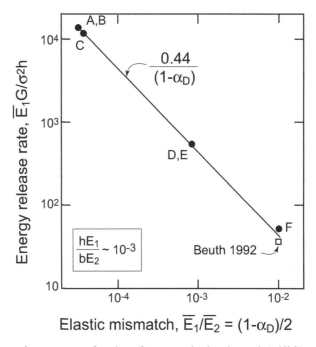

**Figure 6.5** Energy release rate as a function of extreme elastic mismatch (stiff film on a compliant substrate), typical of metals or ceramics on polymers (Begley & Bart-Smith 2005). The material property combinations associated with points labeled A–F are given in Table 6.1. As the ratio $E_1/E_2$ increases, the crack spacing $b/h$ must increase dramatically to obtain the result for an isolated crack.

to reach, such that one must consider the influence of substrate thickness. Similarly, the modulus mismatch implies shear transfer between the film and substrate occurs over very large distances, indicating that crack interactions can be important even when cracks are separated by extremely large distances relative to the film thickness. The paper by Begley & Bart-Smith (2005) provides a complete discussion of behaviors associated with extreme elastic mismatch.

Compliant substrates lead to larger steady-state energy release rates for channeling because they allow for greater crack openings in the wake of the crack. The relationship between energy release rate and crack opening can be understood by considering an alternative approach to computing the steady-state energy release rate. The energy release rate can be calculated as the work done in converting an intact unit slice ahead of the crack into the associated cracked unit slice behind the crack. Denoting $\sigma(x_2)$ as the stress distribution in the film *prior* to cracking, and $\delta(x_2)$ as the crack opening in the film *after* cracking, this implies

$$G = \frac{1}{2h} \int_0^h \sigma(x_2)\delta(x_2)dx_2 \tag{6.4}$$

Thus, all other things being equal, reducing the compliance of the substrate allows for greater crack opening displacements $\delta(x_2)$ and, in turn, increases the energy release rate.

Plasticity in the adjacent layer (beneath the channeling crack) has the same effect; yielding allows larger crack opening displacements and produces larger energy release rates. The paper by Beuth & Klingbeil (1996) provides a complete overview of this effect for stress-free substrates. An important distinction of systems subject to plasticity is that, in addition to the elastic energy in the film, the energy release rate also depends on $\sigma/\sigma_Y$, where $\sigma_Y$ is the yield stress of the adjacent layer. This is shown in Figure 6.6. The bottom curve, corresponding to a stress-free substrate, is typical of those compiled by Beuth & Klingbeil (1996); the energy release doubles relative to that of the elastic system for $\sigma \sim 5\sigma_Y$. Begley and Ambrico studied the effect of residual stress in the adjacent layer, such as might occur for cracks in ceramic passivation layers on top of metallic interconnect in microelectronics (Ambrico & Begley 2002). As one might expect, gross yielding of the adjacent metal layer amplifies the effects of plasticity by promoting even larger crack opening displacements.

The insightful paper by Beuth & Klingbeil (1996) explains that the steady-state energy release rate can still be estimated using two-dimensional simulations of the crack wake, even when plasticity is present. First, they assume that the process of unloading from the stressed, intact film to the cracked film (with stress-free faces) will involve (nearly) proportional stress paths. This implies that the plastic deformation associated with cracking can be accurately modeled using nonlinear elasticity. Second, they noted that, since the system can be treated nonlinear elastic, one can use the definition of complementary energy to obtain the proper result by calculating the response due to a pressure applied to stress-free crack faces. The curves shown in Figure 6.6 were computed by Ambrico & Begley (2002) using Beuth and Klingbeil's insightful method.

The results in Figure 6.6 illustrate that plasticity in the substrate also increases the crack length required to reach steady state; compare, for example, the result for

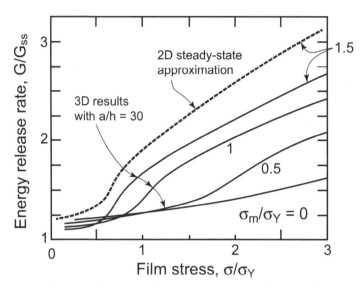

**Figure 6.6** Energy release rate for a channeling crack in an elastic film normalized by the elastic steady-state results, as a function of film stress normalized by the yield stress of the adjacent layer (Ambrico & Begley 2002). Results are shown for both stress-free substrates ($\sigma_m = 0$) and substrates with residual stress ($\sigma_m \neq 0$).

semi-infinite cracks obtained from 2D analysis and the result from 3D analysis with $a/h = 30$. For an elastic system with identical elastic properties, the energy release rate is within 95% of the steady state limit when $a/h = 30$ (see Figure 6.3(b)). The result for same system with plasticity present is at only ~80% of the steady state limit. This is analogous to behavior seen for compliant substrates when $\alpha_D > 0$ (Ambrico & Begley 2002).

## 6.2     A Two-Dimensional Film Cracking Model with Application to Multiple Interacting Cracks

It was possible to compute the steady-state $G$ for the three-dimensional problem in Figure 6.3(a) using two-dimensional solutions owing to the steady state character of the problem. For more complicated film cracking problems that are not steady state, the problems are inherently three-dimensional.

To avoid dealing with fully three-dimensional problems, Xia & Hutchinson (2000) introduced a two-dimensional model which approximates the shear coupling of the film to the substrate by linear springs which, in turn, are calibrated using Beuth's results. Only the essentials of the model will be introduced here, but nevertheless allowing prediction of some useful results for interacting film cracks. The reader is referred to the paper just cited for full details.

A schematic illustration of the model is shown in Figure 6.7. The film is modeled by plane stress theory. At any point, the in-plane components of the restoring force

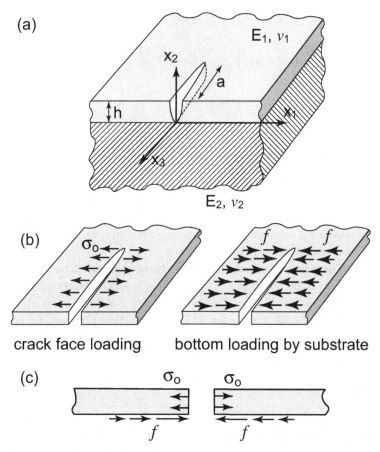

**Figure 6.7** Schematic illustration of the 2D model of film through-cracks illustrated for an isolated mode I channeling crack: (a) the 3D problem. (b) The 2D plane stress model of the film illustrating the two loads associated with the crack process: the loading on the crack faces $\sigma_o$ and the restoring force/area $f$ acting on the underside of the film. (c) Depiction of a typical cross section through the crack.

per unit area exerted on the film by the substrate are modeled by $f_\alpha = -ku_\alpha$, where $u_\alpha$ is the in-plane displacement of the film that occurs due to cracking. Equilibrium requires $h\Delta\sigma_{\alpha\beta,\beta} + f_\alpha = 0$, where $\Delta\sigma_{\alpha\beta}$ is the stress change due to cracking. Along any face the traction change must satisfy $\Delta\sigma_{\alpha\beta}n_\beta = -\sigma^o_{\alpha\beta}n_\beta$, where $\sigma^o_{\alpha\beta}$ is the stress in the uncracked film and $n_\alpha$ is the normal to the crack face. The model preserves the classical mode I and II stress intensity factors, and the energy release rate depends on them in the usual manner for plane stress.

The spring constant $k$ is calibrated by solving the model for a single crack in steady state, that is, the problem in Figure 6.3(a), and then requiring that it reproduces the Beuth result for $G$ in Figure 6.4. This process is straightforward because the model solution for the semi-infinite crack problem in Figure 6.3(a) is exceptionally simple. The result far behind the crack tip for $x_1 \geq 0$ is

$$u_1(x_1) = \sigma_{11}^o \left( \sqrt{\frac{h}{\bar{E}_1 k}} \right) e^{-\sqrt{kx_1/(\bar{E}_1 h)}}; \quad u_3 = 0; \quad G = \sqrt{\frac{h}{\bar{E}_1 k}} \left( \sigma_{11}^o \right)^2 \qquad (6.5)$$

The average crack opening displacement is $2u_1(0)$. The calibration gives

$$\frac{\ell}{h} \equiv \sqrt{\frac{\bar{E}_1}{hk}} = \frac{\pi}{2} g(\alpha_D, \beta_D) \qquad (6.6)$$

with $\ell$ as the decay length of the shear transfer to the film such that $u_1 = (\sigma_{11}^o / \bar{E}_1) \ell e^{-\sqrt{x_1/\ell}}$. The normalized length $\ell/h$ is included in Figure 6.4. In the absence of elastic mismatch between the film and substrate, $\ell/h = 1.98$. However, when the substrate is very compliant compared to the film, the shear decay length $\ell$ can be many times the film thickness.

The distance $\ell$ sets the interaction between cracks. A particularly simple case to analyze with the model is an array of equally spaced cracks advancing together simultaneously in steady state in a film sustaining a uniform stress $\sigma_{11}^o$ (with $\sigma_{12}^o = 0$) in the uncracked state (see the top of Figure 6.8). When the spacing is $b$ the energy release rate at each crack tip is Xia & Hutchinson (2000)

$$G = \frac{\ell \left( \sigma_{11}^o \right)^2}{\bar{E}_1} \tanh \left( \frac{b}{2\ell} \right) \qquad (6.7)$$

This result is plotted in Figure 6.8(a). To our knowledge, there are no results in the literature to check the accuracy of the model prediction (6.7) when the system has elastic mismatch, but the solution for periodic edge cracks in an homogeneous solid in Tada et al. (2000) can be used to show that (6.7) is accurate to within a few percent over the range plotted in Figure 6.8.

Hutchinson & Suo (1992) have noted that the solution for sequential cracking (see the top of Figure 6.8), where a preexisting array with spacing $2b$ experiences sequential cracking with another set of cracks propagating midway between the preexisting cracks, can be solved simply and exactly if the solution for equally spaced arrays is known. The sequential cracking process is a much more realistic model of events leading to the formation of crack arrays such as those shown in Figure 6.1(a). The approach is not limited to the film/substrate spring model and, indeed, applies to the full three-dimensional case. Denote the steady-state energy release rate per crack tip for the equally spaced array with spacing $b$ by $G_o(b)$. The simple energy accounting is as follows. With reference to the top of Figure 6.8, note that far ahead of the advancing array where the cracks have spacing $2b$ the energy/unit thickness that has been released per width $2b$ is $G_o(2b)$. Far behind the advancing crack tips, the spacing is $b$ and the energy/unit thickness that has been released per width $2b$ is $2G_o(b)$. Because there is one advancing crack tip per width $2b$, the energy release rate for each of the sequentially advancing cracks is

$$G = 2G_o(b) - G_o(2b) \qquad (6.8)$$

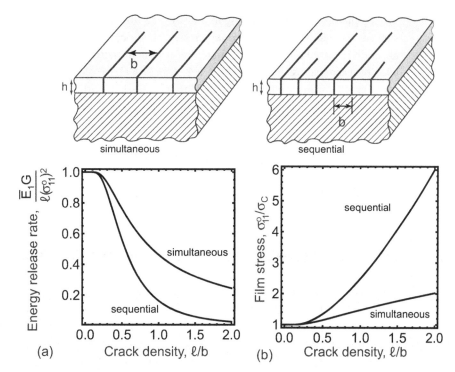

**Figure 6.8** (a) Normalized energy release rate as a function of normalized crack density for channel cracks advancing simultaneously (6.6) and sequentially (6.8). (b) With $\sigma_C = \sqrt{\Gamma_{IC}\bar{E}_1/\ell}$ as the stress at the onset of steady-state crack propagation for an isolated crack, the plot shows the crack density as a function of the film stress for simultaneous (6.9) and sequential (6.10) cracking.

For the film-substrate model with $G_o(b)$ given by (6.7), the result for the sequentially advancing cracks is

$$G = \frac{\ell \left(\sigma_{11}^o\right)^2}{\bar{E}_1} \left[ 2\tanh\left(\frac{b}{2\ell}\right) - \tanh\left(\frac{b}{\ell}\right) \right] \tag{6.9}$$

This result is also plotted in Figure 6.8(a) where it is seen to predict a significantly lower energy release rate than for the simultaneously advancing array. Equation (6.9) is also obtained by directly solving the equations for the film/substrate model.

With $\Gamma_{IC}$ as the mode I film toughness, denote the stress at the onset of an isolated steady-state channel crack from (6.7) or (6.9) with $b/\ell \to \infty$ by $\sigma_C = \sqrt{\Gamma_{IC}\bar{E}_1/\ell}$. Then, (6.7) and (6.9) predict the following relations between the film stress (defined as the stress $\sigma_{11}^o$ in the uncracked film) and the normalized crack density $\ell/b$:

$$\frac{\sigma_{11}^o}{\sigma_C} = \left[\tanh\left(\frac{b}{2\ell}\right)\right]^{-1/2} \quad \text{simultaneous advance} \tag{6.10}$$

$$\frac{\sigma_{11}^o}{\sigma_C} = \left[2\tanh\left(\frac{b}{2\ell}\right) - \tanh\left(\frac{b}{\ell}\right)\right]^{-1/2} \quad \text{sequential advance} \tag{6.11}$$

These relations are plotted in Figure 6.8(b). The film stress $\sigma_{11}^o$ can be expressed as a sum of any residual and/or thermal stresses plus a contribution porportional to overall strain imposed on the fim/substrate. Imposition of increasing overall strain has been used by Delannay & Warren (1991) and Thouless (1990) in their experimental studies of crack density evolution and by Marthelot (2014), with the latter's results shown in Figure 6.1(a) illustrating cracking patterns for several levels of applied strain. The evolution of the channel cracks' density is a sequential process, although the sequential model above is idealized.

The simultaneous results in Figure 6.8(b) significantly overestimate the crack density compared to the sequential results. The reader is also referred to Freund & Suresh (2003) where a presentation on single and multiple film cracks has been given that has much in common with the present one.

## 6.3    Film Cracking Accompanied by Limited Interface Delamination

Under special conditions film cracks can propagate under steady-state conditions across a film with a through-film crack accompanied by limited interface delamination. Figure 6.1(b) shows examples with delamination occurring symmetrically on both sides of the film crack, and Figures 6.1(c, d) show examples with delamination occurring on only one side of the crack giving rise to curved crack trajectories. We begin by considering the symmetric double-sided delamination in Figure 6.9(a), which preserves the mode I behavior of the film crack itself and leads to straight steady-state cracks. Denote the mode I toughness of the film by $\Gamma_{IC}$ and denote the mixed-mode interface toughness relevant to the interface delamination by $\Gamma_{INT}$ (cf. Chapter 3). Continue to assume that $\sigma_{11} \equiv \sigma$ is tensile and uniform.

The existence of a limited delamination trailing behind the advancing tip implies that the interface toughness must satisfy

$$\Gamma_{INT} > \frac{\sigma^2 h}{2\bar{E}_1} \tag{6.12}$$

Otherwise, the film would fully delaminate (cf. Chapter 3). Next, we make use of results presented later in Chapter 9 which reveal that a plane strain interface delamination emanating from a through-crack edge, as in Figure 6.9(a), attains the steady-state energy release rate $\sigma^2 h/(2\bar{E}_1)$ in distances that are a small fraction of the film thickness $h$. This assertion depends somewhat on the elastic mismatch, but only for very compliant substrates is the distance more than the film thickness. Plane strain conditions for delamination are relevant in the steady-state energy accounting. Thus, to an excellent approximation, the energy released per unit film crack advanced by a double-sided delamination of length $c$ on each side of the film crack is $\sigma^2 hc/\bar{E}_1$.

Combining the delamination release with the through-crack release, (6.3), and requiring the total energy released equals the energy dissipated in fracturing the film and the interface, one obtains the energetic propagation condition:

$$\frac{\sigma^2 h}{2\bar{E}_1} (\pi gh + 2c) = h\Gamma_{IC} + 2c\Gamma_{INT} \tag{6.13}$$

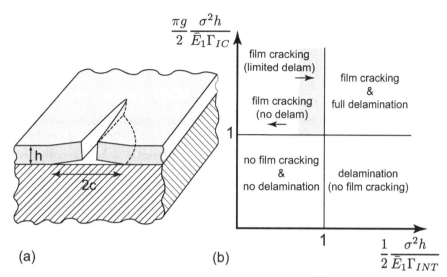

**Figure 6.9** Steady-state film cracking accompanied by limited symmetric double-sided delamination. (a) Notation and geometry. (b) Map depicting the regimes of film cracking with and without delamination.

The left-hand side of (6.13) is the steady-state energy release rate including the effect of delamination, which enhances the driving force for channeling. Equation (6.13) can be rewritten as

$$\frac{\pi g}{2} \frac{\sigma^2 h}{\bar{E}_1} = \Gamma_{IC} + 2\frac{c}{h}\left(\Gamma_{INT} - \frac{\sigma^2 h}{2\bar{E}_1}\right) > \Gamma_{IC} \tag{6.14}$$

where the inequality follows from the condition for limited delamination (6.12). It follows that film cracking with limited delamination requires a film stress in excess of that required for a film crack with no delamination.

The simple regime map in Figure 6.9(b) summarizes the various cracking possibilities. Of special interest here is the regime of film cracking accompanied by limited delamination:

$$\frac{\pi g}{2}\frac{\sigma^2 h}{\bar{E}_1} > \Gamma_{IC} \quad \text{and} \quad \frac{\sigma^2 h}{2\bar{E}_1} < \Gamma_{INT} \tag{6.15}$$

Satisfaction of these two inequalities requires

$$\frac{\Gamma_{INT}}{\Gamma_{IC}} > \frac{1}{\pi g(\alpha_D, \beta_D)} \tag{6.16}$$

With no elastic mismatch between the film and substrate, film cracking with limited delamination will occur only if $\Gamma_{INT}$ is larger than about $\Gamma_{IC}/4$, but it cannot be too large, otherwise delamination will not occur at all. More compliant substrates allow limited delamination at smaller values of $\Gamma_{INT}$ relative to $\Gamma_{IC}$.

The inequalities above are necessary conditions for film cracking accompanied by limited delamination, but they do not ensure that this mode of cracking will occur. Initiation of interface delamination occurs at the propagating front of the crack tip, as is evident in the micrographs given as Figures 6.1(b–d). If $\Gamma_{INT}$ is large enough, one expects that interface delamination will not initiate. Conversely, if $\Gamma_{INT}$ is below a critical value, one expects complete delamination. The precise extent of the regime of limited delamination in Figure 6.9(b) is not yet well defined. The problem governing the film/interface crack front is fully three-dimensional. The expanse of the delamination increases with distance behind the tip until the steady-state width $c$ is attained. Morever, in addition to the 3D geometry of the film/interface crack front the mode mix of the delamination crack near the front is almost certainly not the same as that far behind the propagating tip. To our knowledge, an analysis of this 3D problem has not been addressed, although aspects for special examples have been investigated by Marthelot, Bico, Melo & Roman (2015).

The plane strain analysis of a crack impinging on an interface in Chapter 8 provides further insight into the conditions for arrest at the interface or continued propagation by deflection into the interface followed by delamination. In that chapter, the competition between deflection into the interface and penetration of the interface with cracking of the substrate is also discussed. Thin films on compliant substrates having low toughness are prone to the phenomenon in Figure 6.2(d) where the propagating film crack is accompanied by substrate cracking, even if the substrate is not stressed in the crack state. This phenomenon has been discussed and analyzed by Beuth (1992), Ye, Suo & Evans (1992) and Thouless, Li, Douville & Takayama (2011), while an experimental study of metal films on elastomer substrates has been presented by Douville, Li, Takayama & Thouless (2011) which illustrates this phenomenon.

Yet another possibility exists for film/substrate systems where the substrate is significantly stiffer than the film. Then, as discussed later in Chapter 8, it is likely that the film crack will not extend all the way to the interface with the substrate, and a channeling crack will have a small standoff from the interface. This standoff can be estimated, and its effect on the steady-state energy release rate can be taken into account, as detailed by Beuth (1992).

As noted earlier, there is a rich array of phenomena associated with film cracking, including occurrences in nature such as mud cracking and the crack formations such as the Devil's Post Pile in California and the Giant's Causeway in Ireland. Recent mathematical extensions of brittle fracture mechanics may help explain the evolution of these unusual natural crack patterns, as well as patterns such as those seen in Figures 6.1(c, d), which are presently only qualitatively understood.

## 6.4     Tunnel Cracking in Brittle Buried Layers

There is extensive literature on cracking of brittle layers in multilayers of various kinds, in particular driven by the need to understand cracking as the onset of failure in the brittle matrices of fiber-reinforced composites (including both polymer and ceramic matrices). Early contributions to this problem were made by Gille (1985) and Dvorak & Laws (1986). Many of the basic ideas discussed in connection with the film cracks carry over to cracking of a brittle layer buried within a multilayer and subject to a tensile stress.

In particular, the notion of steady-state cracking retains its relevance, such that for most systems initial flaws on the order of the thickness of the layer have an energy release rate that is nearly as large as the steady-state limit for long cracks, as depicted earlier for films in Figure 6.3(b). Thus, for designing a brittle layer to withstand cracking one may well be able to use the steady-state energy release rate to estimate the crack driving force without being unduly conservative.

The simplest case is a thin layer of thickness $h$ subject to a uniform tension $\sigma$ acting perpendicular to the potential crack plane and sandwiched between two semi-infinite tough half-spaces (see Figure 6.10). If there is no elastic mismatch between the buried layer and the two substrates and if the propagating crack extends all the way across the layer to each interface, then the steady-state energy release rate averaged across the crack front is (Ho & Suo 1993)

$$G_{SS} = \frac{\pi}{4} \frac{\sigma^2 h}{\bar{E}} \tag{6.17}$$

Symmetry dictates a mode I crack tip such that the critical stress for steady-state tunneling is

$$\sigma_C = \sqrt{\frac{4 \bar{E} \Gamma_{IC}}{\pi h}} \tag{6.18}$$

with $\Gamma_{IC}$ as the layer toughness. For reference, a penny-shaped crack extending across the layer with radius $h/2$ has a critical stress given by

$$\sigma_C = \sqrt{\frac{\pi \bar{E} \Gamma_{IC}}{2h}} \tag{6.19}$$

which is only 11% larger than the steady-state critical stress (6.18). This drives home the relevance of the steady-state design approach if flaws on the order of the layer thickness are likely to be present.

The number of parameters characterizing $G_{SS}$ for a brittle layer embedded within a multilayer can be quite large compared to those for a film on a deep substrate. In addition to the properties of the brittle layer itself, one has to consider the thickness of the other layers and any elastic mismatch among the layers. For this reason, the software accompanying this book is an invaluable tool for generating numerical results for specific problems.

An example, originally presented by Ho & Suo (1993), illustrates some of the trends one can expect. The tri-layer in Figure 6.10 has the brittle layer of thickness $h$ sandwiched between two identical layers of thickness $w/2$. The tensile stress in the middle brittle layer is uniform and equal to $\sigma$ – the result presented below is equally valid if this stress is a residual stress or if it is due to a mechanical loading applied to the trilayer. The Young's modulus of the middle layer is $E_a$ while that of the outer layers is $E_s$. In the example in Figure 6.10(b) the Poisson's ratio of each of the three layers is the same and equal to $1/3$, and thus the abscissa is the first Dundurs' parameter. The data shown in Figure 6.10 were computed using the software accompanying the text, but are essentially identical to data presented by Ho & Suo (1993).

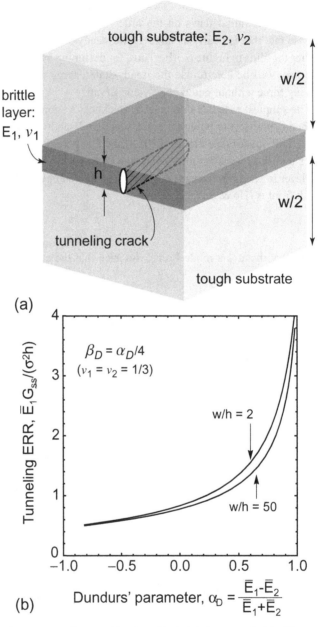

(a)

(b)

**Figure 6.10** Steady-state tunneling cracking in a thin brittle layer sandwiched between two outer layers or substrates, originally studied by Ho & Suo (1993) and computed with the software described in Chapter 16. (a) Notation and geometry. (b) Steady-state energy release rate for an isolated tunneling crack as dependent on elastic mismatch and two relative thicknesses of the substrate. The Poisson's ratios are $v_1 = v_2 = 1/3$, which implies $\beta_D = \alpha_D/4$.

Interestingly, there is little dependence on the thickness of the outer layers for the two cases shown, $w/h = 2$ and $w/h = 50$, except when the outer layers are very compliant compared to the middle layer. Thus, excluding such mismatches, a trilayer with $w/h = 2$ has essentially the same $G_{SS}$ as a middle layer sandwiched by two infinitely deep substrates. In particular, in the absence of any mismatch, (6.17) is applicable. As was seen for thin films, the constraint of the adjoining layers has a strong effect on $G_{SS}$ when those layers are very compliant, significantly enhancing the crack driving force.

# 7 Crack Kinking from an Interface

The interface fracture mechanics framework laid out in Chapter 3 assumes that the path of least resistance for the propagation of the crack is along the interface. When that is indeed the case, the crack propagates in the interface as dictated by the mixed-mode interface toughness, $\Gamma_{INT}(\psi)$, as described in Section 3.3. However, as a preexisting crack is subject to increasing load, a competition necessarily takes place between crack advance within the interface and crack kinking into one of the materials adjoining the interface. This chapter deals with the mechanics of this competition for materials that are sufficiently brittle such that they can be treated as elastic, apart from a small fracture process zone at the crack tip. Whether the crack continues to advance within the interface or kink out of the interface depends on several factors, including the mode mix acting on the preexisting interface crack, the toughness of the adjoining materials relative to that of the interface, the elastic mismatch and the size of flaws in the adjoining material that trigger kinking. This chapter addresses the issues influencing kinking from an interface in four sections organized with the most straightforward aspects addressed first. As will be described throughout the chapter, software is provided to enable calculation of the results presented here using previously published data.

## 7.1  Kinking for Materials Joined at an Interface with No Elastic Mismatch

The mechanics of kinking of an interface crack out of the interface will be described first for the case where there is no elastic mismatch between the two materials joined at the interface. This case exposes the main aspects of the phenomenon without having to deal with the more complex crack solutions required when elastic mismatch exists. The materials, each with Young's modulus $E$ and Poisson's ratio $v$, are joined across a planar interface with plane strain mixed mode toughness $\Gamma_{int}(\psi)$. The mode I toughness of the materials above and below the interface are $\Gamma_{IC}^{(1)}$ and $\Gamma_{IC}^{(2)}$, respectively. Elastic mismatch will be taken into account later in the chapter.

With reference to the interface crack in Figure 7.1(a), denote the mode I and mode II stress intensity factors associated with the loading on the preexisting interface crack by $K_1$ and $K_2$, respectively, anticipating the notation when elastic mismatch will be present. The energy release rate for a crack advancing in the interface under plane strain is (3.15)

$$G_{INT} = \frac{1}{\bar{E}}\left(K_1^2 + K_2^2\right) \tag{7.1}$$

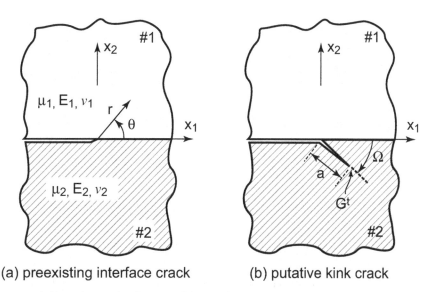

(a) preexisting interface crack  (b) putative kink crack

**Figure 7.1** Kinking of a putative flaw out of the interface between two isotropic elastic materials. (a) Preexisting interface crack. (b) Geometry of the putative flaw emerging from the tip of the preexisting crack.

while its mode mix is $\psi = \tan^{-1} K_2/K_1$. If the crack does not kink out of the interface, the condition for propagation in in the interface is $G_{int} = \Gamma_{INT}(\psi)$.

In addition to the toughness of the adjoining materials, the sign of $K_2$ has a strong influence on the propensity of the crack to kink up or down. With $K_2 > 0$, the mode II shear stress contribution, $\sigma_{12}$, in the region ahead of the crack tip is positive, generating tensile stress at planes oriented at $-45°$ emanating from the tip and compression across planes emanating at $+45°$, as depicted in Figure 7.2. Thus, if the energetic conditions for kinking are favorable, a positive $K_2$ tends to drive a crack downward, while a negative $K_2$ favors upward kinking.

The competition between crack advance in the interface and kinking out of the interface requires the evaluation of the energy release rate for kinking, $G_{KINK}$, and the mode mix, $\psi_{KINK}$, of the putative crack in Figure 7.1(b). Full details of the kinking condition will be discussed below, but, roughly speaking, as the load on the preexisting crack is increased, kinking will occur if $G_{KINK}$ attains the toughness of the adjoining material, $\Gamma_{IC}$, at a load below that at which the condition $G_{INT} = \Gamma_{INT}(\psi)$ is met. The kink angle, $\Omega$, in Figure 7.1(b) is not known in advance; conditions for determining $\Omega$ will be discussed.

As discussed in Section 3.2, the elastic crack problem in Figure 7.1(b) for a putative crack of length $a$ when there is no elastic mismatch has been solved by several methods. For the limit in which $a$ is much less than the length of the preexisting crack, or other relevant in-plane lengths, the stress intensity factors associated with the tip of the putative crack, $K_1^{KINK}$ and $K_2^{KINK}$, depend on the load applied to the preexisting interface crack through $K_1^{INT}$ and $K_2^{INT}$, or equivalently, through $G_{INT}$ and $\psi$, and on the kink

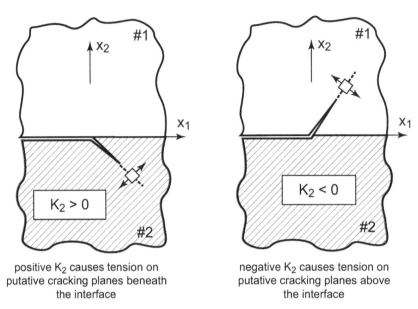

positive $K_2$ causes tension on putative cracking planes beneath the interface

negative $K_2$ causes tension on putative cracking planes above the interface

**Figure 7.2** Positive mode II favors kinking below the interface while negative mode II favors an upward kink.

angle $\Omega$. For plane strain conditions, this can be expressed functionally as

$$K_i^{KINK} = \sqrt{\bar{E} G_{INT}} k_i (\psi, \Omega) \quad i = 1, 2 \tag{7.2}$$

with $k_i$ as dimensionless functions of $\psi$ and $\Omega$.

With $G_{KINK} = [(K_1^{KINK})^2 + (K_2^{KINK})^2]/\bar{E}$ and $\psi_{KINK} = \tan^{-1} K_2^{KINK}/K_1^{KINK}$ as the energy release rate and mode mix of the kink crack tip, an equivalent characterization is

$$G_{KINK} = G_{INT} g(\psi, \Omega); \quad \psi_{KINK} = f(\psi, \Omega) \tag{7.3}$$

with $g$ and $f$ being dimensionless.

With $\psi > 0$, one criterion for kinking down into material 2 is the attainment of $G_{KINK} = \Gamma_{IC}^{(2)}$, with $G_{INT} < \Gamma_{INT} (\psi)$, where $G_{KINK}$ is evaluated for a kink angle $\Omega$, determined such that the putative crack tip is pure mode I:

Criterion (i): $\Omega$ is such that $K_2^{KINK} = 0 \quad (\psi_{KINK} = 0)$

This choice is consistent with the fact that the mode I toughness, $\Gamma_{IC}^{(2)}$, controls propagation in material 2. An alternative criterion which has also been widely employed chooses the kink angle to maximize $G_{KINK}$

Criterion (ii): $\Omega$ maximizes $G_{KINK}$

As discussed in Section 3.1, Criterion (ii) predicts a putative crack which is very nearly mode I. The difference between the two estimates of $\Omega$ is very small, and the difference between the two estimates of $G_{KINK}$ is negligible.

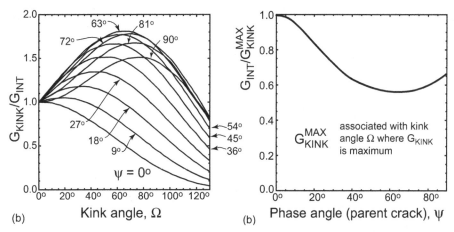

**Figure 7.3** (a) $G_{KINK}/G_{INT}$ as a function of kink angle for several mode mix values for the case with $\alpha_D = \beta_D = 0$ (no elastic mismatch); (b) the ratio of the energy release rates for the parent crack and the kink crack which maximizes $G_{KINK}$ among all kink angles, as a function of mode mix. The kink direction (angle) associated with the maximum values shown in (b) is shown in Figure 3.4(a).

It should be emphasized that the attainment of $G_{KINK} = \Gamma_{IC}^{(2)}$ is a necessary condition for kinking; it is not a sufficient condition for kinking. The criteria above ensure that the energy needed to propagate the kink crack is available, but they do not ensure that a kink crack will be initiated. An interface crack tip that is atomistically sharp can be expected to initiate a kink crack in a brittle adjoining material if either of the above criteria is met. However, for an interface crack tip having a fracture process zone, flaws within the adjoining material at the interface on the order of the size of the process zone will be required to initiate the kink crack. The role of flaws will be taken up in Section 7.4.

The results for $G_{KINK}/G_{INT}$ as a function of mode mix and kink angle are shown in Figure 7.3(a) for the case with no elastic mismatch. This figure is identical to Figure 3.3 presented earlier for a kinked crack in an isotropic material, with different labels to indicate that the parent cracks lies on the interface. For a given mode mix, the energy release rate for the kink crack is a maximum at a particular kink angle; the relationship between this maximum energy release rate and mode mix is shown in Figure 7.3(b). These results are identical to those shown in Figure 3.4(b), again with new labeling to reflect the presence of an interface.

With the maximum for $G_{KINK}$ in hand for a given mixed-mode loading ($\psi > 0$), the necessary criterion for kinking into material 2 versus propagation in the interface is

**Kinking** if $G_{KINK} = \Gamma_{IC}^{(2)}$ is attained with $G_{INT} < \Gamma_{INT}(\psi)$

**Interface propagation** if $G_{INT} = \Gamma_{INT}(\psi)$ is attained with $G_{KINK} < \Gamma_{IC}^{(2)}$    (7.4)

Because $G_{KINK}$ increases in proportion to $G_{INT}$ as the load on the preexisting interface crack is increased, the ratio plotted in Figures 3.4(b) and 7.3(b) is useful because the

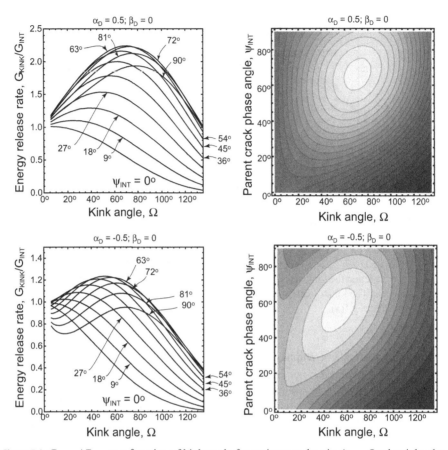

**Figure 7.4** $G_{KINK}/G_{INT}$ as a function of kink angle for various mode mix $\psi_{INT}$. On the right, the lightest tone is associated with the peak values from the figures on the left.

competition is decided by *the toughness ratio*, according to

$$\textbf{Kinking if } \frac{G_{INT}}{G_{KINK}} < \frac{\Gamma_{INT}(\psi)}{\Gamma_{IC}^{(2)}}$$

$$\textbf{Interface propagation if } \frac{G_{INT}}{G_{KINK}} > \frac{\Gamma_{INT}(\psi)}{\Gamma_{IC}^{(2)}} \tag{7.5}$$

The load at which the propagation event occurs requires attainment of the relevant toughness as specified by (7.4). If the load on the preexisting crack has $\psi < 0$, kinking, if it occurs, will be into material 1. The competition can be analyzed with the results presented above by interchanging the roles of the two materials and changing the sign of $\psi$.

## 7.2    Kinking for Materials Joined at an Interface with Elastic Mismatch with $\beta_D = 0$

The competition described above for no material mismatch carries over when the materials joined at the interface have differing elastic properties, but then the solutions for the stress intensity factors of the putative crack also depend on the Dundurs' mismatch parameters $(\alpha_D, \beta_D)$. Specification of the kinking criterion is somewhat more complicated for bimaterial combinations with $\beta_D \neq 0$, due to the weak oscillatory character of the interface crack solution. For most material combinations, the most important effect of the mismatch is generated by the first Dundurs' parameter $(\alpha_D)$. For this reason in this section we consider combinations with $\alpha_D \neq 0$ and $\beta_D = 0$, taking up combinations with $\beta_D \neq 0$ in the next section.

With $\beta_D = 0$, (7.3) for the release rate and mode mix at the tip of the putative crack now becomes

$$G_{KINK} = G_{INT}\, g\,(\psi, \Omega, \alpha_D); \quad \psi_{KINT} = f\,(\psi, \Omega, \alpha_D) \qquad (7.6)$$

The criterion (7.4) governing the competition between kinking and propagation in the interface carries over with no modification because, with few exceptions, it remains true that the kink angle that maximizes $G_{KINK}$ is associated with a putative crack that is nearly mode I. The effect of elastic mismatch on the ratio $G_{KINK}/G_{INT}$ is seen in Figure 7.4. Qualitatively, the trends are very similar to those with no mismatch, but elastic mismatch can be important in the competition between kinking and interface cracking. (It is worth emphasizing that direct, numerical comparisons between specific cases are easily computed with the provided software.) With $\psi > 0$, a more compliant material below the interface $(\alpha_D > 0)$ enhances the likelihood of kinking into that material. These results were generated with the software that accompanies this book; the full prescription for computing this ratio for any mismatch and kink angle is provided in the next section.

The value of the ratio $G_{KINK}/G_{INT}$, maximized with respect to kink angle $\Omega$, is plotted as a function of the interface crack mode for various values of $\alpha_D$ in Figure 7.5. This plot again emphasizes that elastic mismatch enhances the likelihood of kinking when the material into which the putative crack propagates is compliant compared to the material on the other side of the interface. It is worth emphasizing that these plots assume material no. 1 on top, with a downward kink associated with positive $K_2$. Again, the software associated with this book makes the computation of these curves very straightforward.

## 7.3  Kinking for Materials Joined at an Interface with Elastic Mismatch with $\beta_D \neq 0$

The characterization of the interface crack tip and behavior of the kink crack is more complicated than that described above when the second Dundurs' parameter $(\beta_D)$ does not vanish, due to the oscillatory nature of the interface crack tip field. For many bimaterials, these complications can be ignored by taking $\beta_D = 0$ as an approximation. However, the description taking into account $\beta_D$ will be summarized in this section using the results of He & Hutchinson (1989b), and the software accompanying this book takes $\beta_D$ into account in the kinked crack analysis. The criterion for kinking versus continued propagation in the interface, (7.4) or (7.5), continues to apply with two caveats: (1) The interface toughness, $\Gamma_{INT}(\psi)$, depends implicitly on the distance ahead of the interface

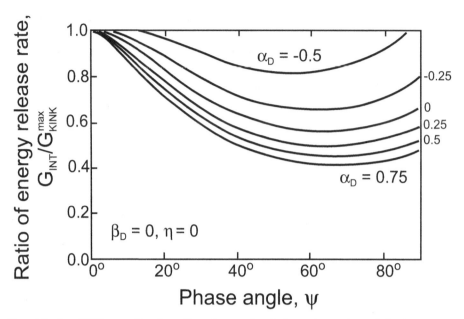

**Figure 7.5** $G_{INT}/G_{KINK}^{max}$ as a function of interface mode mix for various values of elastic mismatch where $G_{KINK}^{max}$ is the maximum over all putative kink angles.

crack tip, $\ell$, used to define the mode mix, $\psi$, as described in Section 3.3. (2) $G_{KINK}$ has a weak dependence on the putative crack length, $a$, as will be detailed below.

As in Section 3.3, it is convenient to make use of complex numbers with $i = \sqrt{-1}$. With $r$ increasing to the right as the distance in the interface ahead of the interface crack tip, the normal and shear stresses on the interface, $\sigma_{22}$ and $\sigma_{12}$, can be expressed from (3.12) as

$$\sigma_{22} + i\sigma_{12} = Kr^{i\epsilon}\frac{1}{\sqrt{2\pi r}} \tag{7.7}$$

where $\epsilon = \frac{1}{2\pi}\ln\frac{1-\beta_D}{1+\beta_D}$. For any bimaterial interface crack problem specified by traction boundary conditions, the complex stress intensity factor will have the following form:

$$K = K_1 + iK_2 = (applied\ stress) \times L^{1/2} \times L^{-i\epsilon} \times F = |K|e^{i\psi}L^{-i\epsilon} \tag{7.8}$$

where $L$ is a relevant in-plane length (e.g., the crack length or a unit of length such as one meter),[1] $F$ is a dimensionless function of the in-plane geometric length parameters and $\alpha_D$ and $\beta_D$, and $\psi$ is the mode mix defined with respect to $L$ (cf. eqn. 3.22). For example, the complex stress intensity factor for an interface crack of length $2L$ between two half spaces and subject to remote normal and shear stresses $\sigma_{22}^{\infty}$ and $\sigma_{12}^{\infty}$ is (England 1965; Rice & Sih 1965; Erdogan 1965)

$$K = \left(\sigma_{22}^{\infty} + i\sigma_{12}^{\infty}\right)(\pi L)^{1/2}(2L)^{-i\epsilon}(1 + 2i\epsilon) \tag{7.9}$$

[1] In the numerical codes accompanying the book, the units of length are those implied by the user's choice when inputing material properties.

The energy release rate and mode mix for propagation within the interface from (3.15) and (3.22) are

$$G_{INT} = \frac{1 - \beta_D^2}{E_*} \left( K_1^2 + K_2^2 \right); \quad \psi_{INT} = \psi \tag{7.10}$$

Here, $E_*$ is defined in (3.14), and $\psi_{INT}$ is defined with respect to $\ell = L$ via (3.22). The mode I and mode II stress intensity factors at the putative crack tip (cf. Figure 7.1(b)) can be expressed as

$$K_I^{KINK} + i K_{II}^{KINK} = cKa^{i\epsilon} + \bar{d}\bar{K}a^{-i\epsilon} \tag{7.11}$$

where $c$ and $d$ are complex valued functions of $\Omega$, $\alpha_D$ and $\beta_D$.

He & Hutchinson (1989b) have computed and tabulated the real and imaginary values of $c$ and $d$ for a range of combinations of $\alpha_D$ and $\beta_D$, for values of $\Omega$ ranging from $0°$ to $135°$ in $1°$ intervals. These tabulated values have been incorporated into the software that comes with this book, along with precoded interpolation functions that makes computations trivial. The kink crack solutions that produced $c$ and $d$ have been recomputed and checked by Noijen, van der Sluis, Timmermans & Zhang (2012), who also developed a scheme for representing the large amount of tabulated data in an efficient accessible form. The work by these authors identified sign errors in some of the tabulated data of He & Hutchinson (1989b), which have been corrected in the software accompanying this book. The present authors are indebted to the authors of the paper by Noijen et al. (2012) for their thorough work and for making their representation available.

The energy release rate of the putative kink crack is given by

$$\begin{aligned} G_{KINK} &= \frac{1}{\bar{E}_2} \left[ \left( K_I^{KINK} \right)^2 + \left( K_{II}^{KINK} \right)^2 \right] \\ &= G_{INT} \left( \frac{1 + \alpha_D}{1 - \beta_D^2} \right) \left[ |c|^2 + |d|^2 + 2Re\left( cde^{2i\tilde{\psi}} \right) \right] \end{aligned} \tag{7.12}$$

with

$$\tilde{\psi} = \psi_{INT} + \epsilon \, ln\frac{a}{L} \tag{7.13}$$

When $\beta_D \neq 0$, the ratio $G_{KINK}/G_{INT}$ depends on the putative crack length through $a/L$ in (7.13), but only weakly for practical purposes, as will be illustrated with an example. Over the entire range, $0 < a/L \leq 1$, the ratio oscillates between

$$\frac{G_{KINK}^{upper}}{G_{INT}} = \left( \frac{1 + \alpha_D}{1 - \beta_D^2} \right) \left[ |c| + |d| \right]^2 \quad and \quad \frac{G_{KINK}^{lower}}{G_{INT}} = \left( \frac{1 + \alpha_D}{1 - \beta_D^2} \right) \left[ |c| - |d| \right]^2 \tag{7.14}$$

and, as $a/L \to 1$, it approaches

$$\frac{G_{KINK}^*}{G_{INT}} = \left( \frac{1 + \alpha_D}{1 - \beta_D^2} \right) \left[ |c|^2 + |d|^2 + 2Re\left( cde^{2i\psi_{int}} \right) \right] \tag{7.15}$$

An example showing the dependence on $a/L$ of this ratio for a bimaterial interface with a relatively large value of $\beta_D$ is shown in Figure 7.6 for $\Omega = 45°$ and $\psi_{INT} = 45°$,

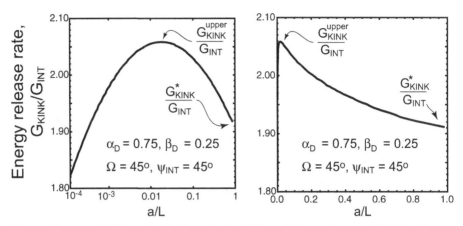

**Figure 7.6** An example illustrating the dependence of $G_{KINK}/G_{INT}$ on the normalized putative crack length for a case with a relatively large value of $\beta_D$. The figure on the left is a log scale for $a/L$, while that on the right is a linear scale for $a/L$.

where the values of $c$ and $d$ can be obtained either from the table in He & Hutchinson (1989b) or the software that comes with this book. In this example, it can be seen that the lower limit is irrelevant if $a/L > 10^{-4}$, and the difference between the upper limit, which is obtained as $a/L = 0.02$, and the value at $a/L = 1$ in (7.15) is only 7%. For most purposes, the competition between kinking and propagation can be adequately assessed using (7.15). Curves of $G^*_{KINK}/G_{INT}$ as a function of $\Omega$ for many $\psi_{INT}$ are shown for a specific bimaterial system in Figure 7.7. Comparison of the first plot in Figures 7.4 (with $\alpha_D = 0.5$ and $\beta_D = 0$) and Figure 7.7 (with $\alpha_D = 0.5$ and $\beta_D = 0.25$) reveal that the trends are identical and the quantitative influence of $\beta_D$ is on the order of only 5%.

For $\beta_D \neq 0$, the software accompanying this book maximizes $G^*_{KINK}/G_{INT}$ with respect to kink angle as a function of $\psi_{INT}$ and the elastic mismatch parameters. If deemed necessary, however, the dependence on $a/L$ for a given kinking problem can be accounted for in a more detailed calculation using the above formulas and tabulated coefficients. If the more detailed approach is taken, realistic choices of the putative crack length $a$ must be motivated by the expected flaw size at the surface of the material into which the crack is expected to kink. The role of surface flaw size is discussed in the next section.

## 7.4     The Roles of Flaw Size and Residual Stress Parallel to the Interface on Kinking

With $\Gamma_{IC}$ as the mode I toughness of the material experiencing kink cracking, the condition $G_{KINK} > \Gamma_{IC}$ is a necessary condition for kinking, but it is not sufficient. A clear illustration of the lack of sufficiency is evident in the mixed-mode toughness data for the epoxy glass interface measured by Liechti & Chai (1991) presented in Figure 3.11. Over the range of mode mix tested, $-90° < \psi < 90°$, the interface crack was observed to propagate within the interface. No kinking into the glass occurred in spite of the fact

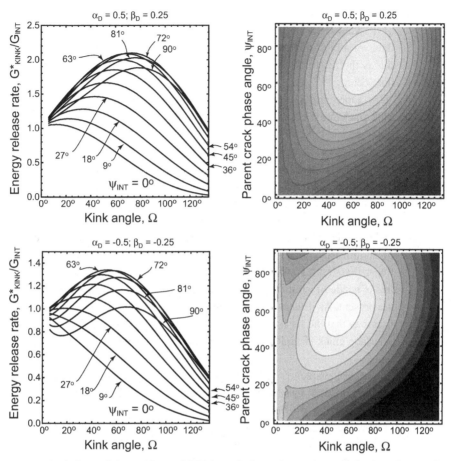

**Figure 7.7** Variations of $G^*_{KINK}/G_{INT}$ with kink angle for various values of mixed-mode $\psi_{INT}$ for two cases of elastic mismatch with relatively large values of $\beta_D$. The scale for the contour plots can be inferred from the plots on the left, with light tone indicating peak value. Compare with Figure 7.4 for plots with the same values of $\alpha_D$ and $\beta_D = 0$.

that the measured interface toughness, $\Gamma_{INT}(\psi)$, exceeded 30 J/m² for the near mode II interface loadings, that is, $\psi \sim \pm 90°$. Although it was not reported, the mode I toughness of the glass substrate is almost certainly below 10 J/m². Thus, there is a range of data for this epoxy-glass interface for which $G_{KINK}$ for kinking into the glass is greater than, say, three times the mode I toughness of the glass. Why did kinking not occur?

The plastic zone size within the epoxy at the interface crack tip is on the order of several microns or more at the onset of crack propagation. Within the plastic zone, the stress on the interface is significantly reduced below the prediction of the elastic crack analysis. If the surface flaw size in the glass is larger than the plastic zone size, the kink crack is likely to be initiated when $G_{KINK}$ exceeds $\Gamma_{IC}$, or soon thereafter, because the flaw experiences the stress predicted by the elastic analysis, to a good approximation. However, if the flaw size is submicron, the plastic zone of the interface crack will subsume the flaw rendering it less potent, and either eliminating or delaying initiation of the kink crack until well beyond the energetic condition, $G_{KINK} = \Gamma_{IC}$. The surface flaws of a high-quality glass substrate can easily be submicron, and it is expected that this is the reason kinking into the glass substrate in the Lechti-Chair experiments did not occur.

To appreciate the distinction between the energetic condition for propagation of a kink crack and the condition for the initiation of a kink crack, it is important to recall that the energetic condition of fracture mechanics, $G_{KINK} = \Gamma_{IC}$, assumes the existence of an initial crack. It seems reasonable to assume that an atomistically sharp interface crack should be capable of initiating a kink crack in the adjoining material when the energetic condition $G_{KINK} = \Gamma_{IC}$ is attained, although atomistic simulations would be required to confirm this, and any anisotropy of the fracture process would come into play, such as specific fracture planes. However, interface cracking in many bimaterial systems of technological interest will be accompanied by a small fracture process zone or zone of nonlinearity, such as the plasticity in the epoxy just discussed. Under these conditions, the stresses attained at the tip of the interface crack are unlikely to attain the levels needed to initiate a kinked crack unless surface flaws in the adjoining material are present. As illustrated by the discussion of the epoxy-glass interface, a rough rule of thumb is that the flaws should be somewhat larger than the fracture process zone of the interface crack if kink cracks are expected to propagate when the energetic condition is met.

Denote the size of the fracture process zone of the interface crack by $\ell$. This length comes into play in several ways in analyzing interface cracks and kinking. First, as just discussed, $\ell$ provides a rough estimate of the minimize size of the putative surface crack, $a$, required to initiate kinking when $G_{KINK} > \Gamma_{IC}$ is attained. Second, for systems with $\beta_D \neq 0$, $r = \ell$ is a naturally choice in (3.22) defining the mode mix, $\psi$, because it then reflects the relative contribution of shear to normal stress in the vicinity where the fracture process occurs. Last, again for systems with $\beta_D \neq 0$, a natural choice for evaluating $G_{KINK}$ in (7.12) is $a = \ell$.

Thus far, the role of the stress parallel to the interface, such as a residual tension or compression in a layer, has been neglected. As will be seen, these stresses are unimportant when the putative crack is sufficiently small, but they can affect both the initiation of kinking from finite-sized flaws and the subsequent extension of the crack. With reference again to Figure 7.1(b), where it is assumed conditions are such (i.e., $\psi > 0$) that kinking occurs below the interface, denote the stress parallel to the crack in Figure 7.1 in material 2 by $\sigma_{11} = \sigma_o$. This is analogous to the 'T-stress' in a homogeneous material discussed in connection with (3.4) and (3.5). The formula for the stress intensity factors of the putative crack tip now becomes

$$K_I^{KINK} + iK_{II}^{KINK} = cKa^{i\epsilon} + \bar{d}\bar{K}a^{-i\epsilon} + \sigma_o ba^{1/2} \tag{7.16}$$

Here, $b$ is a complex valued function of $\alpha_D$, $\beta_D$ and $\Omega$, which has been computed by He et al. (1991). Plots of this coefficient are presented in Figure 7.8, where it is evident that even the dependence on the first mismatch parameter ($\alpha_D$) is relatively small. Included

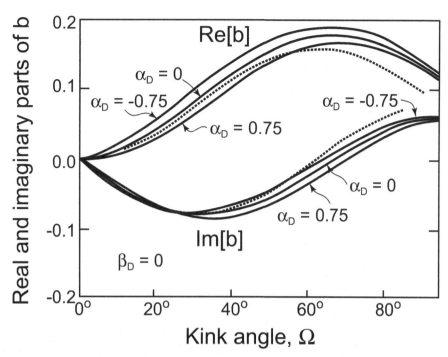

**Figure 7.8** Plots of the real and imaginary parts of $b$ as a function of the kink angle for various values of $\alpha_D$ with $\beta_D = 0$. The dashed curves are given by (7.17).

in Figure 7.8 is a useful approximate result obtained for a homogeneous material by Cotterell & Rice (1980):

$$b = \sqrt{\frac{2}{\pi}} (2 \sin^2 \Omega + i \sin 2\Omega) \tag{7.17}$$

The energy release rate and mode mix of the putative crack can be computed from 7.16 using

$$G_{KINK} = \frac{1}{\bar{E}_2} \left[ \left( K_I^{KINK} \right)^2 + \left( K_{II}^{KINK} \right)^2 \right]$$

$$\psi_{KINK} = tan^{-1} \frac{K_{II}^{KINK}}{K_I^{kink}} \tag{7.18}$$

The dimensionless parameter controlling the effect of the stress parallel to the interface is

$$\eta = \frac{\sigma_o \sqrt{a}}{\sqrt{E^* G_{INT}}} \tag{7.19}$$

As discussed in detail by He et al. (1991), the ratio of $(G_{KINK})_{max}/G_{INT}$ depends on $\psi_{INT}$, $\alpha_D$ and $\eta$ when $\beta_D = 0$. Curves of $(G_{KINK})_{max}/G_{INT}$ for various $\eta$ with no mismatch ($\alpha_D = 0$, $\beta_D = 0$) are shown in Figure 7.9. The important implication is that the

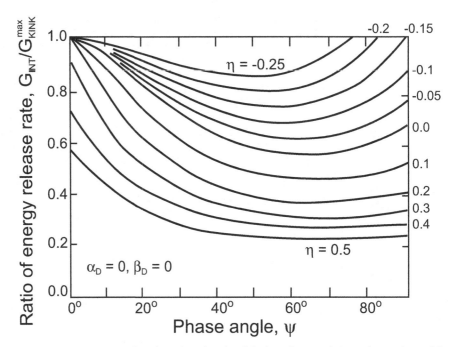

**Figure 7.9** $G_{INT}/G_{KINK}^{max}$ as a function of mode mix of the interface crack for various values of the dimensionless parameter $\eta$ measuring the component of stress parallel to the interface in the material into which crack kinking occurs. $G_{KINK}^{max}$ is the maximum energy release rate over all possible kink angles. There is no elastic mismatch for this figure.

stress parallel to the interface becomes important when $|\eta| > 0.1$. Thus, for example, depending on its sign, stress exceeding $|\sigma_o| > 100$ MPa will either enhance or diminish the likelihood of kinking for a system interface toughness 10 J/m$^2$, modulus 100 GPa and putative flaws on the order of $a \sim 1$ $\mu$m.

# 8     Crack Penetration, Deflection or Arrest?

In Chapter 6, channeling or tunneling of through-cracks in films and layers was considered where one or more of the crack edges arrested at the interface with an adjoining layer, thus confining the crack to the film or layer. This scenario is realistic only if the toughness of the adjoining material and the toughness of the interface are sufficient to suppress penetration of the crack through the interface and prevent propagation within the interface. This chapter lays out the mechanics related to the competition between penetration and deflection of a crack impinging on an interface and the related issue of crack arrest.

We begin by considering the basic mechanics of a plane strain mode I crack within material 2 with the tip at the interface, as shown in Figure 8.1(a). After characterizing the crack tip fields of this preexisting crack, the three possibilities envisioned in Figures 8.1(b–d) are addressed wherein a putative crack of length $a$, which is very short compared to the preexisting 'parent' crack, either penetrates the interface or deflects within the interface. As in the analysis of cracks kinking out of an interface in Chapter 7, the materials are assumed here to be linear elastic. When mismatch is present, these crack problems fall outside the scope of conventional elastic fracture mechanics, as will be discussed. The presentation here follows that of He & Hutchinson (1989a) and He, Evans & Hutchinson (1994).

## 8.1     Penetration and Arrest

The parent crack whose tip is at the interface as in Figure 8.1(a) is assumed to be normal to the interface and loaded symmetrically such that the stress and strain fields are symmetric (mode I) with respect to the crack. When elastic mismatch is present, the stress field at the parent crack tip does not have an inverse square root singularity. Instead, for mode I, the form of the stress field is given by $\sigma_{ij} = k_I (2\pi r)^{-\lambda} \tilde{\sigma}_{ij}(\theta)$, where $r$ is the distance from the tip, $\lambda$ measures the strength of the singularity, and $\tilde{\sigma}_{ij}(\theta)$ is a bounded variation with respect to a polar angle centered at the tip. The nonstandard intensity factor, $k_I$, is normalized such that the normal stress acting on the extended crack plane ahead of the tip in material 1 is

$$\sigma_{11} = \frac{k_I}{(2\pi x_2)^{\lambda}} \tag{8.1}$$

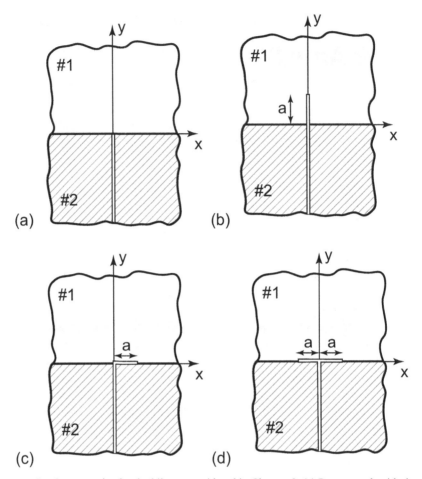

**Figure 8.1** Crack geometries for the bilayer considered in Chapter 8. (a) Parent crack with tip at the interface. (b) Putative crack penetrating the interface. (c) Putative crack deflecting into the interface. (d) Double deflection of putative interface cracks.

The singularity exponent, $\lambda$, is given in terms of the Dundurs' mismatch parameters as the root in the range $0 < \lambda < 1$ of the following equation Zak & Williams (1963):

$$cos\,\pi\lambda = \frac{2\,(\beta_D - \alpha_D)}{1 + \beta_D}\,(1 - \lambda_D)^2 + \frac{\alpha_D + \beta_D^2}{1 - \beta_D^2} \tag{8.2}$$

Curves of $\lambda$ as dependent on the mismatch parameters are displayed in Figure 8.2. If the upper material (no. 1) is compliant relative to the lower material, that is, $\alpha_D < 0$, the singularity is stronger than inverse square root, and the converse is true if the upper material is stiff relative to the lower material. It will be seen below that the compliant material ahead of the crack will be more susceptible to crack penetration than when the material ahead of the parent crack is stiffer. Note that when $\alpha_D < 0$, $\beta_D$ has essentially no influence on the singularity, but it has some effect when $\alpha_D > 0$ and the singularity is weaker than the standard singularity.

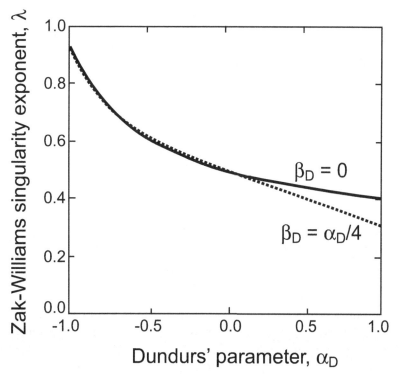

**Figure 8.2** Strength of the singularity of the parent crack in material 2 with the tip at the interface as a function of Dundurs' parameter $\alpha_D$ for $\beta_D = 0$ and $\beta_D = \alpha_D/4$.

The loads applied to the solid in the presence of the parent crack are reflected in the magnitude of $k_I$, which must be obtained from the solution to the problem for the parent crack. The length $a$ of the putative penetrating or deflecting cracks in Figures 8.1(b–d) is assumed to be very small compared to the length of the parent crack and any other relevant in-plane length quantities, such as layer thickness. Under these circumstances, the mode I stress intensity factor of the *penetrating crack* in Figure 8.1(b) can be expressed in terms of $k_I$ by

$$K_I^{penetrate} = c_p(\alpha_D, \beta_D)k_I a^{1/2-\lambda} \tag{8.3}$$

and its energy release rate is

$$G_{penetrate} = \frac{c_p^2 k_I^2}{\bar{E}_1} a^{1-2\lambda} \tag{8.4}$$

If material 1 is less stiff than material 2, that is, $\alpha_D < 0$, the singularity is stronger than that for conventional fracture mechanics. Consequently, the stress intensity and energy release rate of the emerging putative penetration crack are unbounded as $a$ emerges from zero and diminishes with increasing $a$.

For ideally brittle systems, the implication is that material 1 will always experience some penetration if a parent crack reaches the interface, assuming deflection into the interface does not intervene and assuming the parent crack is sufficiently sharp.

Conversely, if material 1 is stiffener than material 2, that is, $\alpha_D > 0$, the intensity and release rate are zero in the limit of $a = 0$ and increasing with $a$. For such material combinations, there is an intrinsic barrier to penetrating the interface which must be overcome by the presence of some type of flaw.

## 8.2     Crack Deflection at an Interface

The stress intensity factors and energy release rate of a putative crack *deflected to the right into the interface* in Figures 8.1(c, d) have the form

$$K_1^{deflect} + iK_2^{deflect} = k_I a^{1/2-\lambda} \left( d_D(\alpha_D, \beta_D) a^{i\epsilon} + e_D(\alpha_D, \beta_D) a^{-i\epsilon} \right) \qquad (8.5)$$

$$
\begin{aligned}
G_{deflect} &= \frac{\left(1 - \beta_D^2\right) k_I^2 a^{1-2\lambda}}{E*} \left[ \left( K_1^{deflect} \right)^2 + \left( K_2^{deflect} \right)^2 \right] \\
&= \frac{\left(1 - \beta_D^2\right) k_I^2 a^{1-2\lambda}}{E*} \left[ |d_D|^2 + |e_D|^2 + 2Re\left( d_D e_D \right) \right]
\end{aligned}
\qquad (8.6)
$$

where as before, $\epsilon = \frac{1}{2\pi} ln \frac{1-\beta}{1+\beta}$. Here, $d_D$ and $e_D$ are complex valued functions of the two mismatch parameters, which must be determined by solving each of the two putative crack problems in Figures 8.1(c, d). He et al. (1994) tabulated $d_D$ and $e_D$ for the two problems for a wide range of $\alpha_D$ with $\beta_D = 0$, and these values can be accessed in the accompanying software. If $\beta_D = 0$, the mode mix of the deflected crack from (8.5) is

$$\psi_{deflect} = \tan^{-1} \frac{K_2^{deflect}}{K_1^{deflect}} = \tan^{-1} \frac{Im\left[d_D(\alpha_D) + e_D(\alpha_D)\right])}{Re\left[d_D(\alpha_D) + e_D(\alpha_D)\right]} \qquad (8.7)$$

The dependence on $a$ (the putative crack length) of the energy release rate of the penetrating and deflecting cracks and on the applied load through $k_I$ are the same. Thus, the ratio

$$\frac{G_{deflect}}{G_{penetrate}} = \frac{1 - \beta_D^2}{1 - \alpha_D} \left( \frac{|d_D|^2 + |e_D|^2 + 2Re\left[d_D e_D\right]}{c_p^2} \right) \qquad (8.8)$$

depends only on the elastic mismatch and on whether single- or double-sided deflection occurs. This ratio, together with the mode mix of the deflected putative cracks, is plotted in Figure 8.3 for mismatch with $\beta_D = 0$.

The penetrating crack is mode I. If there is no elastic mismatch, the energy release rate of the deflected crack is approximately 1/4 that of the penetrating crack. If material 1 is much stiffer than the material containing the parent crack ($\alpha_D \approx 1$), the energy release rate of the deflected crack becomes large compared to that of the penetrating crack, and its mode mix is nearly mode I. Conversely, when material 1 is much more compliant than the material 2 (such that $\alpha_D \approx -1$), the energy release rate of the penetrating crack is roughly twice that of the deflected crack, which has a near mode II mode mix. There is little difference between the energy release rate of the single and doubly deflected

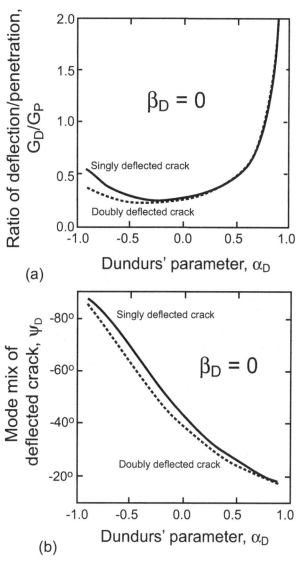

(a)

(b)

**Figure 8.3** $G_{deflect}/G_{penetrate}$ and mode mix of the deflected crack as a function of $\alpha_D$ with $\beta_D = 0$.

cracks except when $\alpha_D \approx -1$, in which case the singly deflected crack has the larger energy release rate.

The competition between penetration and deflection, or whether the loads are such that the parent crack will be arrested at the interface, is complicated by the fact that the crack tips to not have the standard inverse square root singularity. There is a distinction between case for which $\alpha_D < 0$ and $\alpha_D > 0$. If material 1 is more compliant than material 2 (such that $\alpha_D < 0$), the singularity is stronger than the conventional situation, and for ideally brittle systems, there is no barrier to initiating either the penetrating crack or the deflected crack. For elastic mismatch with $\alpha_D < 0$, solutions in the literature for the parent crack with its tip just below the interface reveal that its conventional stress intensity factor increases at the interface is approached. Thus, the tip of the parent crack is 'drawn to' the interface in these circumstances. For an ideally brittle system, the criterion for deflection over penetration or vice versa is

$$\textbf{deflection:}\ \frac{G_{deflect}}{G_{penetrate}} > \frac{\Gamma_{INT}(\psi_{INT})}{\Gamma_{IC}^{(1)}}$$

$$\textbf{penetration:}\ \frac{G_{deflect}}{G_{penetrate}} < \frac{\Gamma_{int}(\psi_{INT})}{\Gamma_{IC}^{(1)}} \tag{8.9}$$

Systems with material 1 as the stiffer material have $\alpha_D > 0$ and weaker crack tip singularities than those of conventional fracture mechanics (cf. Figure 8.2). Several consequences must be considered. First, due to the elastic mismatch, the intensity factor and energy release rate of the crack tip of a parent crack decreases as it approaches the interface from below. There is a barrier to the parent crack even reaching the interface. Many tunneling or channeling cracks in films or layers that are compliant compared to the adjoining layers do not extend all the way to the interface, and thus the issue of deflection or penetration is moot. If the tip of the parent crack does reach the interface, there are additional barriers to deflection and penetration. Even in an ideally brittle system, flaws are required in both cases to initiate the putative cracks. If there is reason to believe that the initial cracklike flaws in the interface and in the surface of material 1 are both of similar size, $a$, then the ratio $G_{deflect}/G_{penetrate}$ is independent of $a$, and the criterion (8.9) governing the competition between deflection and penetration still applies. The critical value of $k_I$ then depends on attaining the critical value of the respective energy release rate, which in turn depends on $a$. If the size of the two sets of flaws are different, say, $a_{penetrate}$ and $a_{deflect}$, then by (8.4) and (8.6), the ratio (8.8) becomes

$$\frac{G_{deflect}}{G_{penetrate}} = \frac{1 - \beta_D^2}{1 - \alpha_D} \left( \frac{|d_D|^2 + |e_D|^2 + 2Re[d_D e_D]}{c_p^2} \right) \left( \frac{a_{deflect}}{a_{penetrate}} \right)^{1-2\lambda} \tag{8.10}$$

Apart from the factor $\left(a_{deflect}/a_{penetrate}\right)^{1-2\lambda}$, this is the ratio in (8.8) which is plotted in Figure 8.3. Criterion (8.9) applies with this modification.

For modest elastic mismatch, for example, in the range $-0.25 < \alpha_D < 0.25$, $\lambda \approx 1/2$ and $G_{deflect}/G_{penetrate} \approx 1/4$. As a useful rule of thumb, it follows that the toughness of material 1 must exceed about four times the toughness of the interface (at a mode mix of $\psi_{int} \approx -45°$) if the objective is to avoid penetration through the interface.

The discussion thus far has assumed ideally brittle systems. The conditions listed above are energetically necessary for propagation of putative cracks, but they are not sufficient conditions. The remarks of Section 7.4 for interface kinking concerning the role of the fracture process zone size, $\ell$, of the parent crack apply as well to the deflection and penetration of putative cracks. In particular, if the flaws in the surface of material

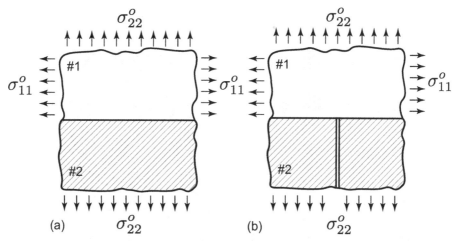

**Figure 8.4** (a) Residual stress in the bilayer system affecting penetration and deflection. (b) Problem for the problems of penetration of material no. 1 or deflection into the interface from the parent crack. Residual stress parallel to the interface in material no. 2 is taken into account in the evaluation of the stress intensity factor of the parent crack.

1 are distinctly smaller than $\ell$ and the crack process zone is confined to the interface, then a putative crack penetrating material 1 is not likely to be initiated. In such cases, deflection into the interface is likely to occur when the load becomes large enough, even if criterion (8.9) indicates otherwise.

As in the interface kinking problem, a prestress, such as a residual stress, in material 1 has an influence on the competition between deflection and penetration. In this competition, the stress parallel to the interface in material 1, denoted as $\sigma_{11}^o$, influences penetration, and a prestress normal to the interfere, denoted as $\sigma_{22}^o$, influences deflection (see Figure 8.4). The prestress component in material 2, denoted as $\sigma_{11}^{(2)}$, influences the loading of the parent crack through $k_I$. The software accompanying the text allows one to consider geometries with initial stresses such as those shown in Figure 8.4, to introduce putative deflected interface cracks or penetrating cracks, and to compute the energy release rate (and mode mix for deflection) of these cracks. In this way, the roles of both prestress components can be taken into account, as can the role of $\beta_D$ if that is deemed important.

The influence of the two residual stress components, $\sigma_{11}^o$ and $\sigma_{22}^o$, on the energy release rate ratio, $G_{deflect}/G_{penetrate}$, in Figure 8.5 have been taken from He et al. (1994). In these figures, the dimensionless residual stress parameters are

$$\eta_1 = \frac{\sigma_{11}^o a^\lambda}{k_I}; \quad \eta_2 = \frac{\sigma_{22}^o a^\lambda}{k_I} \tag{8.11}$$

Assignment of numerical values to these parameters requires knowledge of the putative crack length $a$ and intensity factor $k_I$ of the parent crack. The curves in Figure 8.5 provide quantitative results for the qualitative trends one expects from simple physical

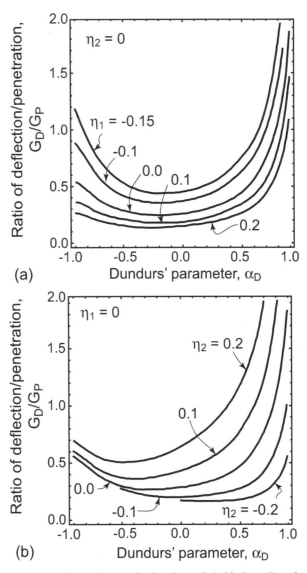

**Figure 8.5** $G_{deflect}/G_{penetrate}$ as dependent on elastic mismatch (with $\beta_D = 0$) and residual stress measured by $\eta_1 = \sigma_{11}^o a^\lambda / k_I$ and $\eta_2 = \sigma_{22}^o a^\lambda / k_I$. (a) Various $\eta_1$ with $\eta_2 = 0$, and (b) various $\eta_2$ with $\eta_1 = 0$.

reasoning. Namely, a tensile stress $\sigma_{11}^o$ promotes penetration while a compressive value of this stress discourages penetration, with no effect on deflection in either case. Similarly, a tensile value of $\sigma_{22}^o$ promotes deflection while a compressive value discourages deflection, with no effect on penetration.

# 9     Edge and Corner Interface Cracks

Chapters 4 and 5 cover delamination in multilayers that have interface crack lengths that are extremely large in comparison to other dimensions, in which case the energy release rate becomes independent of the crack size (i.e., the steady-state limit). However, when the interface crack length is comparable to other dimensions in the problem, the energy release rate generally depends on the interface crack length, and as such, numerical solutions are needed. This chapter describes the relationship between crack length and crack tip parameters, and in so doing provides guidelines that identify crack lengths for which the steady-state solution is accurate.

The focus of coverage is on plane strain (two-dimensional) geometries, with a limited discussion of an interface crack at the corner of a thin film, which is inherently three-dimensional. It will be demonstrated that the two-dimensional results provide the necessary insight for most cases of interest. The two-dimensional results in this chapter were generated using the software described in Chapter 16 and are essentially a reiteration of the results in original papers by Zhuk, Evans, Hutchinson & Whitesides (1998) and Yu, He & Hutchinson (2001). Related coverage of three-dimensional problems is also provided in Chapter 11, which addresses interface cracks between a semi-infinite substrate and patterned lines, that is, thin film strips bonded to very thick substrates.

## 9.1     Interface Edge-Cracks: The Transition to Steady State

Consider the interface cracks located at the edge of a film bonded to a substrate, as illustrated in Figure 9.1. One of the remarkable aspects of thin film mechanics is how quickly a short edge-crack approaches the steady-state limit for a semi-infinite crack. The rapidity of this transition depends on whether the film edge is aligned with the edge of the substrate, as in Figure 9.1(a), or whether the film edge is interior to the edge of the substrate, as in Figure 9.1(b). In this section, we focus on two-dimensional film/substrate systems in plane strain with an infinitely long crack front in the out-of-plane direction. A limited discussion of semicircular crack fronts is provided in the next section, while three-dimensional aspects of debonding of finite width features are considered in Chapter 11.

Suppose the film away from its edges experiences a uniform equi-biaxial residual misfit contraction $\theta$ relative to the substrate, as, for example, due to a thermal strain or a deposition strain as introduced in Chapter 2. The uniform contraction $\theta$ is the strain that

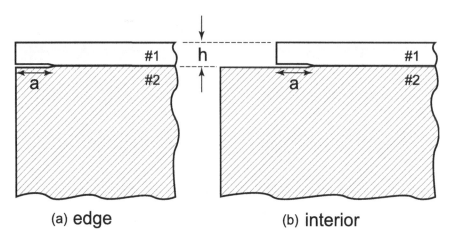

(a) **edge**                                    (b) **interior**

**Figure 9.1** Scenarios in which an interface crack emerge flush with the substrate edge or from a corner of the film lying interior to the substrate.

would occur in the film relative to the substrate if the film were detached from substrate. The stress that develops in the attached film near its edge depends on the details of how it is attached to the substrate and differs in the two cases depicted in Figure 9.1. Nevertheless, the uniform equi-biaxial tensile stress in the film remote from its edge is the same in both cases.

In plane strain, the steady-state energy release rate for the long 'semi-infinite' interface crack derived earlier in Chapter 4 applies to both cases in Figure 9.1:

$$G_{ss} = \frac{\left(1 - v_1^2\right)\sigma^2 h}{2E_1} \tag{9.1}$$

where $\sigma$ is the equi-biaxial stress present in the film far from its edges. The associated steady-state mode mix $\psi_{ss}$ depends on the Dundurs, parameters and is described in Chapter 4.

Results for the two plane strain interface edge crack problems shown in Figure 9.1 are shown in Figure 9.2. These results were generated using the software described in Chapter 16, and they are essentially identical to the data presented by Yu et al. (2001). For the case in which the film edge is aligned with the substrate edge, the energy release rate $G$ is within 10% of its steady-state value $G_{ss}$ when the crack length is greater than 10 times the film thickness ($a/h \gtrsim 10$) when there is no elastic mismatch. The approach to steady state is more rapid for stiffer substrates ($\alpha_D < 0$) and slower for more compliant substrates ($\alpha_D > 0$). The mode mix $\psi$ approach to steady state occurs quite rapidly, attaining steady-state values when $a/h \sim 3$. In all cases, $G$ approaches $G_{ss}$ from below such that, as has been emphasized previously, the use of $G_{ss}$ to approximate $G$ is conservative.

Now consider the case in which the edge crack emerges at the interface of a film whose edge is attached to the interior of the substrate surface, as depicted in Figure 9.1(b). (In some industries, this is referred to as a 'dropped-edge'.) These results

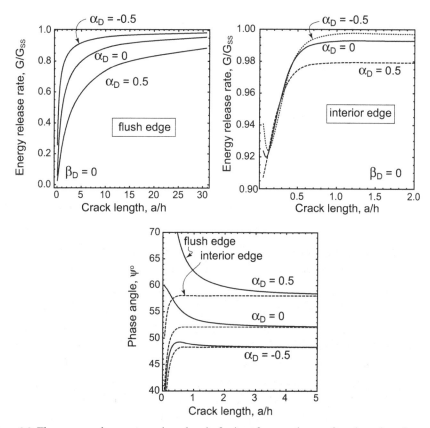

**Figure 9.2** The energy release rate and mode mix for interface cracks as a function of crack length, for several levels of elastic mismatch and two edge conditions. The curves labeled 'flush edge' correspond to the geometry in Figure 9.1(a), while those labeled 'interior edge' refer to the geometry in Figure 9.1(b).

are labeled as 'interior edge' in Figure 9.2. For all practical purposes, the crack behaves as if it were semi-infinite regardless of its length, and when $a/h \gtrsim 0.5$ it is within a percent or so of the steady-state limit with essentially no influence of elasticity mismatch (cf. Figure 9.2). There is an initial singularity when $a = 0$ at the 90° corner, which depends on elastic mismatch, but evidently the mismatch effect on the energy release rate disappears as soon as the edge crack emerges. As shown on the bottom of Figure 9.2, the mode mix $\psi$ attains the steady-state limit almost immediately for the interior edge crack.

There is a relatively simple explanation as to why the delamination from an interior edge attains steady state for interface cracks that are a small fraction of the film thickness. The elastic energy in the separated film behind the crack tip is almost zero at any crack length. The distribution of the stress in the substrate near the tip simply translates with the crack tip, as does the stress distribution in the film. The length of the separated film has essentially no influence on these distributions, and thus, in effect, the interface crack propagates in steady state for all crack lengths.

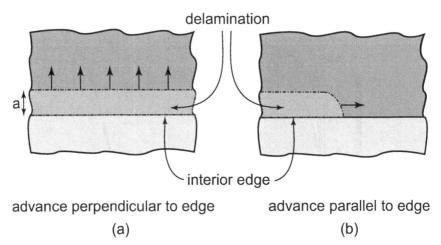

Figure 9.3 Competing modes of crack propagation along a film edge that is interior to the substrate. Normal (a) and parallel (b) crack advance have essentially the same energy release rate.

By contrast, for the case where the film edge is aligned with the substrate edge, the distributions of stress in the substrate and film change as the crack tip advances away from the substrate edge toward the interior. This is because the substrate region over which the crack tip stresses are important scales with the film thickness; for cracks whose crack tip is close to the free edge, these fields are altered by the stress-free surface, and hence the crack tip senses the presence of the surface. Simply put, the environment experienced by the crack tip does not change for delamination from an interior edge, while for an edge aligned with the substrate edge the environment changes significantly as the crack tip moves away from the substrate edge.

Figure 9.2 reveals that a blanket film whose edge aligns with the substrate has built-in protection from edge-crack initiation in the sense that, even if the relevant interface toughness falls slightly below $G_{ss}$, the film will be able to tolerate edge flaws on the order of the film thickness or more. This is especially true for substrates that are compliant compared to the film ($\alpha_D > 0$). By contrast, Figure 9.2 suggests that a film with an interior edge will have no built-in tolerance, and extremely small edge flaws are likely to trigger extensive interface delamination if $G_{ss}$ exceeds the interface toughness.

Another unusual feature of interface cracks located along an interior edge is that it is equally energetically favorable for the crack to advance parallel to the edge as it is for the crack to advance normal to the edge, as sketched in Figure 9.3. Interface crack propagation parallel to the edge is also a steady-state process, and the average energy release rate of the propagating crack front is also $G_{ss}$ in (9.1), to a very accurate approximation. This is due to the fact that the energy release per new area of a separated interface is $(1/a) \int_0^a G(a)da$, where $G(a)$ is the energy release rate plotted in Figure 9.2, which, in turn, is nearly $G_{ss}$. The mode mix $\psi$ for the crack front advancing perpendicular to the edge is in the range $50°$ to $60°$.

Begley & Ambrico (2001) and Ambrico & Begley (2003) have computed the energy release rate along two-dimensional interface flaws located along edges at film corners.

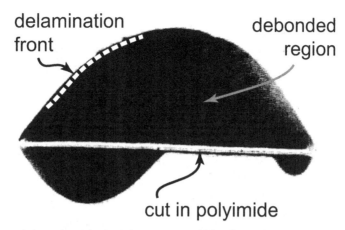

**Figure 9.4** A straight cut is made through a polyimide film bonded to glass substrate (Jensen et al. 1990). The film is under biaxial tension.

In all cases, the curved front of the crack advancing parallel to the edge has a mode III component, and hence mode mix is a combmination of all three modes. Thus, while normal and parallel crack advance along the interior edge are equally favorable energetically, the mode dependence of the interface toughness is expected to play some role, as of now yet not determined, in the manner in which a delamination spreads from an edge flaw. It is likely that a delamination spreading from an edge flaw takes place by a combination of normal and parallel propagation.

A long straight through-crack in a film, extending from the film surface down to the interface, that is under equi-biaxial residual tension $\sigma$ creates an edge condition that is very similar to a film with an interior edge. The so-called 'cut test' (Jensen, Hutchinson & Kim 1990) where a long cut through the film is deliberately introduced has been used to initiate interface crack delamination in films subject to residual tensile stress and, in some cases, to obtain estimates of interface toughness. The outcome of a typical cut test is shown in Figure 9.4. The fact that the interface delamination has spread away from the cut implies $G_{ss} \geq \Gamma_c(\psi_{ss})$ or equivalently from (9.1), $\sigma\sqrt{h} \geq \sqrt{2E_1\Gamma_C(\psi_{ss})/(1-v_1^2)}$. A more definitive relation between the film stress, the interface toughness and the shape of the delamination requires a detailed mixed-mode analysis, including the role of mode III, such as that given by Jensen et al. (1990). On the bottom of the cut in Figure 9.4, one can see that for some reason, perhaps an imperfect cut, delamination has not spread along the entire cut edge.

The discussion above has assumed that the stress in the attached film is tensile. As a consequence, it is evident from the mode mix that the mode I stress intensity factor is positive and a gap opens up between the separated film and the substrate. If the mismatch strain $\theta$ is positive, the biaxial stress $\sigma$ in the blanket film will be compressive. In this case, the separated film remains in contact with the substrate, and, at least approximately, the crack experiences mode II conditions at its tip. This case must be treated on its own merits, as has been done for some special problems Stringfellow & Freund (1993); Balint & Hutchinson (2001); Hutchinson & Hutchinson (2011). If frictional interaction

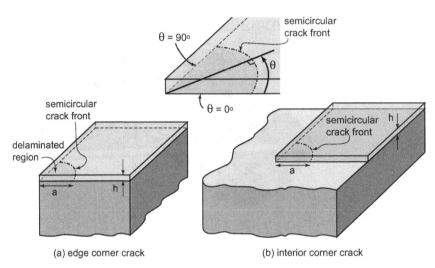

**Figure 9.5** Circular corner cracks: (a) the edge corner crack where the film is flush with the substrate edges; (b) the interior corner crack where the film terminates on the surface of the substrate.

between the film and substrate can be ignored, then (9.1) still stands as the steady-state energy release rate. The compressed film undergoes a mode II edge delamination, which would be resisted by the mode II interface toughness. The approach to steady-state delamination for the two types of edges would be qualitatively similar to those discussed for films under tension in Figure 9.2. In particular, for a film in compression, the attainment of steady state would be almost immediate for interior edge cracks.

## 9.2    Interface Cracks at Thin Film Corners

A natural question is whether delamination is more likely to initiate at flaws along the edge of a film or at a corner feature in the film. Guided by the results discussed for the delamination from an edge, it is clear that one must anticipate distinguishing between film corners that are aligned with the corner of the substrate, as shown in Figure 9.5(a), and corners that terminate at an interior point of the substrate surface, as shown in Figure 9.5(b). Here, the behavior of a semicircular interface crack centered on the corner is considered, as shown in Figure 9.5.

The energy release rate and mode mix vary with position around the crack front with a three-dimensional corner singularity existing at $\theta = 0°$ and $90°$ where the crack meets the edges of the film. The problem for the distribution of $G$ and mode mix is a three-dimensional problem. A numerical finite element solution has been presented by Begley & Ambrico (2001). The energy release rate is relatively uniform around most of the crack front except near the film edges at $\theta = 0°$ and $90°$ where it is largest due to a special singularity at those points. The mode mix involves all three stress intensity factor components, except along the symmetry line on $\theta = 45°$ where $K_{III} = 0$.

The ratio of the energy release rate $G$ along $\theta = 45°$ to the plane strain steady-state energy release rate in (9.1) is plotted in Figure 9.6 as a function of $a/h$ for the corner

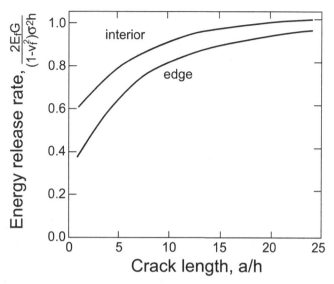

**Figure 9.6** Ratio of the energy release rate at the symmetry point ($\theta = 45°$) to the plane strain steady-state energy release rate, as a function of crack length, for the edge corner crack and interior corner crack. Results are shown for $\alpha_D = 0$ and $\beta_D = 0$, although the results for $-0.3 < \alpha_D < 0.3$ and $\beta_D = 0$ are virtually identical (Begley & Ambrico 2001).

geometries shown in Figure 9.5. The results are taken from Begley & Ambrico (2001), which includes many more results, including energy release rate and mode-mix distributions along the crack front. The energy release rate of the symmetry point along the crack tip is approximately the same as one would obtain for $G$ averaged over the crack front, because there is only a small portion of the crack front near $\theta = 0°$ and $90°$ where $G$ is larger than $G_{ss}$. For quite large delamination cracks, say, $a/h > 20$, $G$ approaches the steady-state energy release rate (9.1) along the vast majority of the crack front. However, for relatively small cracks, $a/h < 2$, the energy release rate of the corner delamination crack is distinctly less than that of an edge crack of the same length, as seen by comparing Figures 9.2 and 9.6 for the same crack length. This comparison holds true for both corner geometries shown in Figure 9.5.

The conclusion to be drawn from these results is that interface delamination starting from a film corner is less likely than delamination initiated at a flaw somewhere along the film edge. (This assumes roughly comparable flaw sizes at each location.) A caveat is that the large, highly localized energy release rate along the edge at a corner may nucleate a small crack in that vicinity, but not enough energy will be available to delaminate the entire corner, assuming $G_{ss}$ in (9.1) is less than the relevant film toughness.

## 9.3    Interface Cracks Approaching Edges: Delamination Arrest

Figure 9.7 illustrates several scenarios in which a semi-infinite interface crack approaches a film edge (or another crack). This leads to a small region of intact interface often referred to as a 'ligament'. When the crack tip is close to either a free surface or

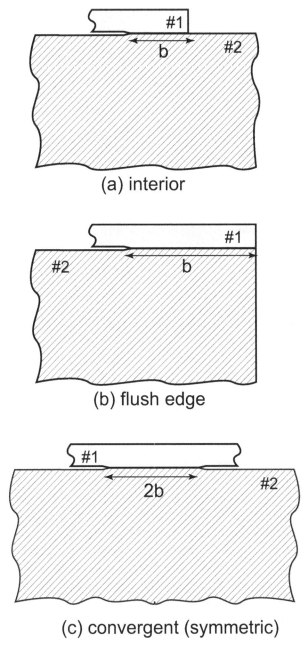

(a) interior

(b) flush edge

(c) convergent (symmetric)

**Figure 9.7** Three different 'convergent' crack geometries, in which the crack tip approaches a ligament of finite width.

another crack, there is a reduction in the energy stored ahead of the crack, and as a result the energy release rate decreases with ligament size. While an interface crack advancing inward from an interior film edge attains steady state almost immediately, a semi-infinite interface crack approaching an interior edge (Figure 9.7(a)) senses the edge at relatively large distances (ligament sizes). This effect explains why large portions of a film may

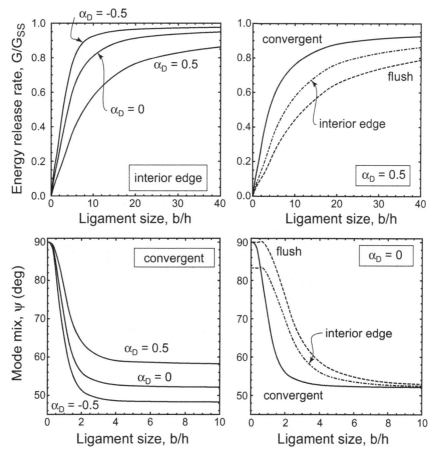

**Figure 9.8** Energy release rate as a function of ligament size, for the cases shown in Figure 9.7. The results were generated using the software LayerSlayer FEA described in Chapter 16 and are virtually identical to those first published by Yu et al. (2001).

delaminate with sections of the film near an edge still attached. By measurement of the remaining attached ligament size, the effect has proved useful in obtaining estimates of the interface toughness (Zhuk et al. 1998).

As in the earlier discussion of edge delamination, consider a thin film on a deep substrate where the film experiences equi-biaxial contractive mismatch $\theta$ relative to the substrate, giving rise to an equi-biaxial tensile stress $\sigma$ in the interior of the film well away from its edges. Imagine, as depicted in Figure 9.7(a), that the film to the left of the interface crack tip has completely separated from the substrate with a ligament of length $b$ still attached. Assume the interface crack separates under plane strain conditions. The same scenario is depicted in Figure 9.7(b), only with the film edge flush with the edge of the substrate. Two 'convergent' interface cracks that approach one another are shown in Figure 9.7(c).

Figure 9.8 shows the dependence of the energy release rate and mode mix on ligament size for these scenarios, for several different levels of elastic mismatch. The results were generated using the software described in Chapter 16 and are identical those first elucidated by Yu et al. (2001). Note that the case with two convergent interface cracks defines the ligament size as $2b$: this implies the only difference between this case (Figure 9.7(c)) and the flush edge case (Figure 9.7(b)) is the boundary condition imposed along the symmetry line. The effects of elastic mismatch are quantitatively and qualitatively similar for each edge condition shown in Figure 9.7.

The distance from the edge where the energy release rate drops well below the steady-state rate depends on the film/substrate elastic mismatch. A film on a compliant substrate (with $\alpha_D = 0.5$ in Figure 9.8) begins to sense the edge when $b \sim 40h$, while the distance for a film on stiff substrate (with $\alpha_D = -0.5$) is about $b \sim 10h$. These are appreciable distances. The steady-state mode mix persists to distances closer to the edge and then transitions to a mode II crack ($\psi = 90°$) for flush edges and convergent cracks. The crack approaching an interior edge is also predominantly mode II, with only a slightly larger mode I contribution than the other cases. The crack will arrest when $G$ drops below $\Gamma_{INT}(\psi)$. For an interface crack with a strong mixed-mode toughness dependence, that is, $\Gamma_{INT}(90°) \gg \Gamma_{INT}(0°)$, both the drop in $G$ and the increasing $\Gamma_{INT}(\psi)$ as the crack tip approaches the edge contribute to arrest.

# 10    Buckling Delamination

For some multilayer systems that experience compression, buckling of the film in an initially debonded region may release energy that becomes available to drive the interface crack and increase the size of the debonded region. This phenomenon, called buckling delamination, occurs in structural composites where a compressed surface layer has debonded from its adjacent layers and buckles when the applied compression is sufficiently large. This surface layer may then delaminate if the interface toughness is not large enough to keep the interface crack from propagating. Buckling delamination is one of the most widely observed failure mechanisms in thin films and coatings on substrates when the stress is compressive. Buckling delamination involves the simultaneous interaction of buckling and interface cracking.

Two of the most commonly observed morphologies of buckling delaminations are seen in Figure 10.1. Both straight-sided and telephone cord delaminations have occurred in Figure 10.1(a) for a multilayered thin film deposited on a glass substrate where the film is subject to equi-biaxial compression. Straight-sided delaminations are rare under equi-biaxial compression. However, they are commonly observed when there is a dominant direction of compression in the film, in which case they propagate perpendicular to the direction of maximum compression. The telephone cord morphology is the mostly commonly observed morphology when the film is under equi-biaxial compression.

The buckle delamination in Figure 10.1(b) has formed under artificially constrained conditions created by patterning on the interface in a long narrowing wedge-shaped region of low adhesion before the film is deposited. The delamination, which takes place only within the low adhesion region, was initiated at the wide end and propagated towards the narrow edge, transitioning from the telephone cord morphology to the straight-sided morphology. Even when the film-substrate interface is not patterned, the delamination arrests along its sides and propagates at the curved front. Figure 10.2 shows a sequence of three closely spaced stages of the front propagation; the upper figures are a numerical solution discussed later, and the lower figures are the experimental observation.

This chapter develops the underlying mechanics of the buckling delamination with particular emphasis on the role of mixed-mode interface toughness, which is essential to understanding the existence, propagation and morphology of the various delamination modes.

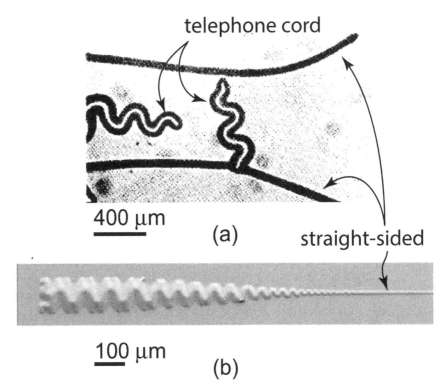

telephone cord

400 µm

(a)    straight-sided

100 µm

(b)

**Figure 10.1** Buckle delaminations. (a) Straight-sided and telephone cord delamination of a multilayer class film under equi-biaxial compression on a glass substrate (picture supplied by M.D. Thouless). (b) A diamondlike carbon film of several hundred nanometers thickness with equi-biaxial compressive stress ~2 $GPa$ deposited on a silicon substrate. Before depositing the film, a long, wedge-shaped region of low adhesion is created on the substrate such that the buckle delamination, which initiates at the left end and propagates to the right, does not spread outside the low adhesion region. In the wide regions the telephone cord morphology appears, while in the narrower region the straight-sided morphology appears (Moon et al. 2004).

## 10.1    A Simple Example: Buckling Delamination of a Symmetric Bilayer

Consider the symmetric clamped bilayer plate of width $2L$ shown in its buckled state in Figure 10.3(a). The ends are constrained from rotation, and a compressive load per unit depth (perpendicular to the page) denoted with $2P$ is applied at the right end, which experiences a horizontal displacement $\Delta$. Each of the two layers has Young's modulus $E$, Poisson's ratio $v$ and thickness $h$. It will be assumed that the plate is constrained against out-of-plane displacement such that plane strain conditions apply with the plane strain modulus $\bar{E} = E/(1 - v^2)$. An initially debonded region of width $2b$ exists in the center of the bilayer. The total width $2L$ is assumed to be below that at which the uncracked bilayer would buckle as a clamped wide Euler column, that is, $L < \pi\sqrt{2\bar{E}h^3/(3P)}$. The following analysis adopts the one-dimensional Kármán-Föppl nonlinear plate theory described in Section 2.2 to model the buckled layers within the region $-b \leq x \leq b$ as

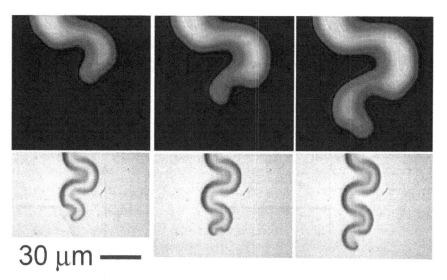

**Figure 10.2** Propagation of a telephone cord buckle delamination of a thin film under equi-biaxial compression (Faou, Parry, Grachev & Barthel 2012). A sequence of three steps reveals how the curved front advances. The upper row is numerical simulation, with light tones indicating larger heights above the substrate. The bottom row is an experimental system: a 120 μm film of molybdenum with equi-biaxial compressive stress of $\sim 2.2\ GPa$ on a silicon substrate.

clamped wide plates buckling in the symmetric manner shown in Figure 10.3(a). The example is amenable to a relatively simple analysis but nevertheless illustrates a number of basic aspects of buckle delamination.

According to Kármán-Föppl theory, the buckling deflection of the upper plate perpendicular to the centerline, $w(x)$, and the buckling load at the onset of buckling, $P_c$, are

$$w(x) = \frac{\delta}{2}\left(1 + \cos\left[\frac{\pi x}{b}\right]\right) \text{ and } P_c = \frac{\pi^2}{12}\frac{\bar{E}h^3}{b^2} \tag{10.1}$$

where $\delta$ is the peak transverse displacement at the center. The displacement at the right end at the onset of buckling is

$$\Delta_c = \frac{P_c L}{\bar{E}h} = \frac{\pi^2}{12}\frac{h^2 L}{b^2} \tag{10.2}$$

The theory predicts that the postbuckling behavior of the clamped plate is such that the buckling deflection increases under constant load, $P = P_c$. Thus, the relationship between $P$ and $\Delta$ has the form indicated in Figures 10.3(b) and 10.3(c). A nonlinear theory more accurate than Kármán-Föppl theory, such as elastica theory, predicts that the postbuckling load increases very slightly after the onset of bucking, but the prediction of Kármán-Föppl theory is accurate and adequate for analyzing buckling delamination as long as the layer is thin, $b/h \gg 1$, and the maximum slope of the buckling deflection is not larger than about $20°$, which, in turn, holds as long as $\delta/b$ is less than about $(2/\pi)\tan 20° = 0.23$.

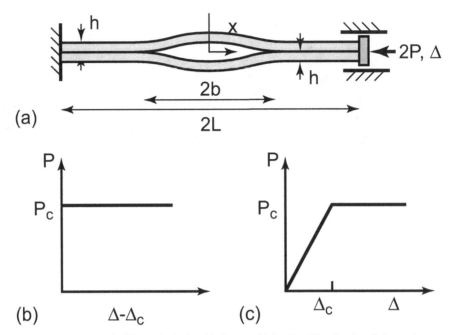

**Figure 10.3** (a) Symmetric bilayer in the buckled state. (b) Postbuckling load-end shortening behavior predicted by nonlinear von Karman plate theory. (c) Load-end shortening behavior.

The nature of this system shows that the loads above $P_c$ cannot be prescribed, and unlimited deflections occur when $P = P_c$. Therefore, rather than prescribing $P$, we prescribe the end shortening $\Delta$. For $\Delta > \Delta_c$, the elastic energy in the bilayer system per depth is the work done to bring it to its current state, that is, the area under the curve in Figure 10.3(c):

$$U = \frac{1}{2}P_c\Delta_c + P_c(\Delta - \Delta_c) = P_c\left(\Delta - \frac{1}{2}\Delta_c\right) \tag{10.3}$$

Noting that both $P_c$ and $\Delta_c$ depend on the interface crack length, $2b$, the energy release rate (per crack tip) is

$$G = -\frac{1}{2}\frac{dU}{db} = \frac{\pi^2}{12}\frac{\bar{E}h^3}{b^3}(\Delta - \Delta_c) \tag{10.4}$$

By symmetry, the interface crack experiences mode I conditions. Denote the mode I toughness by $\Gamma_I$ and assume that the interface is ideally brittle such that the crack advances when $G = \Gamma_I$. Setting $G = \Gamma_I$ in (10.4) and solving for $\Delta$ in terms of $b$ gives

$$\frac{\Delta}{L} = \frac{\pi^2}{12}\left(\frac{h}{b}\right)^2 + \frac{12\Gamma_I}{\pi^2\bar{E}L}\left(\frac{b}{h}\right)^3 = \frac{\Delta_c}{L} + \frac{12\Gamma_I}{\pi^2\bar{E}L}\left(\frac{b}{h}\right)^3 \tag{10.5}$$

This relation gives the applied end shortening required to cause the crack to advance in the interface, that is, the end shortening required for continued buckling delamination given an existing crack length of $2b$. The relation is sketched in Figure 10.4. The end shortening at the onset of buckling is also shown.

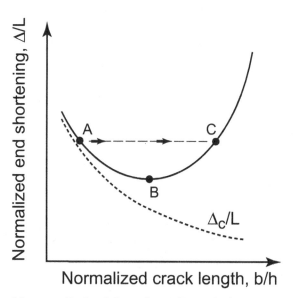

**Figure 10.4** Sketch of the normalized end shortening, $\Delta/L$, required to propagate the interface crack as a function of the existing normalized crack length, $b/h$, from (10.5). The normalized end shortening for the onset of buckling, $\Delta_c/L$ is also shown.

For a preexisting crack with normalized length $b/h$, no buckling occurs until the end shortening attains $\Delta_c$. Then, with imposed end shortening increasing above $\Delta_c$, the energy release rate increases according to (10.4) until the value of $\Delta/L$ attains the value on the upper curve in Figure 10.4. The crack then begins to propagate. For crack lengths greater than that associated with point B in Figure 10.4, crack growth is stable in the sense that $\Delta$ must be increased to continually advance the crack. This is reflected by the fact that, to the right of point B, the curve of $\Delta$ versus $b$ is increasing. This is stable quasi-static crack advance.

On the other hand, for preexisting cracks with lengths less than that associated with point B (such as point A), the curve of $\Delta$ versus $b$ is initially decreasing. If $\Delta$ is imposed and an increment in crack growth occurs, the imposed end shortening exceeds that required to advance the crack. The crack is unstable and will propagate dynamically until it arrests at point C, where the condition for incipient crack growth is again reached and where the crack is stable. This argument ignores the possibility of any 'dynamic overshoot' beyond point C. The crack length associated with the minimum of the curve in Figure 10.4 at point B is

$$\frac{b}{h} = \left( \frac{\pi^4}{216} \frac{\bar{E}L}{\Gamma_I} \right)^{1/5} \tag{10.6}$$

Thus, in this example, by (10.5), increasing the toughness of the interface increases the end displacement necessary to propagate the delamination and, by (10.6), slightly lowers the crack length below which delamination is unstable.

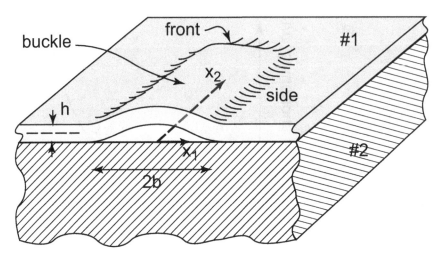

**Figure 10.5** Geometry of the straight-sided buckle delamination.

## 10.2     Straight-Sided Buckling Delamination of Thin Films on Thick Substrates

Buckling delamination is commonly observed for thin films or coatings on thick substrates where stress in the film or coating is compressive arising from a deposition process or from thermal expansion mismatch with the substrate. The mechanics developed in this section also applies to the buckling delamination of a thin surface layer of a composite laminate subjected to overall mechanical compression. The geometry of the system in the buckled state is shown in Figure 10.5 for the case of a straight-sided buckle delamination propagating with a curved front. The substrate is modeled as a semi-infinite elastic half space, and the film layer is again modeled by nonlinear Karman-Föppel plate theory of Section 2.3 under the assumption that the lateral dimension of the buckled region is large compared to the layer thickness and that the magnitude of the buckle slopes do not exceed about $20°$.

The elastic properties of the film and substrate are taken to be isotropic, but differences in their elastic properties are taken into account. The film's Young's modulus and Poisson's ratio are $E_1$ and $v_1$, while those of the substrate are $E_2$ and $v_2$. The film thickness is $h$. The stresses in the film in the unbuckled state, $\sigma_{ij}^\circ$ are assumed to be uniform, and the $(x_1, x_2)$ axes are chosen such that $\sigma_{12}^\circ = 0$ with $\sigma_{11}^\circ < 0$ and $\sigma_{11}^\circ \leq \sigma_{22}^\circ$. Take $\sigma_o = -\sigma_{11}^\circ$ as the maximum compressive stress component in the unbuckled film, which is positive in compression. The results in this section are largely drawn from the article by Hutchinson & Suo (1992).

We begin by analyzing a *section of the straight-sided buckle delamination* shown in Figure 10.5. The section is assumed to lie sufficiently far behind the curved section of the delamination front such that the stresses and strains do not depend on $x_2$. Thus, the one-dimensional 'beam' theory of Section 2.2 is applicable. The information needed to compute the energy release rate and stress intensity factors of the interface crack involves the stress changes between the system in the buckled and unbuckled state. These changes are subject to plane strain. That is, because the substrate is very thick relative to the film (infinitely deep in the model), the substrate constrains the film during buckling such that no changes in the $x_2$-direction occur. It will also turn out that for this part of the analysis the stress component $\sigma_{22}^{\circ}$ has no influence on the results. The buckle is symmetric about $x_1 = 0$, and the half-width of the preexisting interface crack is $b$. Unlike the first example in this chapter, there is no symmetry about the crack plane, and thus the problem must be regarded as a mixed-mode interface crack problem. Mixed-mode behavior of the buckle delamination is an essential element to understanding its behavior, as will be seen.

Analyses which model the elasticity of both the film and substrate have shown that a model which represents the film as a plate clamped at the edge of the delamination is a good approximation, as long as the film modulus is not greater than about three times the substrate modulus, that is, $\alpha_D \leq 1/2$ (Mei, Landis & Huang 2011; Cotterell & Chen 2000; Parry, Colin, Coupeau, Foucher, Cimetière & Grilhé 2005; Yu & Hutchinson 2002). For more compliant substrates, the displacements and rotation at the edges of the delamination become important, and the assumption of clamped conditions must be relaxed. In this chapter, attention will be limited to systems with relatively stiff substrates such that the assumption that the film is clamped at the edge of delamination is a good approximation.

The nonlinear plate equations for the one-dimensional wide plate representation of the film from Section 2.2 reduce to the following equation for the buckling deflection $w(x)$ normal to the substrate and the average compressive stress in the film in the buckled state, $\bar{\sigma}$:

$$D\bar{w}'''' + \bar{\sigma} h w'' = 0 \text{ on } -b \leq x \leq b, \ w(\pm b) = w'(\pm b) = 0 \qquad (10.7)$$

with $x \equiv x_1$, $(\ )' = d(\ )/dx$, $D = \bar{E}_1 h^3/12$, $\bar{E}_1 = E_1/(1 - v_1^2)$ and

$$\bar{\sigma} = \sigma_o - \frac{\bar{E}_1}{4b} \int_{-b}^{b} (w')^2 \, dx \qquad (10.8)$$

According to the nonlinear plate theory, $\bar{\sigma}$ is independent of $x$, but it is reduced below the compressive stress in the unbuckled state, $\sigma_o$. Equation (10.8) arises from the condition derived from (2.25) that there is no change in the displacement $u_1$ at the ends of the buckled region.

The solution to the problem posed by (10.7) and (10.8) is

$$w(x) = \frac{\delta}{2} \left( 1 + \cos \left[ \frac{\pi x}{b} \right] \right) \text{ with } \frac{\delta}{h} = \sqrt{\frac{4}{3} \left( \frac{\sigma_o}{\sigma_c} - 1 \right)} \qquad (10.9)$$

and

$$\bar{\sigma} = \sigma_c = \frac{\pi^2}{12} \bar{E}_1 \left( \frac{h}{b} \right)^2 \qquad (10.10)$$

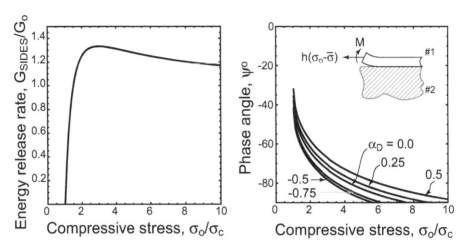

**Figure 10.6** The energy release rate and mode mix along the sides of the straight-sided buckle delamination. The insert shows the load changes at the edge of the delamination used to determine the mode mix. $-90° < \psi < 0$ implies $K_I > 0$ and $K_{II} < 0$.

The moment at the right end of the buckled film is $M = Dw''(b) = \pi^2 D\delta/(2b^2)$, and the compressive force/depth is $\bar{\sigma} h$.

With reference to the insert in Figure 10.6, we make use of the general results in Chapter 4 for the energy release rate, $G$, and the stress intensity factors, $K = K_1 + iK_2$, for an interface crack between a deep substrate and thin layer loaded by $P$ and $M$:

$$G = \frac{1}{2\bar{E}_1}\left(\frac{P^2}{h} + \frac{12M^2}{h^3}\right)$$

$$K = h^{-i\epsilon}\left(\frac{1-\alpha_D}{1-\beta_D^2}\right)^{1/2}\left(\frac{P}{\sqrt{2h}} - i\frac{\sqrt{6}M}{h^{3/2}}\right)e^{i\omega}$$

(10.11)

where $\alpha_D$, $\beta_D$, $\epsilon$ and $\omega$ are defined for plane strain in Chapter 4. The interface crack singularity is produced by the *change* from the unbuckled state and, thus, $P = (\sigma_o - \bar{\sigma})h$. Substitution of $P$ and $M$ into (10.11) for $G$ gives, after some manipulation,

$$\frac{G_{SIDE}}{G_o} = \left(1 - \frac{\sigma_c}{\sigma_o}\right)\left(1 + 3\frac{\sigma_c}{\sigma_o}\right)$$

(10.12)

with $G_o = h\sigma_o^2/(2\bar{E}_1)$ as the elastic energy per unit area stored in the bonded film available for release subject to the plane strain constraint. The notation $G_{SIDE}$ is used to emphasize that this is the energy release rate of the interface crack along the sides well behind the curve section of the delamination front, and the normalized plot of $G_{SIDE}$ is given in Figure 10.6. The stress component $\sigma_{22}^\circ$ in the film contributes to the elastic energy per unit area, but it does not influence the energy available for release under the plane strain constraint, $G_o$, which depends only on $\sigma_o = -\sigma_{11}^\circ$.

The detailed elasticity solution of Chapter 4 is required to obtain the mode mix; (4.11) gives for the crack tip at $x = b$

$$\tan \psi = \frac{Im\left[Kh^{i\epsilon}\right]}{Re\left[Kh^{i\epsilon}\right]} = \frac{\cos \omega + \left(\frac{Ph}{\sqrt{12}M}\right) \sin \omega}{-\sin \omega + \left(\frac{Ph}{\sqrt{12}M}\right) \cos \omega} \qquad (10.13)$$

and from the above solution, $Ph/(\sqrt{12}M) = (1/2)\sqrt{\sigma_o/\sigma_c - 1}$. This expression permits consideration of elastic mismatches for which a nonzero value of the second Dundurs' parameter, $\beta_D$, is taken into account and, thus, $\epsilon \neq 0$. The subtlety associated with $\beta_D \neq 0$ is relatively unimportant for most buckle delamination problems, and we proceed in the remainder of this chapter to take $\epsilon = \beta_D = 0$ such that (11.13) becomes

$$\tan \psi = \frac{K_{II}}{K_I} = \frac{\cos \omega + \frac{1}{2}\left(\sqrt{\frac{\sigma_o}{\sigma_c} - 1}\right) \sin \omega}{-\sin \omega + \frac{1}{2}\left(\sqrt{\frac{\sigma_o}{\sigma_c} - 1}\right) \cos \omega} \qquad (10.14)$$

Recall that $\omega$ is a function of the first Dundurs' elastic mismatch parameter $\alpha_D$ and is plotted in Figure 4.2. For no mismatch, $\omega = 52.07°$.

The mode mix for the straight-sided buckle is plotted in Figure 10.6. For $\sigma_o/\sigma_c$ only slightly greater than one, $\tan \psi = -1/\tan \omega$, or $\psi = \omega - \pi/2$ ($K_I > 0$ and $K_{II} < 0$), which for $\alpha_D = 0$, is $\psi = -37.9°$. The mode mix becomes pure mode II ($K_I = 0$ and $K_{II} < 0$), at $\sigma_o/\sigma_c = 1 + 4\tan^2 \omega$. For no elastic mismatch, mode II is attained when $\sigma_o/\sigma_c = 7.59$ and $\delta/h = 2.96$. For larger values of the applied film stress mode II prevails. In the range $\sigma_o/\sigma_c > 1 + 4\tan^2 \omega$, (10.12) is only approximately valid because the derivation of (10.12) does not account for contact of the crack faces or frictional dissipation due to the sliding of the faces. The phase angle in Figure 10.6(b) has been truncated at $\psi = -90°$; beyond this point, $K_I < 0$ such that the crack tip is presumably closed and pure mode II is a fair approximation.

An alternative calculation of $G_{SIDES}$, which does not rely on the general solution in Chapter 4, can be carried out by calculating the total elastic energy change/depth, $U(b)$, in the unbuckled state minus that in the buckled state, far behind the front, and then evaluating the energy release rate as $G_{SIDES} = (1/2)dU/db$. This alternative procedure gives precisely (10.12).

Now consider the energy released by an advance $\Delta a$ of the crack front in the $x_2$-direction in Figure 10.5. When the length of the straight sides is much greater than the width of the delamination, this is a steady-state problem in the sense that the problem before and after the advance is identical, except shifted by a distance $d$ $\Delta a$ in the $x_2$-direction. The total energy released is the difference between the energy/depth in the unbuckled state and that in the buckled state far behind the front. The energy release rate on the curved front of the interface crack is expected to vary with position, but the energy release rate averaged over the front is precisely

$$\bar{G}_{FRONT} = \frac{1}{b} \int_0^b G_{SIDES}(b) db$$

That is, in steady state the average energy released at the front per unit length of propagation is equal to the energy released far behind the front per unit length of delamination. The two energy release rates are related by

$$\bar{G}_{FRONT} = G_o \left(1 - \frac{\sigma_c}{\sigma_o}\right)^2 = G_{SIDES} \left(\frac{1 - \frac{\sigma_c}{\sigma_o}}{1 + 3\frac{\sigma_c}{\sigma_o}}\right) \tag{10.15}$$

In spite of the fact that $\bar{G}_{FRONT}$ is always less than $G_{SIDES}$, propagation of a straight-sided buckle delamination, once it is established, does occur at the front and not along the sides. To understand why, it is necessary to take into account the difference in the mode mix on the curved front from that on the sides. For this reason, it is useful to consider next circular buckle delaminations before delving into the details of the role played by mixed mode interface toughness.

## 10.3    Circular Buckle Delaminations for Films and Coatings under Equi-biaxial Compression

In this section attention is limited to films or coatings on deep substrates where the stress in the unbuckled layer is uniform equi-biaxial compression, $\sigma_{ij}^o = -\sigma_o \delta_{ij}$. The circular patch of the layer above a circular interface crack of radius, $R$, will buckle away from the substrate in an axisymmetric mode when the combination of $R$ and $\sigma_o$ becomes critical. As in the two previous examples, buckling gives rise to mixed mode stress intensities at the crack edge and releases energy available along the radius of the crack. The results in this section are taken from Hutchinson & Suo (1992), and they are based on the axisymmetric specialization of the nonlinear plate equations given in Section 2.3. More extensive results on circular delaminations, including experiments and a study of the stability of the circular crack front to nonaxisymmetric perturbations, can be found in Hutchinson, Thouless & Liniger (1992).

A circular patch of debonded film or coating of radius $R$ subject to equi-biaxial compression $\sigma_o$ begins to buckle in an axisymmetric mode when $\sigma_o$ attains the critical stress $\sigma_c^*$. If one models the debonded patch of the film as a clamped Karman-Föppel plate, the critical stress and mode shape are given by

$$\sigma_c^* = 1.2235\bar{E}_1 \left(\frac{h}{R}\right)^2$$
$$w(r) = \delta \left(0.2871 + 0.7129 J_o \left(\frac{3.8317r}{R}\right)\right) \tag{10.16}$$

where $r$ is the distance from the center of the patch, $\delta$ is the maximum buckle deflection which occurs at the center, and $J_o$ is the Bessel function of the first kind of order zero. As in the case of the straight-sided buckle, the assumption that the film is clamped at the edge is a good approximation as long as the substrate modulus is not less than about a third of the film modulus, that is, $-1 \leq \alpha_D \leq 1/2$.

Unlike the straight-sided delamination problem, it is not possible to solve the nonlinear plate equations in closed form for the circular configuration at stresses far above the onset of buckling. One must resort to numerical analysis, which was done in the papers cited above. However, it is possible to obtain asymptotic results for the energy release rate and mode mix, which are good approximations for film stresses in the range $1 \leq \frac{\sigma_o}{\sigma_c} < 2$ (Hutchinson & Suo 1992):

$$\frac{G}{G_o^*} = \frac{1}{1 + 0.9021(1 - v_1)} \left[ 1 - \left( \frac{\sigma_c^*}{\sigma_o} \right)^2 \right] \tag{10.17}$$

$$\frac{\delta}{h} = \left[ \frac{1}{0.2473(1 + v_1) + 0.2231(1 - v_1^2)} \left( \frac{\sigma_o}{\sigma_c^*} - 1 \right) \right]^{1/2} \tag{10.18}$$

$$\tan \psi = \frac{\cos \omega + 0.2486(1 + v_1) \left( \frac{\delta}{h} \right) \sin \omega}{-\sin \omega + 0.2468(1 + v_1) \left( \frac{\delta}{h} \right) \cos \omega} \tag{10.19}$$

The total energy/area in the unbuckled film is $G_o^* = (1 - v_1)\sigma_o^2 h / \bar{E}_1 = 2G_o/(1 + v_1)$, where as defined earlier, $G_o = \sigma_o^2/(2\bar{E}_1)$ is the elastic energy/area in the unbuckled film available to be released under plane strain constraint. Accurate numerical results for $G/G_o^*$ and for the dimensionless combination of edge load changes, $Ph/(\sqrt{12}M)$, required to determine the mode mix using (10.13), are shown in Figure 10.7 for equi-biaxial compressive stresses as large as $\sigma_o/\sigma_c^* = 40$.

Based on the plot of $Ph/(\sqrt{12}M)$ in Figure 10.7 and (10.13), the mode mix $\psi$ is presented as a function of $\sigma_o/\sigma_c^*$ in Figure 10.8 for three values of the film Poisson's ratio, $v_1$, for the case of no elastic mismatch between film and substrate. Similar to the behavior on the sides of the straight-sided delamination seen in Figure 10.6, the circular delamination crack is mixed mode with the mode II component becoming more dominant as $\sigma_o/\sigma_c^*$ increases. However, the rate of increase of the ratio of mode II to mode I is not nearly as large for the circular delamination, and, even at very large film stresses, that is, $\sigma_o/\sigma_c^* = 40$, there is still a mode I component. The difference in the mode mix at the curved crack front from that of the straight front is at the heart of why the straight-sided buckle delamination propagates at its curved front and not along its straight sides. It also is central to why telephone cord delaminations tend to be the preferred morphology for films under an equi-biaxial stress state.

## 10.4 Propagation Conditions for the Straight-Sided Delamination

As noted in the introduction to this chapter, delaminations with the telephone cord morphology tend to form for films and coatings subject to equi-biaxial compression, but straight-sided delaminations usually form when the stress state has a dominant compression direction. Straight-sided delaminations propagate in the direction perpendicular to the direction of maximum compression. The energy release rate and mode mix along the sides well behind the curved front are given by (10.12) and (10.14). The average energy release rate along the curved front is given by (10.15), and this will be used to approximate the energy release rate at the propagating front. None of these three quantities are influenced by the stress component $\sigma_{22}^\circ$ acting parallel to the direction of propagation.

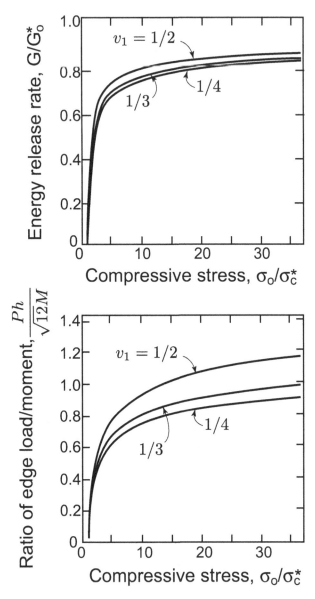

**Figure 10.7** Energy release rate and dimensionless edge loading ratio for a circular delamination of radius $R$ subject to an equi-biaxial compressive stress $\sigma_o$ in the unbuckled state: (a) $G/G_o^*$ with $G_o^* = (1 - v_1)\sigma_o^2 h/E_1$ versus stress, $\sigma_o/\sigma_c^*$ with $\sigma_c^*$ given by (10.16). (b) $Ph/\sqrt{12M}$ versus stress. Results for the circular delamination depend on $v_1$.

The mode mix in Figure 10.8 predicted for a circular blister of radius $R = b$ subject to equi-biaxial stress will be used as an approximation for the mode mix at the curved front of the straight-sided delamination in Figure 10.5. More accurate results for $G$ and $\psi$ at the front would require a full numerical analysis of the detailed geometry such as that carried out for telephone cord delamination discussed in the next section.

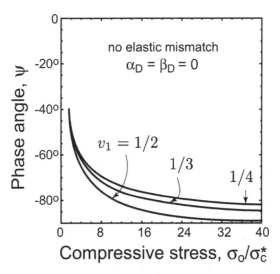

**Figure 10.8** Mode mix of the interface crack of a circular buckle delamination for three values of the film Poisson's ratio. For this example, there is no elastic mismatch between the film and the substrate.

In the following, the compressive stress $\sigma_o = -\sigma_{11}^\circ$ required for steady-state propagation of the straight-sided delamination will be determined together with its half-width $b$. These are the two unknowns, given the elastic properties of the film and the substrate, the thickness of the film (which is assumed to be very small relative to the substrate) and the properties of the interface. In some applications where the compressive stress in the film is regarded as known, the thickness of the film or coating that gives rise to delamination is sought: the critical thickness will also emerge in the analysis. This process will be illustrated for the specific mixed-mode interface toughness displayed in Figure 3.8:

$$\Gamma_{INT} = \Gamma_I f(\lambda, \psi) \text{ with } f(\lambda, \psi) = 1 + \tan^2\left[(1 - \lambda)\,\psi\right] \qquad (10.20)$$

where $\Gamma_I$ and $\lambda$ are regarded as specified and the mode mix $\psi$ differs on the sides and curved front as detailed above. Recall that the ratio of the mode II to mode I toughness, $\Gamma_{II}/\Gamma_I$, depends on $\lambda$, as plotted in Figure 3.8(b). Denote the solution for mode mix along the sides (10.14) by $\psi = \Psi_{SIDES}(\sigma_o/\sigma_c)$ where the dependence on the elastic mismatch is implicit. Similarly, denote the mode mix at the front derived by Figure 10.8 by $\psi = \Psi_{FRONT}(\sigma_o/\sigma_c)$ where the dependence on $v_1$ and elastic mismatch is left implicit. The argument $\sigma_o/\sigma_c$ used in $\Psi_{FRONT}$ is deliberate, and it must be obtained from the plot in Figure 10.8 using $\sigma_o^*/\sigma_c = 1.2235(12/\pi^2) = 1.488$.

The condition for propagation along the sides is $G_{SIDES} = \Gamma_I f(\lambda, \Psi_{SIDES})$ or

$$\frac{G_o}{\Gamma_I} = \frac{f(\lambda, \Psi_{SIDES}(\sigma_o/\sigma_c))}{\left(1 - \frac{\sigma_c}{\sigma_o}\right)\left(1 + 3\frac{\sigma_c}{\sigma_o}\right)} \qquad (10.21)$$

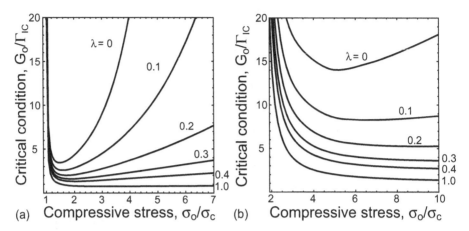

**Figure 10.9** Critical conditions for propagation of a straight-sided buckle delamination for various values of the parameter $\lambda$ in (10.20) which determines the relative importance of mode II to mode I in the interface toughness: (a) along the sides well behind the front; (b) at the curved front. In both plots $G_o = \sigma_o^2 h/(2\bar{E}_1)$ and $\sigma_c = (\pi^2/12)\,\bar{E}_1(h/b)^2$. No elastic mismatch between the film and substrate and in (b) $v_1 = 1/3$.

This relation is plotted for several values of $\lambda$ in Figure 10.9(a) for no elastic mismatch. Note that

$$\frac{\sigma_o}{\sigma_c} = \frac{12\sigma_o}{\pi^2 \bar{E}_1}\left(\frac{b}{h}\right)^2 \tag{10.22}$$

The case for $\lambda = 1$ requires special consideration. It corresponds to the limit in which the interface toughness does not depend on mode mix, that is, $G_{SIDES} = \Gamma_I$. Because the curve in Figure 10.9(a) for $G_o/\Gamma_I$ decreases monotonically for $\sigma_o/\sigma_c > 2.3$, any delamination crack with $\sigma_o/\sigma_c > 2.3$ which begins to propagate will not arrest and will delaminate the entire film. Mixed mode toughness of the interface with $0 \le \lambda < 1$ is required to arrest the delamination as it spreads.

For interfaces with $\lambda < 1$, an interface crack that begins to propagate under conditions lying to the right of the minimum of $G_o/\Gamma_I$ in Figure 10.9(a) will continue to propagate only if $\sigma_o$ is increased. The essential point is this: were it not for an interface that is tougher in mode II than mode I, a delamination crack would completely separate the film from the substrate as soon as it begins to propagate.

The condition for propagation along the curved front of the straight-sided delamination is $G_{FRONT} = \Gamma_I f(\lambda, \Psi_{FRONT})$ or

$$\frac{G_o}{\Gamma_I} = \frac{f(\lambda, \Psi_{FRONT}(\sigma_o/\sigma_c))}{1 - \left(\frac{\sigma_c}{\sigma_o}\right)^2} \tag{10.23}$$

This relation is plotted for several values of $\lambda$ in Figure 10.9(b) for no elastic mismatch. At a value of $\sigma_o/\sigma_c$ around 2, but depending on $\lambda$, $G_o/\Gamma_I$ increases much more slowly with respect to $\sigma_o/\sigma_c$ at the front than on the sides, due to the differences in the mode mix evolution at the respective locations. At some $\sigma_o/\sigma_c$, the value of $G_o/\Gamma_I$ required

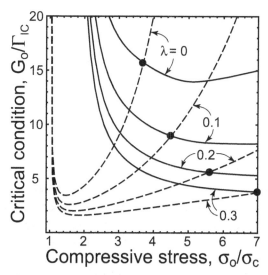

**Figure 10.10** Critical conditions for propagation of a straight-sided buckle delamination on its sides (dashed curves) and front (solid curves) superimposed for various values of the parameter $\lambda$, which determines the mixed mode interface toughness dependence. No elastic mismatch between the film and substrate $v_1 = 1/3$ for the results at the front. The solid points correspond to critical conditions simultaneously satisfied on the sides and the front.

to propagate along the sides exceeds that along the front, assuming $\lambda < 1$. This aspect is seen in Figure 10.10 where the two sets of curves are superimposed. A solid point indicates the combination of $Y \equiv G_o/\Gamma_I$ and $X \equiv \sigma_o/\sigma_c$ for which critical conditions on the front and sides are simultaneously satisfied. We will regard the solid point as characterizing the conditions for steady-state propagation of the straight-sided buckle delamination.

Assuming that the solid point $(X, Y)$ specifies the critical condition for a given $\lambda$, $v_1$ and elastic mismatch, one can solve for the unknowns of interest associated with propagation of the straight-sided buckle delamination:

$$\sigma_o^2 h = 2Y\bar{E}_1\Gamma_I; \quad \frac{b}{h} = \frac{\pi}{2\sqrt{3X}}\left(\frac{\bar{E}_1 h}{2Y\Gamma_I}\right)^{1/4} \tag{10.24}$$

The first of the above formulas combines the critical compressive stress and the thickness: if the thickness is doubled with no other changes, the critical stress decreases by $1/\sqrt{2}$ and so forth. The ratio of the half-width to the thickness depends on the dimensionless combination $\bar{E}_1 h/\Gamma_I$. Increasing $\Gamma_I$ permits larger $\sigma_o^2 h$ and decreases $b/h$, but the dependence of the latter is rather weak. Recalling that smaller values of $\lambda$ correspond to interfaces with larger ratios of mode II to mode I toughness (cf. Figure 3.8(b)), one notes from Figure 10.10 that smaller values of $\lambda$ increase $Y$ and decrease $X$. Thus, enhanced mode II toughness relative to $\Gamma_I$ always increases the critical combination of $\sigma_o^2 h$ and may increase or decrease $b/h$ depending on values of the system parameters.

## 10.5    Telephone Cord Delaminations and Other Aspects of Buckling Delamination

The combination of buckling and fracture is irresistible for many researchers, including the authors of this book, and consequently there is now a fairly large literature on buckling delamination, including both theory and experiments – far too large to cover with any completeness in this chapter. Telephone cord buckling delaminations are particularly intriguing. While considerable insight into the existence of the telephone morphology has appeared in the literature, it is only very recently that detailed and convincing numerical simulations have been published. In this section, background to several morphologies, including the telephone cord, is presented, followed by a brief focus on the most recent work in telephone cord delamination. The reader will have to consult the literature for aspects such as the role of compliant substrates with the references cited in Section 10.2, delamination on curved surfaces and ridge cracks in the film due to tensile bending strains.

Insight into the existence of the telephone cord delamination is gained from considering the buckling pattern of an infinitely long plate of width $2b$ clamped along its edges and subject to equi-biaxial compression, $\sigma_o$, in the unbuckled state. The nonlinear plate equations of Section 2.3 pertain. The onset of buckling occurs when $\sigma_o$ attains $\sigma_c$ given by (10.10) in the wide plate Euler mode. As $\sigma_o$ increases above $\sigma_c$, the amplitude of the Euler mode increases until at some $\sigma_o$, bifurcation occurs into a symmetric mode (the varicose mode) or an antisymmetric mode (akin to a telephone cord mode) with a sinusoidal variation in the long direction of the plate (Jensen 1993; Audoly 1999), illustrated in Figure 10.11(a). Full details are given by Audoly (1999), including the values of $\sigma_o/\sigma_c$ at bifurcation as dependent on $v_1$ and whether the first bifurcation mode is the varicose mode or the telephone mode and the preferred wavelength, $L$.

Moon et al. (2004) computed the difference between the average elastic energy/area in the unbuckled state, $G_o^* = (1 - v_1)\sigma_o^2 h/E_1$, and that of the energy/area in the buckled state, $\bar{U}$. The average energy release rate for delamination of width $2b$ is $\bar{G} = G_o^* - \bar{U}$, where averaging is interpreted as propagation over a full wavelength of the mode. The normalized average energy release rate, $\bar{G}/G_o^*$ as a function of $\sigma_o/\sigma_c$ is plotted in Figure 10.11(b) for $v_1 = 0.3$. For $\sigma_o/\sigma_c < 6.5$, only the Euler mode exists. For $\sigma_o/\sigma_c > 6.5$, the Euler mode becomes unstable. For $\sigma_o/\sigma_c > 6.5$, the telephone cord morphology has the lowest energy in the buckled state and gives rise to the largest average energy release rate. Buckling undulations in the lengthwise direction release some of the compression in that direction, and the telephone cord undulations are more effective than the symmetric varicose undulations when $\sigma_o/\sigma_c$ is large. The buckle delamination that has propagated down the variable width, low adhesion region in Figure 10.1(b) illustrates the transition from the telephone cord mode to the Euler mode when the width is such that $\sigma_o/\sigma_c = (12/\pi^2)\left(\sigma_o/\bar{E}_1\right)(b/h)^2 < 6.5$.

While the results just discussed provide insights into the telephone cord morphology, they have only limited applicability to commonly observed telephone cord delaminations because the width of the delamination in Figure 10.11 has been artificially prescribed. Telephone cord delaminations form on thin film-substrate systems where the

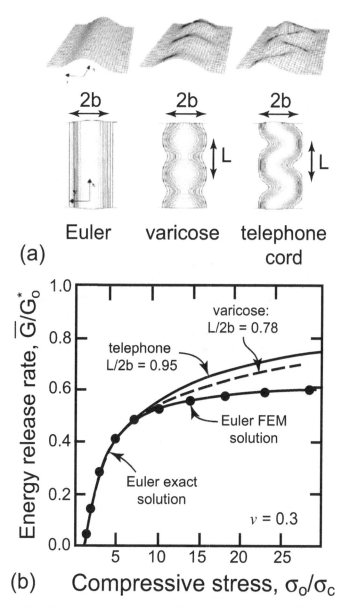

Figure 10.11 (a) Three buckle morphologies for infinitely long plates of width $2b$ clamped along their edges and subject to equi-biaxial compressive stress $\sigma_o$. (b) Ratio of the average energy release rate $\overline{G}$ to the elastic energy per area in the unbuckled state, $G_o^* = (1 - v_1)\sigma_o^2 h/E_1$, as a function of the overstress ratio, $\sigma_o/\sigma_c$, for the three buckle morphologies. At compressive stress above $\sigma_o/\sigma_c = 6.5$ (for $v_1 = 0.3$), the telephone cord mode releases the most energy. Adapted from Moon et al. (2004).

interface properties are nominally the same over large expanses of a film. The boundaries are selected by the delamination, not the analyst!

To our knowledge, the first realistic simulation of telephone cord buckle delamination is that of Faou et al. (2012) shown in the upper half of Figure 10.2. These authors use a finite element plate representation of the film. They employ a mixed-mode cohesive zone with interface toughness of the form (11.20) to model the interface fracture process at the edge of the delamination and link this cohesive zone to the plate displacements at each location along the edge. The calculation is highly nonlinear in several respects, including the fact that as the curved front propagates, oscillating back and forth, portions of the plate that have separated at the interface come back into contact with the substrate as the front advances, as has been observed experimentally.

One of the longstanding objectives of researchers analyzing buckle delamination is to be able to use measured characteristics of a delamination, such as its width, amplitude and, in the case of the telephone cord morphology, the wavelength of the in-plane oscillations to infer properties of the system and, in particular, the interface toughness. In a paper subsequent to that discussed above, Faou, Parry, Grachev & Barthel (2015) employed their computational model for this purpose. These authors developed a relation between the ratio of the wavelength of the telephone cord and its width, $\ell/(2b)$, and the dimensionless combination $(\sigma_o/\sigma_c)\sqrt{G_o^*/\Gamma_I - 1}$. This relation enables one to determine the mode I toughness of the interface, $\Gamma_I$, if the film thickness and the biaxial compressive stress are known and the ratio $\ell/(2b)$ is measured. As long as the ratio of the mode II to mode I toughness of the interface is significant, for example, $\Gamma_{II}/\Gamma_I > 3$, the mix-mode toughness parameter $\lambda$ in (10.20) has a relatively weak effect on the inferred value of $\Gamma_I$ according to their analysis.

# 11 Delamination of Thin Strips (Patterned Lines)

The focus of this book up to this point has been on 'blanket' films, whose in-plane dimensions greatly exceed the film thickness and are largely immaterial. Of course, many applications, particularly microelectronics that involve patterned lines similar to those shown in Figure 11.1, involve features that cannot be idealized as a semi-infinite sheet of material. In this chapter selected topics related to three-dimensional (3D) aspects of thin film delamination are discussed. The subject is quite rich and complex from a mechanics standpoint, and the chapter deals with only some of the most basic issues. Many issues remain open to investigation. In the order addressed, the topics considered here are (1) the effect of the film strip width on the stored elastic energy due to a mismatch strain, (2) the energy release rate for steady-state delamination as dependent on strip width and (3) the energy release rates for short interface cracks for aligned and interior strip ends. The results are largely those first reported in the papers by Zhuk et al. (1998) and Yu & Hutchinson (2003).

## 11.1 Film Strips: The Stored Elastic Energy Due to a Mismatch Strain

The film strip, or line as it is sometimes called, depicted in Figure 11.1 has a thickness $h$, width $w$, Young's modulus $E_f$ and Poisson's ratio $v_f$. The substrate is taken to be infinitely deep with Young's modulus $E_s$ and Poisson's ratio $v_s$. An equi-biaxial in-plane mismatch strain $\theta$ between the bonded film and the substrate is assumed to exist such that in the interior of a blanket film (i.e., in the interior of a very wide, long strip) the stress and elastic energy/area in the film are

$$\sigma_{xx} = \sigma_{yy} = -\frac{E_f\theta}{1-v_f}; \quad \Lambda_B = \frac{E_f h\theta^2}{1-v_f} \tag{11.1}$$

The stress and energy density in a film strip of finite width are not as large for the blanket film due to stress relaxation along the sides. The stress is further relaxed near the strip ends. If the film width is on the order of the thickness, the constraint of the substrate on the film in the width direction (the $y$-direction in Figure 11.1) has almost no influence, and only the stress in the $x$-direction is significant. Well away from the ends, the stress and energy/area in a long narrow strip are

$$\sigma_{xx} = -E_f\theta; \quad \sigma_{yy} \simeq 0; \quad \Lambda_U = \frac{1}{2}E_f h\theta^2 \tag{11.2}$$

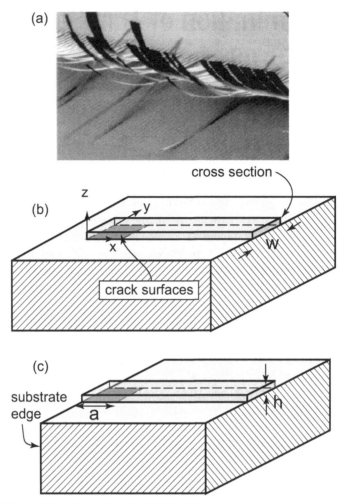

**Figure 11.1** (a) Long film strips undergoing full interface delamination with a few strips that remain bonded or partially bonded (Zhuk et al. 1998). (b) A strip delaminating from an interior end on the substrate surface. (c) A strip delaminating from an end that is aligned with the edge of the substrate. The strip width is $w$, and the interface crack length is $a$.

One can anticipate that (11.1) will apply for very wide strips except near their sides and ends. Similarly, (11.2) will hold for very narrow strips away from the ends. The difference between these two limits depends only on $v_f$, but nevertheless, it is fairly large: $\Lambda_U/\Lambda_B = (1 - v_f)/2$, which is $1/3$ for $v_f = 1/3$.

The stress and energy/area in the film strip vary across the width of the strip and in the $x$-direction near the ends. The energy/area averaged across the width of the strip far from its ends, $\bar{\Lambda}$, has been evaluated by Yu & Hutchinson (2003) as a function of strip width to thickness ratio. It is plotted in Figure 11.2 for $v_f = 0.3$ and for several values of $\alpha_D$ with $\beta_D = 0$. Strips with widths satisfying $w/h \leq 2$ are sufficiently narrow such that $\bar{\Lambda} \simeq \Lambda_U$. Moreover, stiff films on compliant substrates with $\alpha_D \simeq 1$, such as

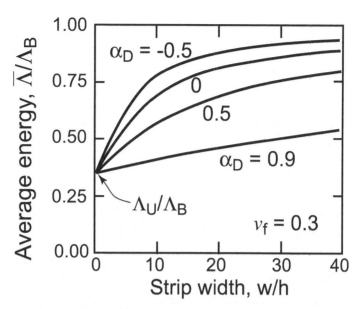

**Figure 11.2** Ratio of the average energy/area in the strip well away from its ends, $\overline{\Lambda}$, to the energy density of a blanket film under equi-biaxial mismatch strain $\Lambda_B$ as a function of the width to thickness ratio of the strip, for several values of $\alpha_D$ with $\beta_D = 0$ and $v_f = 0.3$. Under steady-state propagation of an interface delamination crack, the energy release rate averaged across the crack front is $\overline{G}_{ss} = \overline{\Lambda}$. From Yu & Hutchinson (2003).

metal or ceramic films on polymer substrates, will have an average energy density much closer to the narrow strip limit $\Lambda_U$ than to the blanket film limit $\Lambda_B$ even for quite wide films, that is, for $\sim 10 < w/h \lesssim\sim 40$. When there is no elastic mismatch with $\alpha_D = 0$ in Figure 11.2, $\overline{\Lambda}$ is well below $\Lambda_B$ even when $w/h = 10$ and approaches the blanket film limit $\Lambda_B$ only if $w/h \gtrsim 40$. The case $\alpha_D = -0.5$ corresponding to a substrate that is three times stiffer than the film in Figure 11.2 reveals that a moderately stiff substrate is not much different from a substrate with no elastic mismatch as far as $\overline{\Lambda}$ is concerned.

For delaminations such as those depicted in Figure 11.1, elementary energy accounting provides the steady-state energy release rate averaged over the crack front as $\overline{G}_{ss} = \overline{\Lambda}$ where $\overline{\Lambda}$ is presented in Figure 11.2 for $v_f = 0.3$. This assumes frictional dissipation is neglected if crack face contact occurs. In the next subsection, the interface crack length $a$ required to attain steady state will be discussed. The crack front in steady-state propagation will not necessarily be straight, as depicted in Figure 11.1, but nevertheless the energy released per area of interface separation averaged across the strip will be $\overline{\Lambda}$. If $\theta < 0$, then $\sigma_{xx} > 0$ in the bonded film, and the crack faces will be open; frictional effects do not occur, and the mode mix will be approximately the same as that for the plane strain version of the problem, that is, $\psi_{ss}$ between $50°$ and $60°$ depending on the elastic mismatch. If $\theta > 0$, then $\sigma_{xx} < 0$ in the bonded film, and the crack faces will be in contact over some region behind the crack tip. In this case, the crack tip is approximately mode II ($\psi_{ss} \simeq -90°$), and frictional dissipation may lower the energy/area available to separate the interface. A more precise characterization for $\theta > 0$ accounting for contact and friction requires a detailed analysis such as that given by various authors (Stringfellow & Freund 1993; Balint & Hutchinson 2001; Hutchinson & Hutchinson 2011).

The discussion in the previous subsection regarding the trends in $\bar{\Lambda}$ obviously applies to $\bar{G}_{ss}$. In particular, it is important to note that for many strips, $\bar{G}_{ss}$ will be much closer to the narrow strip limit $\Lambda_U$ than to the blanket film limit $\Lambda_B$. In other words, unless the width to thickness ratio of the strip is fairly large, the steady-state energy release rate driving delamination of the strip will be considerably smaller than that predicted for a blanket film.

## 11.2    Film Strips: Short Interface Cracks and the Transition to Steady-State Energy Release Rate for Delamination

Here, we address the length of the interface crack $a$ for the two cases in Figure 11.1 required to reach steady-state conditions, at least approximately. The discussion is complicated by the fact that there are two factors at play: (1) the role of the substrate compliance at the strip end, and (2) the fact that the separated end of the strip is constrained in the $y$-direction when $a$ is on the order of $w$ or less. Earlier results for blanket films in Chapter 9 provide ample reason to believe that a significant difference can be expected between the stress distribution near the end of a strip whose end aligns with the substrate end, and that experienced by a strip whose end terminates at the interior of the substrate surface. The higher compliance of the former reduces the elastic energy in the strip near its end relative to the other case and consequently gives rise to a lower energy release rate for short interface cracks. The second factor (2) reflects the fact that, for sufficiently small $a$, crack advance takes place under plane strain conditions, except near the strip corners. This constraint is relaxed as $a$ increases.

Film substrate constraint (1) is first addressed by considering the interface delamination of a narrow strip ($w/h = 2$) for which the second factor (2) does not come into play. Figure 11.3 compares the approach to steady state for the two strip end conditions, interior end and aligned end, depicted in Figure 11.1. The qualitative trends in the energy release rate averaged across the crack front $\bar{G}$ are similar to those discussed for the 2D plane strain case in Chapter 9. The strip whose end lies in the interior of the substrate surface attains steady-state conditions at remarkably short crack lengths, $a/h \simeq 1/2$, and the energy release rate is only slightly below the steady state value for all crack lengths computed in Figure 11.3(a). The dependence on the elastic mismatch of the steady state in this case is consistent with that in Figure 11.2 for $w/h = 2$. By contrast, $\bar{G}$ for the film strip whose end is aligned with the edge of the substrate requires much larger crack lengths to attain steady state, with stiff films on compliant substrates requiring the longest cracks. In all these examples, $\bar{G}$ approaches $G_{ss}$ from below.

Wide strips need to be considered to gain some understanding of the transition from plane strain constraint for short interface cracks, for which the misfit strain $\epsilon_{yy}$ parallel to the crack front is not released, to the steady-state limit in which all of the residual strains are released. The elastic energy/area of the blanket film under the equi-biaxial residual

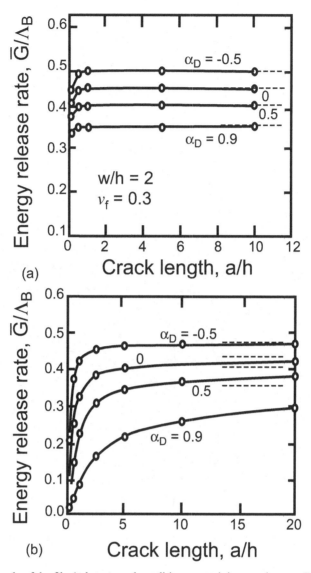

**Figure 11.3** The role of the film/substrate end condition on attaining steady state. Energy release rate averaged across the crack front, $\overline{G}$, as a function of crack length $a/h$ for a narrow strip with $w/h = 2$, $v_f = 0.3$ and various $\alpha_D$ ($\beta_D = 0$). Plot (a) applies for the case of the strip end attached to the interior of the substrate, and plot (b) applies for a strip whose end is aligned with the edge of the substrate. The dashed lines are the steady-state limit from Figure 12.2. From Yu & Hutchinson (2003).

strain $\theta$ is $\Lambda_B$ given by (11.2), and this is the energy/area released for an infinitely wide strip when it is completely separated from the surface. The energy/area of the blanket film released *subject to the plane strain constraint with no change in $\epsilon_{yy}$* is

$$\Lambda_o = \frac{E_f h \theta^2}{2(1 - v_f^2)} \tag{11.3}$$

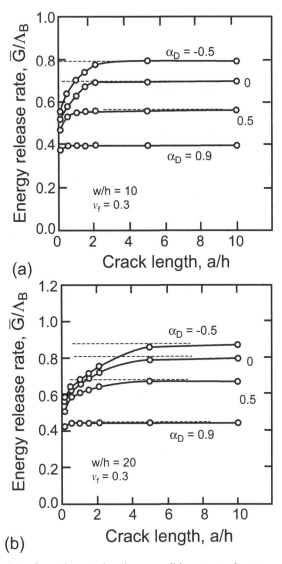

**Figure 11.4** The transition from plane strain release conditions to steady-state propagation for film strips whose ends are attached to an interior point of the substrate surface: (a) for $w/h = 10$ and (b) for $w/h = 20$. Plotted is the energy release rate averaged across the crack front $\overline{G}$ as a function of crack length $a/h$ for $v_f = 0.3$ and various $\alpha_D$ ($\beta_D = 0$). The dashed lines are the steady-state limit from Figure 11.2. From Yu & Hutchinson (2003).

Note that $\Lambda_o/\Lambda_B = 1/[2(1 + v_f)]$, which is $3/8$ for $v_f = 1/3$. The ordering $\Lambda_U < \Lambda_o < \Lambda_B$ holds if $v_f > 0$: in ascending order, this corresponds to the narrow strip limit wherein $\sigma_{yy} \sim 0$ ($\Lambda_U$), the plane strain limit wherein $\epsilon_{yy}$ is the same ahead of and behind the crack tip ($\Lambda_o$), and the blanket film limit wherein equi-biaxial straining occurs with $\epsilon_{yy} = \epsilon_{xx}$ ($\Lambda_B$).

Interface crack propagation from the end of a strip is intrinsically 3D and complicated due to the nonuniform stress distribution in both the $x$- and $y$-directions and the different behavior of the crack front near a corner from that away from the corner. Elastic mismatch between the film and the substrate has a strong influence on the initial stress distribution in the film. Whatever the initial stress distribution, one expects plane strain crack propagation conditions for sufficiently small $a$ compared to $w$ except near the corners. As $a$ increases the plane strain constraint on $\epsilon_{yy}$ of the released film is relaxed and when $a$ attains steady state, all the elastic energy in the strip is released. Thus, as $a$ increases, there will be a transition from the plane strain constraint to full energy release in the strip.

The complicated 3D transition behavior anticipated above is illustrated by the examples in Figure 11.4 for two strip widths, $w/h = 10$ and $w/h = 20$, in both cases for strips whose ends are attached to the interior of the substrate surface and subject to negative equi-biaxial misfit strains $\theta$ such that the bonded film stresses are tensile. The energy release rate $\bar{G}$ averaged across the straight crack front is plotted as a function of $a/h$ for several elastic mismatches. The case of a very stiff film on a compliant substrate, $\alpha_D = 0.9$, attains full release associated with steady state at exceptionally short crack lengths. This is because the initial stress state in the strip is nearly uniaxial (i.e., $\sigma_{yy} \simeq 0$ and see Figure 11.2) due to the high relative compliance of the substrate; accordingly, the plane strain constraint has no effect. On the other hand, the crack in the cases with no elastic mismatch, or with a stiffer substrate, must advance to some fraction of the film width (e.g., a number of film thicknesses) before attaining steady state. Away from the corners, plane strain conditions at the crack tip pertain when the crack is sufficiently short, but $\bar{G} \simeq \Lambda_o$ would pertain only for strips wide enough such that (11.1) is achieved in the central region of the strip.

# 12 Delamination in Multilayers Subject to Steady-State Temperatures

The methods developed in Chapters 4 and 5 enable one to investigate delamination under a variety of conditions including mechanical loading, temperature changes and processing routes. A wide array of conditions fall under the heading of thermal stressing. We have already discussed examples of bilayers subject to uniform temperature changes in Section 4.6. This chapter focuses specifically on delamination due to stresses in multilayers subjected to nonuniform, steady-state (with respect to time) temperature distributions. The main emphasis in this chapter will be on through-thickness temperature gradients generated by an imposed temperature difference across the thickness of the bilayer or multilayer. Through-thickness temperature gradients arise in many technological applications, including thermal barrier coating (TBC) systems such as that shown in Figure 12.1, which play a crucial role in the efficiency of both aircraft engines and gas power turbines. The effect of thermal transients on stress and delamination is left for Chapter 13, which also provides insights into the effects of hot spots and localized thermal shock on cracking phenomena.

## 12.1 Stress Intensity Factors and Energy Release Rate for an Isolated Crack in a Temperature Gradient

To set the stage for delamination of bilayers and multilayers subject to thermal gradients, we begin by presenting a basic two-dimensional, plane strain result for an isolated crack of length $2a$ in an unconstrained infinite solid as depicted in Figure 12.2(a), with Young's modulus $E$, Poisson's ratio $v$, coefficient of thermal expansion $\alpha$ and thermal conductivity $k$. The solid is stress-free at the uniform temperature $T_o$. Imposition of a uniform temperature change $\Delta T$ does not produce any stress in the plane of Figure 12.2(a), but a temperature gradient perpendicular to the crack plane does give rise to stress intensity factors if the crack impedes the heat flow.

Consider an imposed temperature change $\Delta T(x_1, x_2)$ such that far from the crack the constant temperature gradient is $\Delta T' \equiv \partial(\Delta T)/\partial x_2$. The heat flow far from the crack is $q = -k\partial(\Delta T)/\partial x_2$. If the crack does not interrupt the heat flow, then the imposed temperature gradient produces no in-plane stress in the solid and no stress intensity factors. Denote the heat transfer coefficient (assumed to be uniform) across the crack by $h_c$ such that the heat flow across the crack at any point is $q = -h_c(T^+ - T^-)$ where $T^+$ and $T^-$ are the temperatures on the top and bottom crack faces. Define the dimensionless Biot

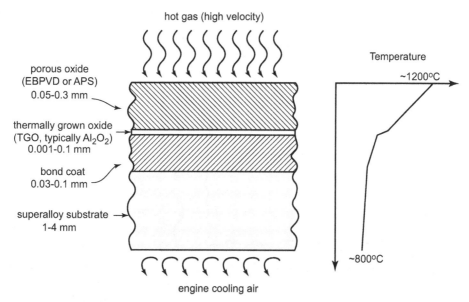

**Figure 12.1** A thermal barrier coating system on a superalloy substrate; the top coat is typically a low thermal conductivity material such as yttria-stabilized zirconia (YSZ). The bond coat is a metallic layer that serves as as aluminum reservoir to form a protective oxide. The purpose of the multilayer is to allow higher gas temperatures while maintaining the superalloy substrate at a temperature where creep is diminished.

number characterizing thermal conductivity across the crack by $B_c = h_c a / k$. The crack experiences pure mode II intensities ($K_I = 0$, $K_{II} / \Delta T' > 0$) with the energy release rate given by

$$ G = \frac{(1 - v^2) K_{II}^2}{E} = \frac{\pi}{16} \frac{(1 + v) E a \, (\alpha a \Delta T')^2}{(1 - v)} \frac{1}{\left(1 + 3.709 B_c + \left(\frac{\pi B_c}{2}\right)^2\right)} \quad (12.1) $$

This result is exact for the limit with no heat transfer across the crack, $B_c = 0$ Florence & Goodier (1960); Sih (1962), and approximate but accurate for $B_c \neq 0$ Kuo (1990); Xue et al. (2009). The energy release rate is a strong function of the crack length, increasing in proportion to $a^3$ in the limit with no heat transfer across the crack.

The above result for the isolated crack remote from any boundaries has been extended to cracks near the surface in coatings subject to high heat flux by Xue et al. (2009). The energy release rate and the mode mix depend on the depth of the crack below the surface as well as its length relative to the depth. These cracks, like the isolated crack, are buried cracks in the sense that neither end of the crack intersects a free surface. When one end of the crack is not constrained such as when the crack intersects a free edge of the coating or at an open vertical gap in the coating, more energy can be released in the cracking process. Edge-delamination problems such as those shown in Figures 12.2(b) and 12.2(c) will be considered next.

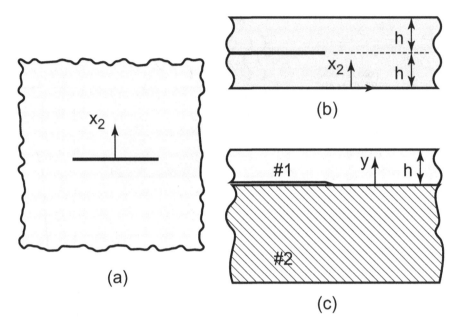

**Figure 12.2** (a) A two-dimensional crack under plane strain conditions in a homogeneous body under a uniform temperature gradient $\partial \Delta T / \partial x_2$ remote from the crack. (b) A bialyer layer with equal thickness layers and uniform elastic properties with thermal expansion mismatch subject to linear through-thickness temperature change and undergoing an edge delamination. (c) A thin film or coating subject to a linear temperature change $\Delta T(y)$ on a deep substrate with high thermal conductivity undergoing an edge delamination. Both $y$ and $x_2$ will be used for the vertical coordinate in this chapter.

## 12.2    Edge Delamination in a Bilayer Due to a Steady-State Temperature Distribution

The notation used in the general analysis of a bilayer in Chapter 4 is adopted here: $E_1$, $v_1$, $\alpha_1$ and $h_1 = h$ characterizing the upper layer, and $E_2$, $v_2$, $\alpha_2$ and $h_2 = H$ characterizing the lower layer. In this section, only stresses due to temperature changes are considered. A reference temperature distribution $T_o(x_2)$ at which there are no stresses must be defined. Under a steady-state through-thickness temperature distribution in the intact bilayer, the temperature $T(x_2)$ varies linearly within each layer, assuming the thermal conductivity is uniform within each layer. In the same notation employed in Chapter 5, denote the misfit strains due to thermal expansion as $\epsilon_{11} = \epsilon_{33} = \theta(x_2)$, and assume the misfit varies linearly within each layer such that $\theta_b^1$ and $\theta_t^1$ are the values at the bottom and top of the top layer with $\theta_b^2$ and $\theta_t^2$ as the corresponding values of the bottom layer.

The misfit strains depend on the coefficients of thermal expansion and the solution to the steady-state thermal problem for changes from the reference state, which, in turn, depends on the temperatures imposed above and below the bilayer, the heat transfer coefficients at the top and bottom surfaces, and the (uniform) thermal conductivity within each layer. The governing equations for such temperature distributions are described in Section 2.6, while details of the associated calculations for general multilayers are provided in Chapter 14, which describes software associated with this chapter.

The analysis of the stress distribution through the intact bilayer and the delamination analysis are similar to analyses in Section 4.6 for the case of uniform temperature changes, and it is a special case of the general multilayer analysis spelled out in Section 5.3. The thermal misfit strain in each layer relative to the reference temperature is $\theta_i = \alpha_i [T(x_2) - T_o(x_2)]$. The stress in the 1-direction is

$$\sigma_{11} = \bar{E}_1 \left( \epsilon_o + \kappa x_2 - \bar{c}_1 \left( \theta_b^1 + \frac{(\theta_t^1 - \theta_b^1)}{h_1} (x_2 - h_2) \right) \right), h_2 < x_2 \leq h_1 + h_2$$

$$\sigma_{11} = \bar{E}_2 \left( \epsilon_o + \kappa x_2 - \bar{c}_2 \left( \theta_b^2 + \frac{(\theta_t^2 - \theta_b^2)}{h_2} x_2 \right) \right), 0 \leq x_2 \leq h_2 \qquad (12.2)$$

where $\epsilon_o$ is the strain at the bottom of the bilayer, $\kappa$ is its curvature, and the moduli quantities $\bar{E}_i$ and $\bar{c}_i$ apply to plane stress, plane strain or equi-biaxial stressing depending on their definition as given in Table 2.1. The conditions for overall equilibrium are

$$\int_0^{h_1+h_2} \sigma_{11} dx_2 = 0; \quad \int_0^{h_1+h_2} \sigma_{11} x_2 dx_2 = 0$$

which are the same as (5.2) with a simple change in notation. They yield (5.6) and (5.7) for the elongation $\epsilon_o$ and curvature $\kappa$, which are repeated here for the two-layer system as a matter of convenience:

$$\begin{aligned} a_{11}\epsilon_o + a_{12}\kappa &= b_1 \\ a_{21}\epsilon_o + a_{22}\kappa &= b_2 \end{aligned} \qquad (12.3)$$

with

$$y_b^2 = 0; \quad y_t^2 = h_2; \quad y_b^1 = h_2; \quad y_t^1 = h_1 + h_2$$

$$a_{11} = \bar{E}_1 h_1 + \bar{E}_2 h_2; \quad a_{12} = a_{21} = -\sum_{i=1}^2 \bar{E}_i \left( \frac{(y_t^i)^2 - (y_b^i)^2}{2} \right)$$

$$a_{22} = \sum_{i=1}^2 \bar{E}_i \left( \frac{(y_t^i)^3 - (y_b^i)^3}{3} \right)$$

$$b_1 = \frac{1}{2} \sum_{i=1}^2 \bar{c}_i \bar{E}_i h_i \left( \theta_b^i + \theta_t^i \right)$$

$$b_2 = -\sum_{i=1}^2 \bar{c}_i \bar{E}_i h_i \left[ \frac{h_i}{3} \left( \theta_t^i - \theta_b^i \right) + \frac{1}{2} \left( \theta_t^i y_b^i + \theta_b^i y_t^i \right) \right]$$

With reference to Figure 4.6 for the definition of $P$ and $M$ used in the interface delamination calculation for the bilayer, one has

$$P = -\int_0^{h_2} \sigma_{11} dx_2$$

$$M^* = M + \frac{P}{2}(h_1 + h_2) = -\int_0^{h_2} \sigma_{11}\left(x_2 - \frac{h_2}{2}\right) dx_2 \qquad (12.4)$$

The stress intensity factors and energy release rate of the steady-state interface delamination crack are given in terms of $P$ and $M$ by (4.7). While the algebraic expressions for all the quantities involved in computing the stress intensity factors and energy release rates are explicit, for most cases the expressions are too lengthy to provide insights into trends without computing curves. Thus, for most cases one should either use the software accompanying this book to generate numerical trend curves or write one's own small code for that purpose.

Two exceptional cases which do reduce to relatively simple and revealing algebraic formulas are (1) equal thickness layers with no elastic mismatch subject to a linear temperature distribution, and (2) a thin film or coating on a deep substrate with high thermal conductivity.

### Bilayer with Equal Thickness, No Elastic Mismatch and $\alpha_1 \neq \alpha_2$, Subject to a Constant Temperature Gradient

In this example, illustrated in Figure 12.2(b), $h_1 = h_2 = h$ and there is a uniform reference temperature $T_o$ at which there is no stress in the layers. Denote the temperature change from $T_o$ within the bilayer by $\Delta T(x_2)$. Assume the thermal conductivity is uniform throughout the intact bilayer such that $\Delta T(x_2)$ is linear and given by

$$\Delta T(x_2) = \Delta T_o + (x_2 - h)\Delta T' \quad 0 \le x_2 \le 2h \qquad (12.5)$$

where $\Delta T' = [\Delta T(2h) - \Delta T(0)]/(2h)$ is the constant gradient. The misfit strains are defined by

$$\begin{aligned} \theta_b^1 &= \alpha_1 \Delta T_o; \quad \theta_t^1 = \alpha_1(\Delta T_o + h\Delta T') \\ \theta_b^2 &= \alpha_2(\Delta T_o - h\Delta T'); \quad \theta_t^2 = \alpha_2 \Delta T_o \end{aligned} \qquad (12.6)$$

These can be used with the equations (12.3) to obtain

$$\begin{aligned} h\kappa &= -\frac{3\bar{c}}{4}(\alpha_1 - \alpha_2)\Delta T_o - \frac{\bar{c}}{2}(\alpha_1 + \alpha_2)h\Delta T' \\ \epsilon_o - h\kappa &= \frac{\bar{c}}{2}(\alpha_1 + \alpha_2)\Delta T_o - \frac{\bar{c}}{4}(\alpha_1 - \alpha_2)h\Delta T' \end{aligned} \qquad (12.7)$$

and the stress in the lower layer of the intact bilayer is then found to be

$$\sigma_{11} = \frac{\bar{c}\bar{E}}{4}(\alpha_1 - \alpha_2)\left[2\Delta T_o + h\Delta T' + \frac{x_2 - h}{h}(3\Delta T_o + 2h\Delta T')\right] \qquad (12.8)$$

The factor $\bar{c}\bar{E} = E/(1 - v)$ is the same for both equi-biaxial strain and plane strain, and thus $\sigma_{11}$ is the same for these two conditions. The in-plane stress vanishes if there is no thermal expansion mismatch between the layers, consistent with the fact that a homogeneous bilayer undergoes pure stretch and bending subject to a change in temperature that varies linearly through the thickness.

Equations (12.4) give

$$P = -\frac{\bar{c}\bar{E}h}{8}(\alpha_1 - \alpha_2)\Delta T_o; \quad M = -\frac{\bar{c}\bar{E}h^2}{48}(\alpha_1 - \alpha_2)(3\Delta T_o - 2h\Delta T') \quad (12.9)$$

By (4.7), the stress intensity factors are

$$K_I = -\frac{\bar{c}\bar{E}\sqrt{h}}{4\sqrt{3}}(\alpha_1 - \alpha_2)h\Delta T'; \quad K_{II} = -\frac{\bar{c}\bar{E}\sqrt{h}}{4}(\alpha_1 - \alpha_2)\Delta T_o \quad (12.10)$$

where use has been made of the fact that $\cos\omega = \sqrt{3/7}$ and $\sin(\gamma + \omega) = 1$ for this case. For the case with no temperature gradient, (12.10) predicts a pure mode II response in agreement with the result derived earlier in (4.27). For the case with an imposed temperature gradient and no average temperature change ($\Delta T_o = 0$), the delamination is pure mode I. The crack faces open ($K_I > 0$) if $(\alpha_1 - \alpha_2)\Delta T' < 0$, and (12.10) is valid only under these conditions. Assuming this is to be the case, the energy release rate under plane strain delamination is

$$G = \frac{K_I^2 + K_{II}^2}{\bar{E}} = \frac{[\bar{c}\bar{E}(\alpha_1 - \alpha_2)]^2 h}{16\bar{E}}\left(\Delta T_o^2 + \frac{1}{3}(h\Delta T')^2\right) \quad (12.11)$$

As noted in Chapter 4, for prestressing under either equi-biaxial straining or plane strain followed by plane strain delamination, the factor $(\bar{c}\bar{E})^2/\bar{E} = E(1 + v)/(1 - v)$ is the same. For plane stress this factor is simply $E$.

If $(\alpha_1 - \alpha_2)\Delta T' > 0$, crack face contact will occur, and neither (12.10) or (12.11) is valid. However, if frictional dissipation between the crack faces can be ignored, (12.11) can be used an approximation for $G$. The result (12.11) is simply the statement that the edge delamination releases all the elastic energy stored in the bilayer subject to the plane strain constraint for that case. Even if there is crack face contact, nearly all this energy will be released assuming friction does not play a dominant role. The crack is nearly mode II, but it may have a small mode I component with the faces near the tip propped open.

The above analysis ignores the fact that the delamination crack at the interface may impede the heat flow through the interface region in the manner considered in Section 12.1 and thereby change the temperature distribution near the crack tip and in the delaminated region. In turn, this change in the temperature distribution may alter the stress intensity factors and the energy release rate. However, if the delamination crack advances rapidly, as would be expected if the energy release rate exceeds the critical interface toughness, there would not be time for the redistribution to take place, and the results derived above would be valid.

## A Thin Film on a Thick Substrate, with a Constant Thermal Gradient in the Film

In this example, we model aspects of delamination of a thermal barrier coating (TBC) accounting for the temperature gradient in the coating as depcted in Figure 12.2(c). TBCs are thin ceramic coatings with low thermal conductivity bonded to a relatively thick metal alloy substrate. As depicted in Figure 12.1, the substrate is cooled from below or through internal channels while the surface of the coating is exposed to high temperature of the burning fuel. A significant temperature drop across the layered coating is created by these conditions, and the TBC provides a thermal insulation 'barrier' for the underlying substrate. Under steady-state temperature conditions, the high thermal conductivity of the metal results in a much lower temperature gradient in the substrate than the coating.

Here, as an approximation, it will be assumed that the temperature change in the substrate from the zero-stress reference condition, $\Delta T_{sub}$, is uniform such that the temperature change of the intact interface is also $\Delta T_{sub}$. The temperature change at the top surface of the coating from the reference temperature is denoted $\Delta T_{sur}$ such that the temperature change in the upper layer is

$$\Delta T = \Delta T_{sub} + y\Delta T'; \quad \Delta T' = \frac{\Delta T_{sur} - \Delta T_{sub}}{h} \tag{12.12}$$

The bilayer in Figure 12.2(c) has both elastic and thermal expansion mismatch, and the substrate is taken to be infinitely deep. The misfit strain in the substrate is $\theta_2 = \alpha_2 \Delta T_{sub}$, while that in the coating is $\theta_1 = \alpha_1 (\Delta T_{sub} + y\Delta T')$. The system of equations (12.3) for $\kappa$ and $\epsilon_o$ can be used, but it is more instructive to proceed directly by noting the substrate imposes its deformation ($\kappa = 0$ and $\epsilon_o = \bar{c}_2\alpha_2\Delta T_{sub}$) on the coating such that the stress in the coating is given by

$$\sigma_{11} = \bar{E}_1[(\bar{c}_2\alpha_2 - \bar{c}_1\alpha_1)\Delta T_{sub} - \bar{c}_1\alpha_1 y\Delta T'] \tag{12.13}$$

The calculation of $P$ and $M$ in Figure 4.3 defined in (4.10) leads directly, via (4.11), to the following energy release rate and stress intensity factors for the steady-state delamination crack:

$$G = \frac{\bar{E}_1^2 h}{2\tilde{E}_1}\left(X^2 - XY + \frac{1}{3}Y^2\right) \tag{12.14}$$

$$K_I = \sqrt{\frac{\tilde{E}_2}{\tilde{E}_1 + \tilde{E}_2}}\bar{E}_1\sqrt{h}\left[\left(X - \frac{1}{2}Y\right)\cos\omega - \frac{1}{2\sqrt{3}}Y\sin\omega\right]$$

$$\tag{12.15}$$

$$K_{II} = \sqrt{\frac{\tilde{E}_2}{\tilde{E}_1 + \tilde{E}_2}}\bar{E}_1\sqrt{h}\left[\left(X - \frac{1}{2}Y\right)\sin\omega + \frac{1}{2\sqrt{3}}Y\cos\omega\right]$$

where

$$X = (\bar{c}_2\alpha_2 - \bar{c}_1\alpha_1)\Delta T_{sub}; \quad Y = \bar{c}_1\alpha_1 h\Delta T' = \bar{c}_1\alpha_1(\Delta T_{sur} - \Delta T_{sub}) \tag{12.16}$$

The above result covers predelamination stresses generated either in plane strain or in equi-biaxial strain (with $\bar{E}$ and $\bar{c}$ defined in Table 2.1) followed by delamination in plane strain with $\tilde{E} = E/(1 - v^2)$. In this derivation, it is assumed that $\beta_D = 0$, and $\omega(\alpha_D)$ is plotted in Figure 4.2. The condition for an open crack, $K_I > 0$, is

$$X - \frac{1}{2}\left(1 + \frac{1}{\sqrt{3}}\tan\omega\right)Y > 0 \tag{12.17}$$

The conditions for the delamination to be an open pure mode I crack are (12.17) plus

$$X - \frac{1}{2}\left(1 - \frac{1}{\sqrt{3}}\cos\omega\right)Y = 0 \tag{12.18}$$

It is important to note that influence on $G$ of the temperature change of the substrate is proportional to the thermal mismatch, $\bar{c}_2\alpha_2 - \bar{c}_1\alpha_1$, while the influence of the gradient change through the coating is proportional to the full thermal expansion coefficient of the coating, $\bar{c}_1\alpha_1$. For this reason, the effect of the temperature gradient can be more important than otherwise might be expected. These results are specialized further with the TBC application in mind.

A common assumption made in the stress and delamination analysis of TBCs is that the stress in the coating relaxes to zero at the highest operating temperature by creep mechanisms. Stresses develop in the coating as the bilayer cools, and this process is assumed to be sufficiently rapid that it can be treated as being elastic. Thus, in applying the results above, $\Delta T_{sub}$ and $\Delta T_{sur} - \Delta T_{sub}$ are negative if the surface of the coating is hotter than the substrate at the start of cool-down. Assume the stresses are generated under conditions for equi-biaxial stressing followed by plane strain delamination such that $\bar{c}_1 = \bar{c}_2 = 1$ and $\bar{E}_1^2/\tilde{E}_1 = E_1(1 + v_1)/(1 - v_1)$. With these assumptions, (12.14) and (12.16) become (Evans & Hutchinson 2007)

$$G = \frac{(1 + v_1)\bar{E}_1 h}{2(1 - v_1)}\left(X^2 - XY + \frac{1}{3}Y^2\right) \tag{12.19}$$

$$X = (\alpha_2 - \alpha_1)\Delta T_{sub}; \quad Y = \alpha_1(\Delta T_{sur} - \Delta T_{sub}) \tag{12.20}$$

The condition for an open crack is $X - 0.870Y > 0$ and, assuming this holds, the requirement that the crack is mode I is $X - 0.275Y = 0$, both computed using $\omega = 52.07^{\circ}$. For a ceramic coating on a metal substrate, $(\alpha_2 - \alpha_1) > 0$, and, for temperature changes due to cooling, $X < 0$ and $Y < 0$. A contour plot of constant values for the dimensionless energy release rate relevant to the TBC model is given in Figure 12.3 with a dashed line marking the regime with an open crack (mixed mode) and that with a closed crack (approximately mode II), together with another dashed like indicating the condition for an open mode I crack. In the regime in which the crack is closed, $G$ has been approximated by (12.19) with the justification that this expression is the energy/area stored in the intact coating and that almost all of this energy will be released by the delamination even in mode II provided frictional effects are not overly significant.

The contour plot in Figure 12.3 reveals that the energy release rate under a combination of reduction in substrate temperature, $-\Delta T_{sub}$, and a reduction in the coating

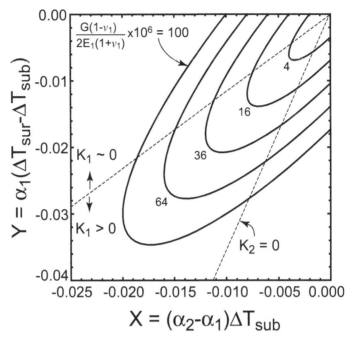

**Figure 12.3** Contours of constant energy release rate for edge delamination of a thin coating/deep substrate bilayer during cool-down. The bilayer is subject to combinations of substrate temperature change, $\Delta T_{sub}$, and change in the coating gradient, $\Delta T_{sur} - \Delta T_{sub}$, assuming no residual stress when $\Delta T_{sub} = 0$ and $\Delta T_{sur} = 0$.

gradient, $-(\Delta T_{sur} - \Delta T_{sub})$, can be significantly less than either reduction acting alone. This is readily understood by noting that the temperature reduction of the substrate produces compression in the coating (cf. (12.13)), while the reduction in the coating gradient produces tension in the coating. The trajectory of the changes in $\Delta T_{sub}$ and $\Delta T_{sur}$ in Figure 12.3 during cool-down can have a significant effect on the energy release rate experienced by a potential delamination crack. Various scenarios such as rapid and slow cool-down have been discussed by various authors (Sundaram, Lipkin, Johnson & Hutchinson 2013; Evans & Hutchinson 2007; Jackson & Begley 2014).

Another important aspect of TBC delamination evident from either (12.19) or Figure 12.3 is the fact that $G$ is proportional to the coating modulus $E_1$. Low modulus ceramic coatings are achieved by deposition processes which generate porous microstructures and coatings with an effective modulus that is a small fraction of the fully dense material. One challenge facing airlines and aircraft engine manufacturers is the increasingly common occurrence wherein engines ingest high-altitude airborne dust generated by either volcanos or desert wind storms which melts and penetrates into the TBC. The generic term for this dust is 'CMAS', which strictly speaking stands for calcium-magnesium-alumino-silicate, though CMAS is used for a broad range of mineral compositions. The unfortunate effect is to increase the effective modulus of the coating, in some cases to essentially that of the fully dense ceramic, such that the driving force for delamination is dramatically increased. This is a relatively recent phenomenon triggered by many successful advances in TBC systems, which allow high enough operating surface temperatures to melt the dust. An example illustrating this phenomena is analyzed in the next section using the numerical tools described later in the book.

## 12.3    Edge Delamination Due to Steady-State Thermal Gradients in Multilayers

To provide an illustration of the effect of temperature gradients on delamination, the system shown schematically in Figure 12.1 was analyzed using the software described in Chapter 14. The analysis includes a portion of the top coat that has been penetrated by CMAS, as shown in Figure 12.4(a). Experiments have revealed that CMAS penetration and solidification increases the modulus of the coating, with little apparent effect on its thermal conductivity Jackson et al. (2015). The properties needed for the analysis can be found in the software files associated with this chapter.

In the present analysis, typical surface temperatures during operation are prescribed on the top and bottom. The resulting piecewise linear steady-state temperature distribution is calculated using the theory described in Section 2.6 and software described in Chapter 14. In the mechanical analysis, it is assumed that the steady-state temperature distribution shown in Figure 12.4(b) defines the stress-free reference state; cooling from this state to room temperature generates these stresses shown in Figure 12.4(c). The energy release rates shown in Figure 12.4(d) are those associated with the stress state after cooling.

The temperature profile shown in Figure 12.4(b) illustrates that the bond coat and thermally grown oxide have little impact on the resulting temperature profile. This is because the thermal conductivity of the bond coat is similar to that of the substrate, which implies the change in temperature gradient between these two layers will be negligible. With regard to the oxide, it has negligible impact on the thermal problem because it has a much lower thermal resistance due to the fact it is much thinner than the other layers. The presence of CMAS penetration in the YSZ top coat has no effect on the temperature distribution, since in the present case the thermal conductivity of this layer is identical to the original top coat. In contrast to the thermal problem, the presence of the bond coat and oxide are clearly evident in the stress distributions shown in Figure 12.4(c). Similarly, the CMAS-infiltrated layer does not affect the thermal problem, but its higher modulus has a dramatic effect on stresses, as shown in Figure 12.4(c).

Figure 12.4(d) shows the energy release rate for a delamination crack at the position indicated on the ordinate axis; the position in the multilayer is measured from the bottom of the substrate. A comparison of the two solid lines (one for a system with CMAS and one without) illustrates that the energy release rate for delamination is higher throughout the entire multilayer whenever CMAS is present. The sharp spike in ERR seen at the position ~3.0 mm corresponds to the location of the thermally grown oxide. Despite the thinness of this layer, the driving force for delamination is substantially higher underneath the oxide, as it is heavily stressed and contains significant strain energy that is released by delamination. The kink seen in the upper curve at ~3.2 mm is associated

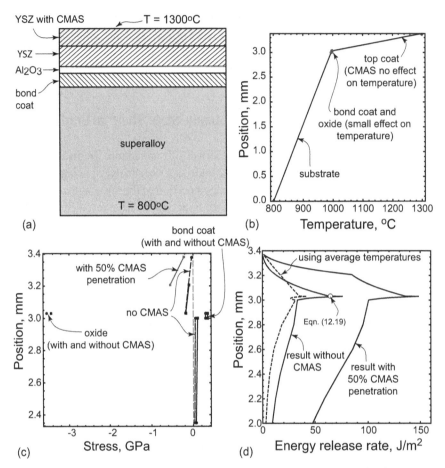

**Figure 12.4** Analysis of a thermal barrier coating system such as the one shown in Figure 12.1, produced using the software described in Chapter 14: (a) a schematic of the system showing prescribed surface temperatures, (b) the resulting temperature distribution in the multilayers, (c) the resulting stress distribution after cool-down, (d) the resulting energy release rates for a delamination crack placed at the position shown, comparing various approximations and illustrating the dramatic impact of CMAS penetration.

with traversing the boundary between the CMAS-infiltrated region and the original top coat.

The single open data point in Figure 12.4(d) corresponds to the predictions based on (12.19), which assumes a uniform temperature change in the substrate. To make this prediction, the uniform change in substrate temperature defined in (12.19) is taken the change in temperature at the top of the substrate as determined via the thermal analysis, $\Delta T_{sub} \simeq -995°C$. Meanwhile, the change in the top surface temperature is simply $\Delta T_{sur} = -1300°C$. This point, which imposes uniform temperature changes to the substrate, differs from the full analysis that includes gradients in all layers by less than 1%. One can conclude that for systems dominated by thermal gradients in the top layer, (12.19) provides a reasonable estimate for the energy release rate even when gradients in the substrate exist, as shown in Figure 12.4(b).

To illustrate the importance of temperature gradients on the energy release rate (or more generally, gradients in misfit strains), the energy release rate was computed assuming constant temperatures in each layer, for the system without CMAS. For each layer, the uniform temperature was prescribed to be the average of the full linear distribution shown in Figure 12.4(b). The resulting energy release rate is shown as the dashed line in Figure 12.4(d). While the same trend is evident, the calculation using average temperatures significantly underestimates the peak driving force for delamination near the TGO. This comparison emphasizes the importance of accounting for misfit gradients when they are a significant fraction of the average temperature in a given layer.

# 13 Cracking under Transient Temperature Distributions

The previous chapter dealt with steady-state, time-independent temperature distributions which induce stress in a body due, for example, to interruption of heat flow across a preexisting crack or to thermal expansion and/or conductivity mismatches across material interfaces. This chapter deals with problems where transient (time-dependent) temperature distributions induce stresses in a body and, in turn, how these stresses can cause cracking or interface delamination.

A simply connected solid with uniform properties which is unconstrained at its boundaries experiences no stress under temperature distributions imposed on its surface once the temperature distribution ceases to change and steady-state conditions are attained. However, during the transient period while the temperature is changing, stresses will generally be induced, and the solid may be susceptible to cracking.

The first set of examples considered in this chapter deals with an unconstrained uniform semi-infinite solid having an initial uniform temperature which is suddenly subject to a temperature change imposed on its entire free surface. Cold shock, with cooling imposed, induces transient tensile stresses acting parallel to the surface with the potential to cause mode I cracking perpendicular to the surface, as well as mixed-mode delamination cracking on a weak interface parallel to the surface. When a sudden increase in surface temperature is imposed, the resulting hot shock produces compressive stresses parallel to the surface such that subsurface mixed-mode delamination is again a possibility if weak interfaces exist parallel to the surface.

The second example again considers a uniform semi-infinite solid at an initial uniform temperature, but in this case *localized* hot shock is considered with a higher temperature suddenly imposed over a local region on its surface. The stress field is more complicated in this case with both tensile and compressive components such that various modes of cracking must be considered.

## 13.1  Cold and Hot Shock of an Entire Surface of a Uniform Half-Space

With Figure 13.1 as a reference, consider a semi-infinite block of material with Young's modulus $E$, Poisson's ratio $v$, thermal expansion coefficient $\alpha$ and thermal diffusivity $\kappa$, defined as $\kappa \equiv \rho c_p / k$, where $\rho$ is the density, $c_p$ is the specific heat capacity of the material and $k$ is the thermal conductivity. The block is unconstrained and is at uniform temperature $T_o$ at time $t = 0$. For $t \geq 0$ a uniform temperature change $\Delta T_o$ is imposed

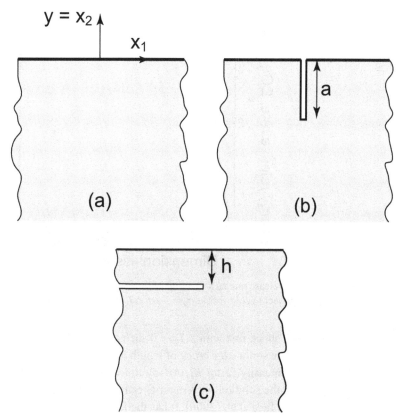

**Figure 13.1** (a) Notation for the half-space. (b) A plane strain edge crack in a half-space subject to a cold shock. (c) A delamination crack in a half-space subject to either cold shock or hot shock.

on the surface $y = 0$. The diffusion equation governing the resulting temperature $T(y, t)$ for $t \geq 0$ is (see Section 2.6)

$$\nabla^2 T = \frac{1}{\kappa} \frac{\partial T}{\partial t} \qquad (13.1)$$

For the semi-infinite solid subject to the sudden uniform temperature change at its surface, the solution to (13.1) is well known and given by

$$T(y, t) = T_o + \Delta T_o \text{erfc}\left(-\frac{y}{2\sqrt{\kappa t}}\right) \quad \text{with} \quad \text{erfc}(x) = 1 - \frac{2}{\sqrt{\pi}} \int_0^x e^{-\eta^2} d\eta \qquad (13.2)$$

The dependence based on the single similarity variable arises due to the fact there is no physical length in the problem with $y/\sqrt{\kappa t}$ as the only possible dimensionless variable. A transient equi-biaxial stress state acting parallel to the surface is induced in the solid:

$$\sigma_{11} = \sigma_{33} \equiv \sigma(y, t) = \sigma_o \text{erfc}\left(-\frac{y}{\sqrt{\kappa t}}\right) \quad \text{with} \quad \sigma_o = -\frac{E\alpha\Delta T_o}{(1 - v)} \qquad (13.3)$$

The temperature change and the stress progress into the half-space in proportion to $\sqrt{\kappa t}$.

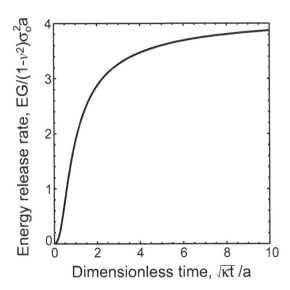

**Figure 13.2** Energy release rate as a function of time for an edge crack of length $a$ at the surface of a half-space subject to cold shock. $\sigma_o = -\alpha E \Delta T_o / (1 - v)$.

Consider cold shock first with $\Delta T_o < 0$ such that the stress $\sigma(y, t)$ is tensile. A two-dimensional, plane strain *edge crack* of length $a$ as depicted in Figure 12.1(b) will have a mode I stress intensity factor $K_I$. At any time $t$, with stresses given by (13.3) in the uncracked body, the solution for $K_I$ and $G$ can be obtained using the basic edge crack solution given by Tada et al. (2000). It has the form:

$$K_1 = \sigma_o \sqrt{a} f\left(\frac{\sqrt{\kappa t}}{a}\right) \quad or \quad G = \frac{(1 - v^2)K_I^2}{E} = \frac{(1 - v^2)\sigma_o^2 a}{E} f^2\left(\frac{\sqrt{\kappa t}}{a}\right) \tag{13.4}$$

with the dimensionless energy release rate plotted in Figure 13.2. The energy release rate approaches the limit for an edge crack subject to a uniform stress $\sigma_o$ at times $t \sim 100 a^2 / \kappa$, but $G$ is already about one-half this limit when $t \sim a^2 / \kappa$. Heat transfer across the crack plays no role in this problem because there is no lateral gradient in temperature and therefore no lateral heat transfer.

The potential for an edge crack to become critical due to *cold shock* far exceeds the likelihood of delamination parallel to the surface unless a weak interface exists below the surface. To see this, consider a plane strain edge delamination at $y = -h$ below the surface, and assume at any time $t$ the stress (13.3) due to the cold shock acts on the uncracked half-space. Further, as we have done in other problems, assume that delamination occurs sufficiently rapidly so that the temperature distribution at $t$ is not affected by the delamination crack. If this were not the case, the effect of heat transfer across the crack would have to be taken into account in determining $T(\underline{x}, t)$.

The problem ignoring the effect of the crack on the temperature distribution was analzyed by Hutchinson & Suo (1992). It can be solved in terms of the resultant force/depth $P$ and moment/depth $M$ introduced in Section 4.3, which are equal in magnitude and opposite in sign to the resultants associated with $\sigma(y, t)$ in (13.3).

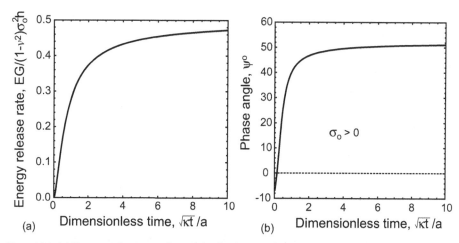

**Figure 13.3** (a) Energy release rate for a delamination crack at depth $h$ below the surface as a function of time for both cold shock and hot shock. (b) Mode mix for cold shock. For hot shock with $\sigma_o < 0$, the crack is mode II, or nearly mode II, and the energy release rate in (a) is approximate. $\sigma_o = -\alpha E \Delta T_o / (1 - v)$.

With $\xi = h/\sqrt{\kappa t}$,

$$P = \sigma_o h I(\xi) \quad and \quad M = \sigma_o h^2 \left[\frac{1}{2} I(\xi) - J(\xi)\right] \tag{13.5}$$

with

$$I(\xi) = \frac{1}{\xi} \int_0^\xi \operatorname{erfc}\left(\frac{\eta}{2}\right) d\eta \quad and \quad J(\xi) = \frac{1}{\xi^2} \int_0^\xi \operatorname{erfc}\left(\frac{\eta}{2}\right) \eta d\eta \tag{13.6}$$

By (4.7), the dimensionless energy release rate for the delamination crack is

$$\frac{EG}{(1 - v^2)\sigma_o^2 h} = \frac{1}{2}\left(I(\xi)^2 + 12\left(\frac{I(\xi)}{2} - J(\xi)\right)^2\right) \tag{13.7}$$

This requires the crack to be open, which is the case for all $\xi$ if $\sigma_o > 0$. This result is plotted in Figure 13.3(a) where it can be seen it approaches the limit of a delamination crack for a layer subject to uniform $\sigma_o$ for sufficiently large $t$. By (4.9), the mode mix is given by

$$\tan \psi = \frac{I(\xi)\tan \omega - 2\sqrt{3}\left[\frac{1}{2}I(\xi) - J(\xi)\right]}{I(\xi) + 2\sqrt{3}\left[\frac{1}{2}I(\xi) - J(\xi)\right]\tan \omega} \tag{13.8}$$

with $\omega = 52.07°$, which is plotted in Figure 13.3(b).

For $\sqrt{\kappa t} > 10$, the energy release rate approaches the classical result for a layer subject to uniform stress $\sigma_o$ on a deep subtrate, and for cold shock with $\sigma_o > 0$ the mode mix in this limit is $\psi \to \omega$. There exists a time, $\sqrt{\kappa t} = 0.147$, when the combination of $P$ and $M$ produces a mode I crack ($\psi = 0$) having $\bar{E}G/(\sigma_o^2 h) = 0.036$ and $K_1/(\sigma_o \sqrt{h}) = 0.19$. In principle, this suggests that mode I propagation parallel to the surface would be possible in a homogeneous material at this time, but the driving force is so low compared to those at larger times that such occurrences seem highly unlikely. In fact, if one entertains a competition between cracking due to a surface edge crack and delamination parallel to the interface with $h = a$, one sees that the energy release rate for edge cracking is an order of magnitude larger than that for delamination. In cold shock, delamination should be expected only if a weak interface parallel to the surface exists with a toughness that is one tenth or less than that of the bulk material. Otherwise, edge cracking from the surface is most likely.

*Hot shock* of the half-space produces compressive stress parallel to the surface, $\sigma_o < 0$, rendering any surface edge cracks harmless. However, delamination is still possible. In this case the combination of $P$ and $M$ keeps the crack closed such that mode II delamination occurs, at least approximately. The result for $G$ in (13.6) and Figure 13.3(a) can be used as a approximation for the energy release rate if friction effects can be ignored. The result is expected to be relevant only for delamination along a weak interface at a depth $h$ below the surface or possibly by materials with highly anisotropic fracture resistance whose planes parallel to the surface are the most brittle. If the body has isotropic fracture resistance, the delamination path does not meet the condition for mode I advance. For hot shock, it is important to consider the finite size of the body because then the compressive stress at the surface are offset by tensile stresses in the interior. These tensile stresses may trigger cracking from internal flaws as evidenced by an ice cube dropped into a glass of warm water.

## 13.2    Cracking Due to a Sudden Localized Temperature Increase – Localized Hot Spots

Two additional problems illustrate basic aspects of transient temperature distributions on thermal cracking. They correspond to cases in which the surface of the half-space is subject to a suddenly applied localized hot spot. A half-space at uniform temperature at $t = 0$ is subject to an abrupt local temperature change $\Delta T$ on its surface starting at $t = 0$, as depicted in Figure 13.4. Once applied, $\Delta T$ at the surface is held constant. The cracking analysis involves three steps: solving for the time-dependent temperature distribution $T(x, y, t)$, solving for the stress distribution $\sigma_{ij}(x, y, t)$ due to the temperature change, and finally, solving for the stress intensity factors of potential cracks as a function of time due to $\sigma_{ij}(x, y, t)$. As in the previous section, the effect of the crack on the temperature change will not be taking into account. The results presented below have been extracted from Tvergaard et al. (1993).

The time-dependent temperature and stress distributions for the two-dimensional problem in Figure 13.4(a) can be solved in closed form, although in terms of integrals which require numerical integration. With $\sigma_{ij}(x, y, t)$ in hand, solutions for the stress intensity factors and energy release rates for two-dimensional (plane strain) cracks of various lengths, locations and orientations within the half-space can be obtained using formulas from Tada et al. (2000). Time-dependent stress fields and stress intensity factors for various crack locations and orientations have been presented by Tvergaard et al. (1993). Some of the most important aspects from the point of view of basic understanding can be summarized as follows.

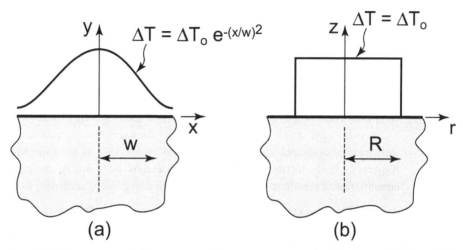

**Figure 13.4** Homogeneous half-space at a uniform temperature and subject to a suddenly applied temperature increase on its surface starting at $t = 0$. (a) Two-dimensional problem, and (b) axisymmetric problem.

Immediately after the temperature change is imposed on the surface, the stress at the surface at the center of the hot spot is compressive with

$$\frac{(1-v)\sigma_{xx}}{E\alpha\Delta T_o} = -1; \quad \sigma_{yy} = 0; \quad \sigma_{33} = \sigma_{xx} \ at \ x = y = t = 0. \tag{13.9}$$

As $t$ increases, the stress distribution evolves with tensile components occurring in some regions of the half-space. For sufficiently large $t$, the stresses decay to zero everywhere because the temperature distribution becomes time-independent. Cracking, if it occurs, will do so within the window of time and space in which tensile stress components attain their maximum. The largest tensile components of stress occur below the peak of the temperature distribution imposed on the surface along $x = 0$. The maximum of all tensile components is

$$\frac{(1-v)\sigma_{xx}}{E\alpha\Delta T_o} = 0.0951 \ at \ \frac{y}{W} = -0.505 \ and \ \frac{\sqrt{\kappa t}}{w} = 0.15 \tag{13.10}$$

This tensile stress is about one-tenth the maximum compressive stress (13.9). If this stress causes cracking, it would propagate mode I subsurface cracks aligned perpendicular to the $x$-axis. Estimates of the stress intensity factors of sufficiently short cracks (compared to $w$) can be obtained using (13.10), and such detailed results are presented by Tvergaard et al. (1993). Because these cracks occur along the symmetry plane where $\partial T/\partial x = 0$, they would not alter the temperature distribution. If small subsurface flaws are present in the half-space which become critical at a tensile stress $\sigma_c$, then it follows from (13.10) that the half-space will be able to withstand hot spots without cracking if

$$\Delta T_o < 10.5 \frac{(1-v)\sigma_c}{\alpha E} \tag{13.11}$$

For comparison, the maximum tensile component $\sigma_{yy}$ also occurs along $x = 0$ and it is approximately one-half of the maximum $\sigma_{xx}$ in (13.10):

$$\frac{(1 - v)\sigma_{yy}}{E\alpha\Delta T_o} = 0.0438 \text{ at } \frac{y}{W} = -1.23 \text{ and } \frac{\sqrt{\kappa t}}{w} = 0.31 \qquad (13.12)$$

It follows that vertical cracking is more likely than horizontal cracking unless there is significant fracture anisotropy or weak horizontal interfaces lie beneath the surface.

The maximum tensile stresses that occur in the axisymmetric problem with surface temperature conditions imposed according to Figure 13.4(b) are quite similar in most respects to those for the two-dimensional problem. Specifically, the peak radial and circumferential tensile stress components occur along $r = 0$ according to

$$\frac{(1 - v)\sigma_{rr}}{E\alpha\Delta T_o} = 0.085 \text{ at } \frac{z}{R} = -0.46 \text{ and } \frac{\sqrt{\kappa t}}{R} = 0.13 \qquad (13.13)$$

$$\frac{(1 - v)\sigma_{zz}}{E\alpha\Delta T_o} = 0.070 \text{ at } \frac{z}{R} = -0.95 \text{ and } \frac{\sqrt{\kappa t}}{R} = 0.25 \qquad (13.14)$$

In addition, a tensile circumferential stress develops at the surface of the half-space ($z = 0$) outside the hot spot. Its maximum value and place and time of occurrence are

$$\frac{(1 - v)\sigma_{\theta\theta}}{E\alpha\Delta T_o} = 0.035 \text{ at } \frac{r}{R} = 1.63 \text{ and } \frac{\sqrt{\kappa t}}{R} = 0.25 \qquad (13.15)$$

Unless there is strong fracture anisotropic or weak subsurface interfaces, the conclusion for the axisymmetric hot spot is similar to that for the two-dimensional hot spot: vertical subsurface cracks are the most likely to become critical.

## 13.3 The Effect of Temperature Transients on Delamination in a Single Layer with Finite Thickness

In this section, we consider the transient response of a layer with finite thickness $H$, as shown in Figure 13.5(a). The initial temperature distribution is linear through the layer, defined by $T_0$ at the bottom of the layer ($y = 0$) and $T_0 + \Delta T_o$ at the top of the layer ($y = H$). The layer is then subjected to convective heat transfer at the top surface driven by the ambient temperature $T_o$, while the bottom of the layer is held fixed at $T_o$. This is a somewhat limited example, given that the ambient temperature driving cooling at the top of the layer is not generally equal to the ambient temperature bottom of layer. However, it provides significant insight into the important transient effect of differential cooling, wherein the temperature at the hot surface cools faster than the interior of the layer. This is sometimes referred to as a temperature 'inversion', because the temperature gradient near the surface changes sign. We will see that such localized cooling near a surface has important implications for delamination. The example also provides a means to benchmark the numerical procedure detailed in Chapter 15.

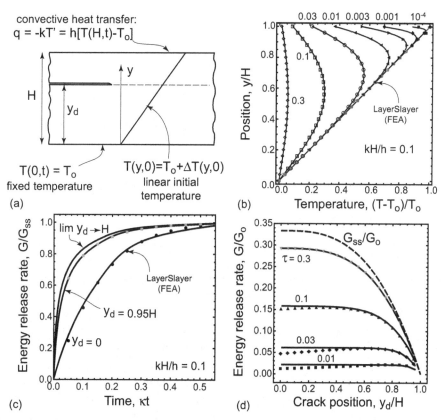

**Figure 13.5** (a) Schematic of the problem of a layer subjected to fixed temperature on the bottom and convective heat transfer on the top. (b) Temperature profiles through the layer at several times. (c) Energy release rate near the top and at the bottom of the layer, normalized by the steady-state result. (d) Energy release rates as a function of crack position for various times after cooling of the top surface has been turned on: $G_o = (1 + v)EH\alpha\Delta T_o^2/[2(1 - v)]$. In (b)–(d), the solid lines are generated from the series solution, (13.19) or (13.21), while the data points are the FEA-based results from the software described in Chapter 15.

The problem can be described in terms of the temperature change relative to the bottom of the layer, given by

$$\Delta T(y, t) = T(y, t) - T_o \qquad (13.16)$$

Because $T_\infty$ is constant, (13.1) holds with $\Delta T$ replacing $T$. Similarly, from (2.48) and (2.49), the convective heat transfer condition on the top surface is

$$q = h \cdot \Delta T(H, t) = -k\left(\frac{\partial \Delta T}{\partial y}\right)_{y=H} \qquad (13.17)$$

The problem and solution can be stated compactly by defining $0 \leq \bar{y} = y/H \leq 1$ and $\tau = \kappa t$ where $\kappa \equiv \rho c_p/k$ as before. Let $\tilde{T} = \Delta T(\bar{y}, \tau)/\Delta T_o$, where $\Delta T_o = \Delta T(1, 0)$ is the initial temperature change at the top of the layer. The complete problem statement in dimensionless variables is therefore

$$\dot{\tilde{T}}(\tilde{y}, \tau) = \tilde{T}''(\tilde{y}, \tau); \quad \tilde{T}(0, \tau) = 0; \quad \tilde{T}(1, \tau) = -\tilde{k}\tilde{T}'(1, \tau); \quad \tilde{T}(\tilde{y}, 0) = \tilde{y} \qquad (13.18)$$

where $(\dot{\ })$ denotes $\partial(\ )/\partial t$ and $\tilde{k} = kH/h$. This can be solved via separation of variables, with the solution given by

$$\tilde{T}(\tilde{y}, \tau) = \sum_{n=1}^{\infty} B_n \sin(\lambda_n \tilde{y}) e^{-\lambda_n^2 \tau}$$

$$B_n = \frac{4 \sin \lambda_n - 4\lambda_n \cos \lambda_n}{\lambda_n (\sin 2\lambda_n - 2\lambda_n)} \qquad (13.19)$$

with $\lambda_n$ defined by the roots to the characteristic equation

$$\tan \lambda_n = -\tilde{k}\lambda_n \qquad (13.20)$$

Temperature distributions based on this solution are shown in Figure 13.5(b). The solution has been coded in the software accompanying this chapter.

The dip in temperature near the surface that occurs immediately after the start of cooling is of central interest. A large number of terms in the series solution are required to accurately capture the initial condition near the top surface $\tilde{y} = 1$. While the coefficients $B_n$ decay quickly with increasing $n$, the product of $B_n e^{-\lambda_n^2 \tau}$ converges slowly for small values of $\tau$. As such, a significant number of terms are required for an accurate solution. Fortunately, modern software such as *Mathematica* makes this root finding and series evaluation trivial. Figure 13.5(b) illustrates the temperature distributions predicted with the first 200 terms from the series, for $\tilde{k} = kH/h = 0.1$ at several different times. Though mostly obscured in Figure 13.5(b), there is a barely perceptible squiggle near the edge $\tilde{y} = 1$ for very small times, a consequence of truncating the series. Nevertheless, the error is very small: with 200 terms, the maximum error in the initial condition is less than 1%.

The implications of this solution for a plane strain delamination crack within the layer are as follows. We assume that the layer is confined against bending or elongation, which is consistent with the layer being bonded to a semi-infinite substrate. As before, we assume that delamination cracking occurs so fast as to not alter the temperature distribution through the layer. If one assumes the elevated temperature condition is stress-free, the misfit strain is given by $\theta(\tilde{y}, \tau) = \alpha \Delta T_o[\tilde{T}(\tilde{y}, \tau) - \tilde{y}]$. In the present treatment, we assume that the substrate does not change temperature and therefore experiences no strain and does not influence the results. For plane strain conditions, the stress ahead of any delamination crack and beneath any delaminated segment is equi-biaxial and given by $\sigma_x = \sigma_z = -E\theta(\tilde{y}, \tau)/(1 - v)$. In a delaminated section of the layer, $\sigma_x = 0$ and $\sigma_z(\tilde{y}, \tau) = -E\theta(\tilde{y}, \tau)$.

The energy release rate for a delamination crack that lies at $\tilde{y}_d$ is therefore given by

$$G(\tilde{y}_d, \tau) = \frac{(1 + v)EH\alpha \Delta T_o^2}{2(1 - v)} \int_{\tilde{y}_d}^{1} [\tilde{T}(\tilde{y}, \tau) - \tilde{y}]^2 d\tilde{y} \qquad (13.21)$$

where $\Delta T_o = \Delta T(1, 0)$ is the initial temperature rise from top to bottom in the layer. The solution at infinite time, when $\tilde{T}(\bar{y}, \infty) = 0$, is given by

$$G_{ss} = \frac{(1+v)EH\alpha\Delta T_o^2}{2(1-v)} \left( \frac{1}{3} - \frac{\bar{y}_d^3}{3} \right) \qquad (13.22)$$

Figure 13.5(c) plots the energy release rate near the top of the layer and at the bottom as a function of time, for the dimensionless heat transfer coefficient $\tilde{k} = kH/h = 0.1$. Results are shown based on both the series solution (13.19) and the finite element approach used in Chapter 15.

The finite element results prevent curling of the strip shown in Figure 13.5(a) by explicitly modeling the substrate layer, with an exaggerated thickness and thermal conductivity. The latter promotes a nearly uniform temperature in the substrate. Note that cracks near the surface reach the their temporal steady state much more quickly than a crack at the bottom of the layer (i.e., the layer/substrate interface). For a crack near the top surface ($y_d = 0.95H$), the finite element results compute the energy release rate using a single element in the debonded layer; nevertheless, the results are quite accurate. (There are 20 elements in the calculations for Figure 13.5.)

Figure 13.5(d) plots the energy release rate (normalized by the prefactor in 13.21) for all crack locations accessible through the finite element calculation, and compares the results with those obtained via the series solution. As seen in both Figure 13.5(c) and Figure 13.5(d), the FEA results near the bottom of the strip at short times are slightly different than the analytical solution. This discrepancy arises because the FEA solution explicitly models the substrate, such that a small amount of heat is conducted out of the layer due to the substrate's finite thermal conductivity. Put another way, in the FEA the temperature at the bottom of the layer (top of the substrate) is enforced only approximately and is not strictly held constant, as imposed in the analytical solution.

In this simple example, the maximum energy release rate occurs at the bottom of the layer and at infinite time (after the layer has been fully cooled). However, this is because the CTE *mismatch* with the substrate plays no role because it is constrained to have zero deformation. In multilayer problems subject to rapid cooling, as first elucidated by Sundaram et al. (2013) and detailed in the next section, the factor controlling the energy release rate switches from the CTE of the coating at small times to the CTE mismatch with the substrate at large times. The critical implication of this switch is that the peak energy release for the system can be dictated by transients, with values larger than the fully cooled state.

## 13.4     The Effects of Temperature Transients on Delamination in Multilayers

In multilayers, the effect of transient temperature distributions on cracking can be critical, as first elucidated for bilayers by Sundaram et al. (2013). In the previous section, heat transfer through the substrate was effectively suppressed, such that the substrate experienced no thermal strain and did not influence the results. Here, that assumption is relaxed to consider the more realistic scenario where heat flows out of the substrate and

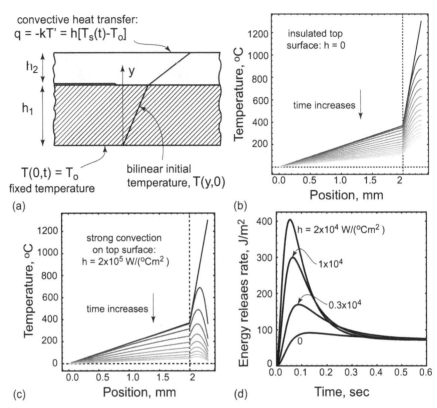

**Figure 13.6** (a) Schematic of the problem of a bilayer subjected to fixed temperature on the bottom and convective heat transfer on the top. (b) Temperature profiles through the bilayer at several times assuming no convection on the top surface, $h = 0$; curves span from $t = 0$ sec (black) to $t = 0.375$ sec in increments of $0.025$ sec. (c) Temperature profiles through the bilayer at several times assuming strong convection on the top surface, $h = 2 \cdot 10^5 \; W/(°Cm^2)$; curves span from $t = 0$ sec (black) to $t = 0.25$ sec in increments of $0.025$ sec. (d) Energy release rates as a function of time for several levels of convection. All results have been generated with LayerSlayer, described in Chapter 15.

substrate contraction can occur. Consider the bilayer shown in Figure 13.6(a). Properties for these calculations are included in the software files associated with this chapter: they are reasonable for a porous thermal barrier coating on a superalloy substrate, though the thickness values were chosen to elucidate the results and are somewhat atypical. Typically, the substrate would be slightly thicker than used for Figure 13.6, while the coating would be slightly thinner. The top surface is subject to convective heat transfer, while the bottom-layer has fixed temperature. At time $t = 0$, the ambient temperature surrounding the top surface is switched to zero.

The temperature distributions at constant time intervals are shown in Figure 13.6(b), for the case where the heat transfer coefficient of the top surface is zero ($h = 0$). This corresponds to the top surface being insulated. Note that this implies $q = -kT' = 0$ for the top surface, so the spatial derivative of the temperature is therefore zero as well.

Despite the fact that the temperature of the bottom of the substrate is fixed, heat flows out through the bottom, and the temperature drops with time. Figure 13.6(c) shows the results of the calculation when the heat transfer coefficient at the top surface is not zero. In this case, heat flows out of the system on both the top and bottom surfaces at the same time. Figure 13.6(c) clearly shows that switching on strong convection ($h \uparrow$) leads to rapid cooling of the top surface: the second curve corresponds to $t = 0.025$ sec, by which point the surface temperature has dropped more than 800°C while the interface temperature has not changed significantly.

This temperature 'inversion' at small times in Figure 13.6(c) implies differential cooling, wherein the coating cools much faster than the substrate. In this time regime, the substrate has not changed temperature, such that the stresses in the coating scale with $\alpha_2$ directly, as opposed to the mismatch $\alpha_1 - \alpha_2$. The impact of this on delamination is shown in Figure 13.6(d), which plots the plane strain energy release rate at the substrate/coating interface as a function of time. Differential cooling of the coating that is promoted by strong convection produces a large spike at small times. The size of this spike scales the heat transfer coefficient. As the rate at which heat is pulled from the surface increases, the coating experiences sudden contraction by the amount $\alpha_2 \Delta T$, which generates significant strain energy and drives cracking. At larger times, the substrate also contracts, switching the scaling on misfit strains to $\alpha_2 - \alpha_1$, lowering the stresses and the energy release rate. At very large times, the energy release rate asymptotes to the values predicted by the steady-state calculations described in Chapter 5.

To make matters worse, the mode mix for a delamination crack that is controlled by transients such as those in Figure 13.6 is typically closer to mode I than the steady-state result at long times. This implies that the relevant toughness controlling delamination during the transients is lower than those associated with steady state, which are typically closer to mode II. Additional details regarding the time dependence of the phase angle are given by Sundaram et al. (2013) and Jackson & Begley (2014).

Given that the asymptote at large time depends on the thermoelastic properties and is independent of strictly thermal properties, it is evident that the transient spike in energy release rate illustrated in Figure 13.6(d) is not necessarily higher than the asymptotic result associated with the full cooled state. An important question therefore arises: when is cracking more likely during transient cooling than the fully cooled state?

To address this question, Jackson & Begley (2014) conducted a parametric study that imposed temperatures on the top and the bottom of a bilayer that decay with a known time constant. That is, they defined

$$T_s = T_o + (T_s^o - T_o)e^{-t/\tau}; \quad T_b = T_o + (T_b^o - T_o)e^{-t/\tau} \quad (13.23)$$

where $T_s$ is the temperature of the top surface of the coating, and $T_b$ is the temperature of the bottom surface of the substrate. $T_o$ is the ambient temperature prescribed to be 25°C, while $T_s^o$ and $T_b^o$ are the initial temperatures of the top surface and bottom surface, respectively. The parameter $\tau$ is a constant (with units of time) that controls the rate of cooling. As $\tau$ increases, the rate of cooling decreases. This approach is motivated by the fact that heat transfer coefficients vary widely with application, and the fact that one can utilize experimental measurements of top and bottom surface temperatures. Laser

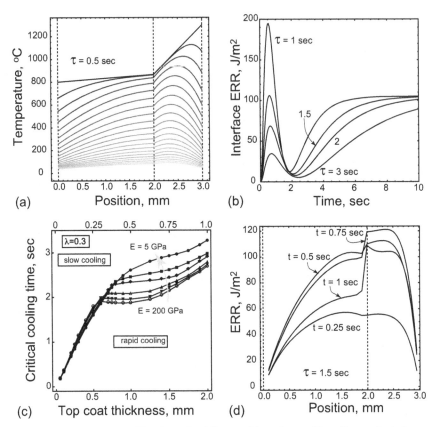

**Figure 13.7** (a)Temperature profiles through a bilayer subjected to rapid cooling on both faces; curves span from $t = 0$ sec (to) to $t = 2$ sec (bottom) in $0.1$ sec intervals. (b) Energy release rate at the bilayer interface as a function of time, for several time constants that control the rate of cooling. (c) Critical cooling time at which the transient ERR becomes larger than the cooled state as a function of various bilayer parameters; $\lambda = 0.3$ defines the mode mix toughness parameter, which is factored into the calculations. (d) Energy release rate distributions throughout the bilayer for one cooling rate and various times: note that the peak ERR does not occur at the interface under certain cooling conditions. All results have been generated using *LayerSlayer*, the software described in Chapter 15.

experiments have illustrated that (13.23) is a reasonable approximate form for the time dependence of surface temperatures when the heating laser is switched off and ambient temperature air is blown across the surface.

Figure 13.7 details the behaviors observed in a bilayer subjected to this type of cooling. The temperature profiles in Figure 13.7(a) illustrate the consequences of heat flowing out of both faces, that is, large drops in temperatures near the faces while the interior remains hot. The energy release rate at the interface of the bilayer is shown in Figure 13.7(b) as a function of time, with sharp peaks at short times arising from the differential cooling that takes place near the top coat surface. This behavior was first shown by Sundaram et al. (2013) for convective heat transfer but is qualitatively similar.

The cooling rate (here described by the time constant used in the exponential cooling law, $\tau$) dictates whether or not the transient response leads to higher crack driving forces; for slow enough cooling, the worst case scenario is the fully cooled state. Jackson & Begley (2014) conduct a broad parameter study for the bilayer problem and used the software accompanying this book to tabulate critical cooling times $\tau$ associated with the case where the peak ERR in the transient becomes larger than the fully cooled state; Figure 13.7(c) shows this critical time as a function of top coat thickness and modulus.

The results in Figure 13.7(c) are based on a slightly modified form of the results shown in Figure 13.7(b) which takes into account mode mix. As the temperature profile evolves in time, so too does the residual stress and in turn the resultant edge force and moment; hence, the phase angle evolves in time as well. Jackson and Begley used the framework described in Chapter 5 to compute the resultants and inserted these results into those in Chapter 4 to compute the mode mix. The energy release rates (such as those shown in Figure 13.7(b)) were then scaled by the mixed-mode toughness, as described in Chapter 3; the critical times shown in Figure 13.7(c) reflect the results from using this scaled ERR, and hence account for mode mix.

Finally, Figure 13.7(d) shows the energy release rates at various locations in the bilayer, for several times. (These results have not been adjusted to account for mode mix.) One notes that the location with the maximum energy release rate is not necessarily at the interface; in this example, at time $t = 0.25$ sec, the maximum energy release rate is in the substrate, while at time $t = 0.75$ sec, the energy release rate is maximum close to the midpoint the coating. This is of no practical consequence if the interface toughness is lower than the bulk toughness of the layers (e.g., they are obviously irrelevant for ductile substrates); however, for some systems with very brittle layers, such behavior may imply midlayer cracking will occur during transients. The paper by Jackson & Begley (2014) has further details regarding the system parameters that lead to such behaviors.

# 14 Software for Semi-Infinite Multilayers: Steady-State Delamination

In this chapter, we describe software that analyzes multilayers that are at steady-state conditions with respect to time; in order to be applicable, the multilayers' width must be much larger than the total stack thickness, such that one can neglect lateral gradients in the direction of the layers. The software can be used to determine the temperature distribution in the layers, the deformation (elongation and stretch) and stress distribution in the layers of the multilayer, or any submultilayers formed by delamination cracks. The software is based on one-dimensional heat transfer described in Section 2.6 and general multilayer mechanics framework outlined in Chapter 5. The software is applicable to any number of layers, and using this framework, the results can be used to compute the energy release rates for semi-infinite delamination cracks located at any location in the stack. The code associated with this chapter has been dubbed *LayerSlayer (LS)*.

Since the energy release rate associated with a semi-infinite crack in a semi-infinite multilayer is independent of the crack length, the results are also at steady state with respect to crack length, which almost always represents the maximum possible value for any crack length. Hence, in this chapter, steady state refers to both time and crack length. One should recall that the analysis in Chapter 5 does *not* yield the mode mix, which requires full solution of the associated elasticity problem. This is covered in Chapter 16.

A complementary code that solves for time-dependent (transient) behavior under the same geometric restrictions as imposed in this chapter is described in the next chapter and is called *LayerSlayer Transient (LST)*. It may be appreciated that the mechanical analysis conducted in *LayerSlayer Transient* is essentially identical to that in *LayerSlayer*, only with each physical layer broken into through-thickness elements that describe nonlinear temperature distributions as a series of piecewise linear segments. As such, the interfaces for both *LS* and *LST* are remarkably similar. While both are completely self-contained (i.e., the reader can use either code without needing the other), the description of *LST* builds off this chapter, and hence it is recommended that the reader become familiar with *LS* first, even if transient solutions are sought.

The codes are currently provided in the commercial multipurpose code *Mathematica* (Wolfram Research). While there are many reasonable choices (and versions of *LayerSlayer* for other software platforms can be anticipated in the future), *Mathematica* was chosen because it runs on most operating systems, e.g., Windows, Unix, Mac OS) and is very consistent from one version to the next, so that users do not need to worry about version compatibility (at least so far) or pesky differences in compilers. Perhaps most importantly, *Mathematica* includes prepackaged commands to handle a variety of high-level mathematical functions, most notably:

- Linear equation solvers, nonlinear solvers (including root-finding) and differential equation solvers that make solving for deformation variables trivial, even when time-dependent or nonlinear problems are being tackled.
- Powerful, easy-to-use plotting routines, including both *x-y* plots and contour plots when there is more than one independent variable.
- The ability to easily define new functions that include new dependencies without changing the structure of the underlying code.

The last feature is particularly powerful, because it means that one can define new functions to describe properties as a function of another variable, such as temperature. For example, we can define the modulus of different layers as Em[T], a function that describes the moduli as a function of temperature, and then use those functions to define a multilayer whose definition depends on temperature, as in Multilayer[T]; one can then use the provided modules without further alteration, generating results as a function of temperature. Examples of this 'variable' property definition are sprinkled through the analysis files associated with each chapter.

Naturally, one could use *any* programming code to achieve the same functionalities described in this chapter, with the only change being in syntax and structure. It is our intention that the preceding chapters in this book provide a complete set of equations to be coded. We offer the codes in this chapter and the next simply as a convenience, recognizing that there will likely be scenarios where 'home-grown' codes are to be preferred.

## 14.1    Overview of *LayerSlayer*'s Assumptions, Capabilities and Interface

*LayerSlayer* (LS) can be used to analyze a multilayer stack comprising any number of semi-infinite layers, as shown in Figure 14.1. The layers are assumed to have much larger in-plane dimensions than their through thickness dimension, and one can assume various behavior with regard to out-of-plane behavior (e.g., plane strain $\epsilon_z = 0$ or equibiaxial straining with $\epsilon_z = \epsilon_x$). The code assumes that the constitutive behavior is thermoelastic and the response is independent of time; that is, steady state has been reached with respect to temperatures. A spatially uniform misfit strain can be prescribed in addition to thermal strains, which can represent deposition/growth strains or swelling strains and so forth. The key assumption is that the strain distribution in bonded layers is given by

$$\epsilon_x(y) = \epsilon_o^x - \kappa_z y + \alpha \left[ T(y) - T_o(y) \right] + \epsilon_g, \tag{14.1}$$

where $\epsilon_o^x$ is the elongation in the *x*-direction at the bottom of the stack, $\kappa_z$ is the curvature of the stack about the *z*-axis, $\alpha$ is the coefficient of thermal expansion in the layer, $T_o(y)$ is a reference temperature that has a linear distribution in a given layer, and $T(y)$ is the linear temperature distribution dictated by the temperature at the top and bottom of a

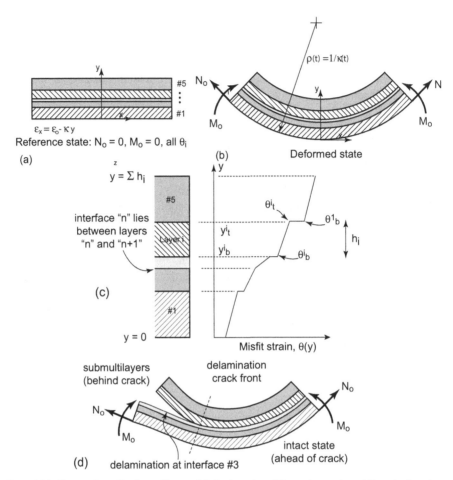

**Figure 14.1** Conventions for *LayerSlayer*: (a) the intact multilayer is numbered from bottom to top, (b) deformation modules report elongation and curvature of intact multilayer (or submultilayer), (c) misfit strains are linear functions of position and written as superposition of thermal strains and 'growth strains', (d) the energy release rate for an interface crack is analyzed using the multilayer ahead of the crack and submultilayers formed by the delamination.

given layer. The specified parameter $\epsilon_g$ represents a spatially uniform misfit strain that varies from layer to layer and arises from sources other than thermal expansion.

*LS* allows one to prescribe linear temperature distributions in each layer by prescribing the temperature at the top and bottom of each layer. Alternatively, one may use *LS* to solve for them directly using the conditions of steady-state heat transfer. The code allows for jumps in temperatures (eigenstrains) across interfaces, as might arise due to finite thermal conductance of the interfaces or swelling strains that are different from layer to layer.

The framework is general for an arbitrary number of layers, which are assumed to be elastic, isotropic layers described by a modulus, coefficient of thermal expansion, thermal conductivity, a reference (strain-free) temperature, growth strain, and the temperatures at the top and bottom of each layer (which can be solved for using steady-state thermal analysis). The code predicts the following:

- steady-state temperature distribution through the layers (for either two prescribed temperatures or a prescribed heat flux with one temperature)
- deformation of the multilayer as described by its elongation and curvature, assuming various stressing conditions such as plane strain or equi-biaxial stressing.
- energy release rate for debonding, assuming various stressing conditions ahead of and behind the delamination crack, such as equi-biaxial stressing followed by plane strain delamination.

The analysis is conducted for a stack of blanket films, as shown in Figure 14.1. The films are assumed to be infinite in the $(x, z)$-plane, with a semi-infinite crack that grows along an interface that lies on a specific $(x, z)$-plane, with a crack front that is aligned with the $z$-direction; energy release rates are computed for the steady-state growth scenario, that is, when the crack is long enough so as to not influence the results. The problem is essentially one-dimensional, such that behavior in the $(x, y)$-plane is analyzed: in the third dimension (out-of-plane to the analysis), one can either assume plane strain deformation (i.e., $\epsilon_z = 0$) or biaxial deformation (i.e., $\epsilon_z = \epsilon_x$).

Temperatures are assumed to be uniform in the plane of the blanket, that is, in the $(x, z)$-plane, with gradients occurring only in the through-thickness direction, that is, in the $y$-direction. The thermal analysis solves for the steady-state distribution through the stack, which yields a piecewise linear temperature profile: if finite conductance is specified at the interfaces, the temperature is not continuous across the interface but rather experiences jumps proportional to the inverse of the interface conductance: hence, the temperatures on either side of the interface (i.e., bottom of one layer and top of layer beneath it) are retained as possibly independent variables. In the limit of extremely high interface conductance, the temperatures on either side of the interface become identical.

When analyzing cracking, it is assumed that the crack has no influence on the temperature distribution, that is, the intact stack is analyzed. While cracking would naturally alter the heat flow in the cracked region, it is assumed that cracking events are sufficiently fast so as to not allow for temperature changes associated with debonding. That is, the scenario being analyzed consists of a steady-state crack growing at speeds faster than thermal transport. (The analysis of initiation of crack growth, involving a different temperature profile behind the crack due to debonding, requires a 2D thermal analysis to account for spatial gradients in the $x$-direction.) A variety of scenarios can be addressed with regard to the behavior parallel to the crack front, including plane strain deformation ($\epsilon_z = 0$) or biaxial deformation ($\epsilon_z = \epsilon_x$).

A given layer is defined by the list

$$Layer = \{h, E, v, \alpha, \{T_o^b, T_o^t\}, \epsilon_g, k, \{T^b, T^t\}\} \tag{14.2}$$

where

- $h$ is the layer thickness
- $E$ is the Young's modulus of the layer

- $\alpha$ is the coefficient of thermal expansion of the layer
- $\{T_o^b, T_o^t\}$ are the reference temperatures at the bottom and top of the layer
- $\epsilon_g$ is a growth or deposition strain (defined below)
- $k$ is the thermal conductivity of the layer
- $T^b$, $T^t$ are the current temperatures at the the bottom and top of the layer

Naturally, the order of the properties in the multilayer is fixed, and one must always be sure that the properties are properly loaded. Note that the code assumes that the reference temperature and current temperature distributions are linear, defined by the temperature at the bottom and top of the layer, as described in Chapter 5.

A multilayer is defined by forming a list of layers:

$$Multilayer = \{Layer1, Layer2, \ldots, LayerN\} \tag{14.3}$$

Once the multilayer is defined as a list of layers, with each layer represented as a list of properties, the problem is completely defined and ready for subsequent analysis.

In the software implementation, modules are created to calculate the response of the multilayer, with each module producing quantities of interested. Examples of these modules are

$$\{\epsilon_o, \kappa\} = \texttt{BiaxialDeformation}[Multilayer]$$

$$stress\ list = \texttt{GetStress}[Multilayer, iflag1, iflag2]$$

$$G = \texttt{PlaneStrainERR}[Multilayer, iflag1, iflag2]$$

where *iflag1*, *iflag2* is a list of additional information required to define the desired analysis, such as the deformation constraints imposed in the analysis, stress component of interest, location of the delamination crack for which the energy release rate is to be computed and so forth. Here, anything written in `Courier font` identifies the text as a *Mathematica* module (or command); this includes predefined modules in *LayerSlayer*. That is, the modules in *LS* are written to be used as *Mathematica* functions.

## 14.2    Steady-State Temperature Distributions

### Numerical Formulation for Steady-State Temperatures

In this section, we describe the numerical framework used to solve for the steady-state temperature distribution a semi-infinite multilayer; this temperature distribution serves as input to the mechanical analysis described subsequent sections. Naturally, for cases where the temperatures in each layer are known, one can skip the thermal analysis and directly specify the temperatures in the mechanical analysis. The reader is referred to Section 2.6 for a description of one-dimensional heat transfer that provides the theoretical foundation for the framework that follows.

When solving a multilayer problem, one must take care to define heat flux across boundaries appropriately: the heat flux coming into a given layer must have a different sign than that leaving the layer. The heat flux through layer $i$ (with positive heat flux occurring in the positive $y$-direction) can be written as

$$\begin{pmatrix} q_{in} \\ q_{out} \end{pmatrix} = \frac{k_i}{h_i} \begin{bmatrix} -1 & 1 \\ 1 & -1 \end{bmatrix} \begin{pmatrix} T_{2i-1} \\ T_{2i} \end{pmatrix} \tag{14.4}$$

where $h_i$ is the thickness of the layer, $T_{2i-1}$ is the temperature at the bottom of layer $i$, and $T_{2i}$ is the temperature at the top of layer $i$. It is clear from the above equations that we have defined $q_{in} = -q_{out}$ (i.e., if you multiply the first equation by $-1$ you get the second equation). This is a statement that total heat energy in the layer is conserved, because the heat flux into and out of the layer is the same. The negative sign on $q_{in}$ arises because the surface normal points in the negative $y$-direction, that is, positive heat flux is defined as that leaving the film and moving in the direction of a positive surface normal.

Using the same approach, we can write the equation governing the the jump condition across the interfaces, which arises due to finite interface conductance. To account for the jump condition, the parameter $k_i/h_i$ is replaced with $k_{INT}$, the conductance of the interface, and the temperatures are replaced with those on either side of the interface. That is, both individual layers and interfaces act as resistive elements characterized by either $k_i/h_i$ or $k_{INT}$: both have units of thermal conductivity per unit length. The total heat flux at any *internal* boundary, represented as the sum of the heat flux leaving one domain (i.e., film) and entering another domain (i.e., film), must be zero such that the boundary does not gain or lose thermal energy.

In the following, the nodal temperatures refer to the temperatures at the top and bottom of the film: these are numbered sequentially starting with $T_1$ at the bottom of the layer at the bottom of the stack and ending at $T_{2N}$ at the top of the stack, where there are $N$ layers. That is, there are two temperatures associated with each layer, one at the bottom of the layer and one at the top. Thus, the layer numbered $p$ is defined by nodes numbered $T_{2p-1}$ (at the bottom) and $T_{2p}$ (at the top).

With this in mind, we can sum the heat flux at all nodes (i.e., $\sum q_j = q_{out}^j - q_{in}^{j+1} = 0$, where $j$ is the node number and superscripts denote element number) and use (14.4) to derive a set of $2N$ coupled equations:

$$[K]_{2N \times 2N} \{T\}_{2N \times 1} = \begin{bmatrix} -q_{out} \\ 0 \\ \vdots \\ 0 \\ 0 \\ q_{out} \end{bmatrix}_{2N \times 1} \tag{14.5}$$

where $[K]$ is a global thermal conductivity matrix, involving properties for the layers $k_i/h_i$ and the interfaces $k_{INT}$. In (14.5), $q_{out}$ defines the condition governing heat flux coming out of the system, and $q_{in} = -q_{out}$ defines the flux coming into the system. The right-hand side reflects the sum of heat fluxes from connected elements at that location: it is zero for all interior nodes because net heat flux summed over element boundaries is zero for internal elements. It should be kept in mind that global assembly includes "elements" representing the interface, which behave as artificial layers with zero thickness (wherein $k/h$ is replaced with $k_{INT}$). It may be apparent to some readers that (14.4) and (14.5) are equivalent to a first-order finite difference scheme in one dimension.

The governing equations given as (14.5) are singular until appropriate boundary conditions are specified. To enable a solution, any two of the unknown variables (nodal temperatures and heat flux) can be specified. Given that $q_{in} = -q_{out}$, we see that if heat flux is a specified in the analysis, we can prescribe only one temperature. For example, if heat flux going into the stack is defined, the output conditions must impose the condition $T_{out} = -(q_{in}/h_t) + T_\infty$, where $h_t$ is the heat transfer coefficient and $T_\infty$ is the external temperature driving heat flux. Alternatively, one can specify both surface temperatures and eliminate $q_{in}$ and $q_{out}$ as variables. In the software implementation, the specified conditions are imposed using Lagrange multipliers, such that for the case of two specified temperatures, the software computes the associated heat flux as value of the Lagrange multiplier.

The output of the temperature analysis is the nodal temperatures, keeping in mind that there are two nodes associated with each interface (to allow for finite temperature jumps across an interface associated with finite interface conductance). That is, the thermal analysis module produces

$$Temperature = \{\{y_1^b, T_1^b\}, \{y_1^t, T_1^t\}, \{y_2^b, T_2^b\}, \{y_2^t, T_2^t\}, \dots, \{y_N^b, T_N^b\}, \{y_N^t, T_N^t\}\}$$

(14.6)

where $y_j^b$ is the location of the bottom of layer $j$, $y_j^t$ is the location of the top of layer $j$, and similarly for the temperatures. These quantities are precisely those required to compute the mechanical response of the multilayers, as described in Chapter 5.

## Thermal Analysis Modules

For thermal analysis, one must specify a list of layer properties as in (14.2), and an order list of layers as (14.3), where the layer numbering is from bottom to top, as in Figure 14.1. The user must also provide a list of *interface* conductances, again numbered from bottom to top, as in

$$ifaces = \{k_{INT}^{12}, k_{INT}^{23}, \dots, k_{INT}^{(N-1)(N)}\},$$

(14.7)

where *ifaces* is a user-defined name of the list containing interface properties, $k_{INT}^{12}$ is the interface conductance between layers 1 and 2, $k_{INT}^{23}$ is the interface conductance between layers 2 and 3 and so forth. Setting the interface conductances to a large number (in comparison to the effective layer conductances, $k_i/h_i$) results in nearly perfect conduction, with (piecewise) continuous temperatures throughout the stack. Two modules for steady-state thermal analysis are included, distinguished by different types of boundary conditions.

The module TempSol[] solves for the temperature distribution in the multilayer: it requires two temperatures to be specified within the entire multilayer stack. One can specify any two temperatures in the stack. These boundary conditions are specified as

$$TempBC = \{\{Layer\ \#,\ location,\ T_1\},\ \{Layer\ \#,\ location,\ T_2\}\}, \qquad (14.8)$$

where *TempBC* is a user-defined name of the list containing boundary conditions; *location* $= 0$ for temperature specified at the bottom of the layer, and *location* $= 1$ for temperature specified at the top of the layer. When interface conductance is essentially infinite, it does not matter if you specify the temperature at the top of one layer (i.e., the bottom of the interface) or the temperature at the bottom of the adjacent layer (i.e., at the top of the interface). That is, for very large, interface conductance, specifying either $\{1, 1, T_1\}$ or $\{2, 0, T_1\}$ yields the same result. When interface conductance is finite, such that there is a temperature drop across the interface, you must specify the temperature at the proper side of the interface.

Once the interface properties and temperature boundary conditions have been specified, the steady-state temperatures at the bottom and top of each large are computed by the module

$$\text{TempSol}\,[Multilayer,\ ifaces,\ TempBCs,\ iflag] \qquad (14.9)$$

where *Multilayer*, *InterfaceProps* and *TempBCs* are the names of the lists defined earlier. The parameter *iflag* defines whether or not the temperatures should be used to redefine the layer temperatures: for *iflag=0*, the original definition of the reference temperatures (at the top and bottom of each layer) and current temperatures (at the top and bottom of each layer) in *Multilayer* are left unchanged. For *iflag=1*, the reference temperatures in *Multilayer* are redefined to be the solution temperatures from the thermal analysis; this corresponds to the assumption that the stress-free condition is associated with the temperatures arising from the thermal analysis. For *iflag=2*, the current temperatures in *Multilayer* are redefined to be the solution temperatures from the thermal analysis.

The module returns a list of position/temperature pairs, running from the bottom to the top of the stack. For example, for a two-layer system, the output is the following:

$$\{\{0,\ T_{bot}^1\},\ \{y_1,\ T_{top}^1\},\ \{y_1,\ T_{bot}^2\},\ \{y_2,\ T_{top}^2\}\} \qquad (14.10)$$

That is, a list of absolute global positions in the stack and temperatures at those positions is returned; positions are repeated at interfaces, such that there are different pairs associated with the top and the bottom of the interface. One can easily plot the temperature distribution using this list, illustrating any temperature jumps at the interfaces that are a result of finite interface conductance. It must be kept in mind that thermal analysis will rewrite the definition of the multilayer for *iflag* $= 1$ or *iflag* $= 2$.

An alternative analysis can be run where the heat flux and one temperature are specified (as opposed to two temperatures); this case is handled by the module $\text{TempSolQ[]}$. One can specify the temperature either at the bottom or the top of a given layer. For this module, the boundary conditions are specified as

$$\text{HeatFluxBCs} = \{\{Layer\ \#,\ location,\ T_1\},\ Q\}, \qquad (14.11)$$

where *location* = 0 for temperature specified at the bottom of the layer, and *location* = 1 for temperature specified at the top of the layer. $Q$ is the heat flux in the layer, defined as positive from bottom to top. (If the top temperature is higher than the bottom temperature, then $Q < 0$.) Again, when interface conductance is set to a very large number (compared to the values of $k/h$ for the adjacent layers), it does not matter whether the temperature at the top of one layer is specified or the temperature at the bottom of the adjacent layer.

The module input for the heat flux scenario is

$$\text{TempSolQ}\,[\textit{Multilayer, InterfaceProps, HeatFluxBCs, iflag}] \tag{14.12}$$

where *Multilayer*, *InterfaceProps* and *HeatFluxBCs* are the names of the lists defined earlier. These commands return a list of position/temperature pairs, corresponding from bottom to top of the stack, identical to the output from the other boundary condition module, e.g., that given by (15.14). Again, the parameter *iflag* defines whether or not the temperatures should be used to redefine the layer temperatures: for *iflag* = 0, the original definition of the reference temperatures (at the top and bottom of each layer) and current temperatures (at the top and bottom of each layer) in *Multilayer* are left unchanged. For *iflag* = 1, the reference temperatures in *Multilayer* are redefined to be the solution temperatures from the thermal analysis; this corresponds to the assumption that the stress-free condition is associated with the temperatures arising from the thermal analysis. For *iflag* = 2, the current temperatures in *Multilayer* are redefined to be the solution temperatures from the thermal analysis.

## 14.3    Deformation and Stresses

Once the temperatures throughout the multilayer are known (being either specified by the user directly or computed from the thermal analysis described above), the deformation and stress in the layers can be computed following the framework described in Chapter 5. (It should be noted that the deformation and stress do *not* need to be computed a priori to compute the energy release rate, which has its own self-contained module.)

There are two modules to calculate the elongation and curvature of a multilayer: one that assumes equi-biaxial deformation with no externally applied loads, and one that assumes plane strain deformation with externally applied moment and axial force. In this calculation, one can specify a uniform strain parallel to the crack front, that is, $\epsilon_z = constant$, which is referred to as generalized plane strain.

The biaxial deformation is found with

$$\text{BiaxialDeformation}\,[\textit{Multilayer}] \tag{14.13}$$

where *Multilayer* is the name of the stack that was defined previously by creating a list of layers. This module returns the elongation of the bottom of the stack, $\epsilon_o$, and the curvature of the stack, $\kappa$, in a list ordered $\{\epsilon_o, \kappa\}$. One can substitute an explicit list of the layers desired for analysis, as in $\text{BiaxialDeformation}\,[\{\textit{Substrate, Coating}\}]$.

That is, you can define the multilayer in the function call itself and avoid having to define additional multilayers if you want to consider cases that do not include all layers.

Similarly, there is a separate module for generalized plane strain deformation, in which one prescribes the applied moment, the applied axial resultant and the value of the out-of-plane strain. This type of deformation is found with

$$\texttt{PlaneDeformation}[Multilayer, M_a, N_a, \epsilon_z] \qquad (14.14)$$

where *Multilayer* is the name of the stack currently being analyzed, $M_a$ is the value of the applied moment, $N_a$ is the value of the applied axial force, and $\epsilon_z$ is the constant value of the strain in the $z$-direction. Again, this module returns the elongation of the bottom of the stack, $\epsilon_o$, and the curvature of the stack, $\kappa$, in a list ordered $\{\epsilon_o, \kappa\}$.

Once the deformation variables are found, stresses are computed using a module that is supplied with the elongation and curvature of the stack in the $x$- and $z$-directions. That is, one provides $\{\epsilon_o, \kappa\}_x$ and $\{\epsilon_o, \kappa\}_z$. The module that computes stress is

$$\texttt{GetStress}[\{\epsilon_o, \kappa\}_x, \{\epsilon_o, \kappa\}_z, Multilayer, dir, scale] \qquad (14.15)$$

where $\{\epsilon_o, \kappa\}_x$ define the total strain in the $x$-direction via (14.1), while $\{\epsilon_o, \kappa\}_z$ define the total strain in the $z$-direction via the analogous version of (14.1) for that direction. *Multilayer* is the list of layer properties defined by (14.3) and (14.2), and $dir = 1$ for $\sigma_x$ (and $dir = 3$ for $\sigma_z$). The variable *scale* is simply a scale factor that allows you to choose convenient units. For example, $scale = 10^6$ yields stresses in MPa if all the unit quantities are in base SI units.

The command returns an ordered list, with each entry being a list of the position and stress at the bottom and top of the layers:

$$\{ ...., \{\{y_{bot}^i, S_{bot}\}, \{y_{top}^i, S_{top}\}\}, ... \} \qquad (14.16)$$

where the positions (e.g., $y_{bot}^i$) are measured from the bottom of the stack. Thus, using `ListPlot` with the output of `GetStress` produces a plot of the stresses in the layers, with a different (linear) line representing the stress in each layer.

## 14.4 Energy Release Rates

Once temperatures at the top and bottom of each layer have been defined (either directly or computed using the temperature module), the energy release rate for delamination can be computed using the framework described in Chapter 5. The deformation module and stress module are not needed to compute the energy release rate; one can proceed directly to computing the energy release rate once the temperatures have been determined.

There are three commands to calculate energy release rates for debonding, each corresponding to a different assumption regarding the stress states ahead of and behind the interface crack:

$$\texttt{ClassicERR}[Multilayer, \#]$$

$$\texttt{BiaxialERR}[Multilayer, \#]$$

$$\texttt{PlaneStrainERR}[Multilayer, \#, M_a]$$

In all commands, *Multilayer* is the name of the list that has the properties of all layers in the stack, and the entry # denotes the interface number where debonding occurs. The interfaces are numbered from the bottom of the stack, such that interface $i$ lies between the $i$ and $i + 1$ layers.

In the *ClassicERR* command, the stress state ahead of the crack is equi-biaxial, and delamination is assumed to occur subject to plane strain constraints. In the *BiaxialERR* command, the stress state is assumed to be equi-biaxial both before and after debonding. In the *PlaneStrainERR* command, plane strain conditions are enforced (i.e., $\epsilon_z = 0$) for all sections (the intact section ahead of the crack and the two sublayers formed by the delamination crack). The quantity $M_a$ refers to the applied moment applied to both ends of the multilayer: this moment is applied to the whole stack ahead of the crack, and only to the bottom stack (i.e., the bottom stack formed by the debond) behind the crack front.

## 14.5 A Basic Example with Three Layers

In this example, the basic analysis steps of *LayerSlayer* are illustrated for a three-layer system with the properties listed in Table 14.1; the properties are fictitious and have been chosen simply to produce plots that clearly illustrate the nature of the results. It should be emphasized that there are no units embedded in *LS*; the units listed in Table 14.1 are relevant only to the intrepetation of the inputs and outputs. The steps taken to analyze more sophisticated multilayers with many more layers are identical.

### Layer Definition
The temperatures at the top and bottom of the layer are set to zero since they will be calculated by the thermal analysis. In the above, the reference temperatures are uniform and equal to 70°C. The multilayer is defined as follows:

$$\text{Layer1} = \{0.003, 200.x10^9, 0.3, 15x10^{-6}, \{70, 70\}, 0, 25, \{0., 0.\}\};$$

$$\text{Layer2} = \{0.002, 40.x10^9, 0.2, 11x10^{-6}, \{70, 70\}, 0, 1, \{0., 0.\}\};$$

$$\text{Layer3} = \{0.001, 100.x10^9, 0.2, 11x10^{-6}, \{70, 70\}, 0, 2, \{0., 0.\}\};$$

$$\text{Multilayer} = \{\text{Layer1}, \text{Layer2}, \text{Layer3}\}$$

Thus, the layers are named appropriately from bottom to top, that is, Layer1 is on the bottom, and Layer3 is on the top.

**Table 14.1** List of Properties Used in the Case Study. Growth Strains are Zero, and the Temperatures are Determined Via Steady-State Thermal Analysis

| Layer | Thickness, m | $E$, Pa | $\upsilon$ | $\alpha$, 1/°C | $\kappa$, W/(m²K) |
|---|---|---|---|---|---|
| Layer1 | 0.003 | $200 \times 10^9$ | 0.3 | $15 \times 10^{-6}$ | 25 |
| Layer2 | 0.002 | $40 \times 10^9$ | 0.2 | $11 \times 10^{-6}$ | 1 |
| Layer3 | 0.001 | $100 \times 10^9$ | 0.2 | $11 \times 10^{-6}$ | 2 |

**Thermal Analyses**

First, consider the case where the temperature at the bottom of the stack is fixed to be 800°, while the temperature at the top of the stack is fixed to be 1300°. Let's assume that the interface between Layer1 and Layer2 has finite conductance, such that there is a temperature drop across that interface: the interface between Layer2 and Layer3 has a large conductance (nearly perfectly conductive), such that there will be no temperature drop. This is stated as

$$\text{TwoTbcs} = \{\{1, 0, 800\}, \{3, 1, 1300.\}\}$$

$$\text{ifc} = \{10^4, 10^7\}$$

Second, consider the case where the top surface temperature is known to be 1300°C and the heat flux (from top to bottom) with is known to be $Q = 169{,}244$, then we would specify

$$\text{TQbcs} = \{\{3, 1, 1300\}, -169{,}244\}$$

$$\text{ifc} = \{10^4, 10^7\}$$

where a negative heat flux is prescribed between the top temperature is known to be higher than the bottom temperature (i.e., heat flows from top to bottom, which is negative by convention). The thermal analysis for the 'two-temperature boundary condition' is conducted by executing:

$$\text{Tout} = \text{TempSol[Multilayer, ifc, TwoTbcs, 2]}$$

Note that since *iflag* = 2, the solution from the temperature analysis has been used to redefine the current temperature distribution in the multilayer; hence, these stresses are those associated with heating a multilayer that is stress-free at room temperature. Executing the module returns the following:

Heat flux: -169,294.

$$\{\{0., 800.\}, \{0.003, 820.315\}, ...., \{0.005, 1215.35\}, \{0.006, 1300.\}\}$$

Figure 14.2 plots the results. Since the heat flux is defined as positive from bottom to top, and the top temperature is highest, the heat flows in this case from top to bottom, which is a negative heat flux according to the adopted convention. Note the $\sim 56°C$ temperature drop between Layer1 and Layer2.

Note that, if the temperature is known at the bottom of Layer2, we could get the same results by running the following boundary conditions:

$$\text{twoTbcs} = \{\{2, 0, 876.8\}, \{3, 1, 1300.\}\}$$

$$\text{ifc} = \{10^4, 10^7\}$$

That is, one can prescribe two temperatures at any locations within the stack, including those on *either* side of an interface with finite conductance. Conversely, if we knew the top surface temperature was 1300°C and the heat flux was from top to bottom with

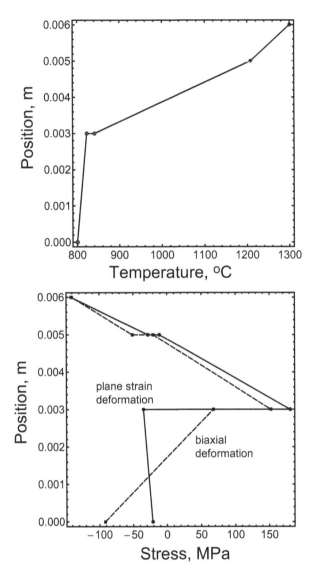

**Figure 14.2** Temperature profiles and stresses profiles through a three-layer stack, for the properties given in Table 14.1.

magnitude $Q = 169, 294$, we could run the 'one temp and heat flux bc' analysis, as in

```
Tout2 = TempSolQ[Multilayerex, ifc, {{3, 1, 1300}, -169, 294}, 2]
```

with the output given by

$$\{\{0., 800.001\}, \{0.003, 820.317\}, \ldots, \{0.006, 1300.\}\}$$

This result is identical to that obtained for the other boundary condition, as expected. Again, *iflag* = 2, so the solution from the temperature analysis has been used to redefine the current temperature distribution in the multilayer.

**Mechanical Analysis**

To get stresses in the layer, we first run the analysis that gives us the total elongation and curvature of the stack. Examples are:

$$\{\texttt{e1p, k1p}\} = \texttt{PlaneDeformation[Multilayer, 0, 0, 0]} \quad (14.17)$$

$$\{\texttt{e1b, k1b}\} = \texttt{BiaxialDeformation[Multilayer]} \quad (14.18)$$

where the first command gets the elongation and curvature for the plane-strain case, and the second command gets the elongation and curvature for the biaxial case. These results are then used to compute the associated stresses, in this case in the $x$-direction (scaled to report results in units of MPa, as implied by the units given in Table 14.1): the stresses for plane strain deformation and equi-biaxial deformation are computed from

```
Pstresses = GetStress[{e1p,k1p},{0,0}, Multilayer, 1, 10^6]

Bstresses = GetStress[{e1b,k1b},{e1b,k1b}, Multilayer, 1, 10^6]
```

with the output given as
$\{\{0, -28.7655\}, \{0.003, -18.7446\}\}, \ldots, \{\{0.005, -28.9895\}, \{0.006, -130.041\}\}\}$
$\{\{\{0, -99.3313\}, \{0.003, 84.399\}\}, \ldots, \{\{0.005, -52.112\}, \{0.006, -129.011\}\}\}$
where the first data set are the stresses assuming plane-strain deformation, and the second data set are those assuming biaxial deformation. The stresses in the layers for the present example are shown in Figure 14.2.

The calculation of energy release rates are simply function calls to the commands outlined above; for the interface between Layer1 and Layer2, one uses

```
G=ClassicERR[Multilayer, 1]
Gb=BiaxialERR[Multilayer,1]
Gps=PlaneStrainERR[Multilayer, 1, 0]
```

where the zero in the last command indicates no moment is applied to the multilayer. Note that the energy release rate modules require specification of the interface number; one can easily use the commands in *Mathematica* to loop through all interfaces in a stack. Figure 14.3(a) presents a bar chart of the energy release rates at two interfaces in the three-layer system, for the three conditions describing the stress states.

One can easily compute the energy release rate for any location within the stack by splitting a given layer into two identical sublayers, with an artificial interface defined between the sublayers. Figure 14.3(b) shows the outcome of this procedure; the example file accompanying this chapter details the trivial commands needed to split the layers. Setting a high interface conductivity for the artificial sublayer interfaces, but retaining the same true interface conductances, one obtains identical temperature and stress distributions as in Figure 14.2 (only with the code returning more points). In Figure 14.3(b), the plane strain ERR is computed at all interfaces, both real and artificial: we see that the maximum ERR does indeed occur at the first interface for this problem.

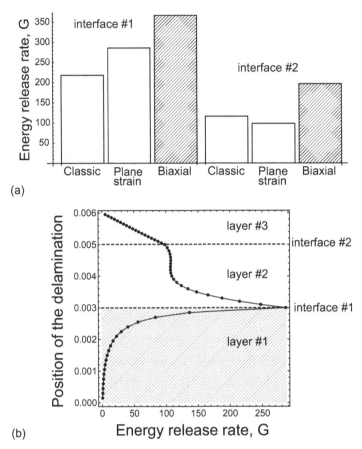

**Figure 14.3** (a) The energy release rates at the two interfaces in the three-layer example specified in Table 14.1, showing the impact of various assumptions regarding the constraints experienced by the delaminated submultilayers. (b) The plane strain energy release rate as a function of the position of a delamination crack for the three-layer example specified in Table 14.1; such plots can be generated by dividing a single layer into multiple, identical sublayers.

## 14.6    An Advanced Example: An Environmental Barrier Coating (EBC) System

*LayerSlayer* is a powerful tool to aid in the design of complex multilayers; here, we consider the performance of a candidate environmental barrier coating system intended to protect silicon carbide (SiC) composite substrates, shown in Figure 14.4. A summary of the system is given in Table 14.2. In order to prevent oxidation and volatilization of the substrate, rare earth silicates such as $Yb_2Si_2O_5$ (YbMS) and $Yb_2Si_2O_7$ (YbDS) are utilized. Both ytterbium silicates may be needed; the mono-silicate is an effective oxygen barrier, but has a high CTE and so needs to be used sparingly. The Si bond coat will be prone to oxidation (as the diffusion barriers are not perfect), hence a $SiO_2$ layer above the bond coat is included.

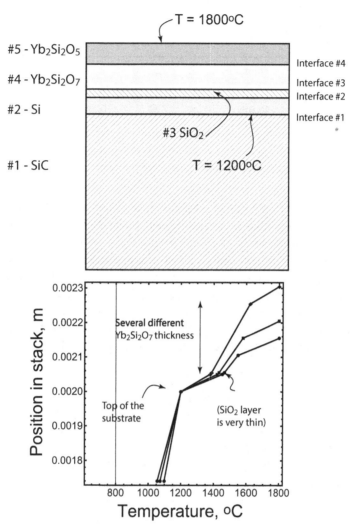

**Figure 14.4** An environmental barrier coating system designed to protect *SiC* composites from oxidation and volatilization with properties given in Table 14.2; the temperature profiles shown are steady-state results generated with *LayerSlayer*, assuming a top surface temperature of 1800°C and that the top of the substrate is held at 1200°C. These temperatures are used as the reference temperatures to define the stress-free state.

An important aspect of the development of such systems is that the properties of the layers are often uncertain; while the properties listed in Table 14.2 are reasonable estimates, literature values span quite a range due to different materials-processing conditions, temperature-dependent properties and the inherent variability in measurements. Further complicating the analysis, $SiO_2$ can experience phase changes with large volumetric strains that increase driving force, while creep in the layers may alter the response. In this example, it is assumed that the system relaxes at high

**Table 14.2** List of Properties Used in the Design Study of an Environmental Barrier Coating System. Growth Strains are Zero, and the Temperatures are Determined Via Steady-State Thermal Analysis

| Layer | Thickness | $E$, GPa | $v$ | $\alpha$, ppm/$^{\circ}$C | $\kappa$, W/(m$^2$K) |
|-------|-----------|----------|-----|---------------------------|----------------------|
| SiC | 2 mm | 300 | 0.16 | 4.67 | 9 |
| Si | 50 μm | 163 | 0.22 | 4.1 | 1 |
| SiO$_2$ | 5 μm | 70 | 0.18 | 3.1 | 2 |
| Yb$_2$Si$_2$O$_7$ | 50, 100, 200 μm | 180 | 0.27 | 3.1 | 3 |
| Yb$_2$Si$_2$O$_5$ | 50 μm | 172 | 0.27 | 7.1 | 1 |

temperature, such that the steady-state temperature distribution through the stack is utilized as the stress-free reference temperature. Plastic flow or creep during cool-down is neglected.

Figure 14.4 shows the steady-state temperature distributions in three different multilayers distinguished by three different Yb$_2$Si$_2$O$_7$ layer thicknesses. Note that all curves past through the same point at the top of the substrate, since the temperature is prescribed at this location. The TempSol command was used with *iflag* = *1*, which implies the steady-state temperature distribution is then used as the reference temperature distribution. Therefore, the stresses are zero at elevated temperature. Upon cooling to a uniform temperature of 20$^{\circ}$C, stresses develop in the layers, as shown in Figure 14.5(a). The CTE of the top YbMS is largest; since the substrate is thickest, it provides constraint against contraction during cooling and develops tension in the top layer. The curvature is positive, indicating the stack shown in Figure 14.4 will curl upwards; the elongation $\epsilon_o$ is negative, such that the substrate is almost entirely in compression (due to the contraction of the top layers).

Figure 14.5(b) shows the plane strain energy release rates for delamination cracks placed at various locations through the multilayer. The main figure covers only the layers above the substrate; the inset shows the entire distribution, illustrating that the energy release rate approaches zero as crack location approaches the top and bottom surfaces. Figure 14.5(b) reveals that the maximum energy release rate occurs at the YbDS/YbMS interface; however, it should be kept in mind that the probable location of a delamination crack is determined by comparison of the values in Figure 14.5(b) and the relevant interface toughness – the bulk layer toughness for cracks that occur away from an interface, and the interface toughness associated with each interface.

Two questions frequently arise in the design of such multilayers. First, how sensitive is the energy release rate to the myriad of required input properties? This is particularly important when during the devleopment of new material systems, as the measurement of properties can be expensive and time-consuming, and ideally one would like to limit such measurements to only those properties that significantly influence the response. Second, can cracking be avoided through modification of the layers' thickness?

Naturally, one can clearly address these questions by making suitable changes to the individual properties of interest; however, the automation of *LayerSlayer* enables a global view of the system's performance. For example, let us define the sensitivity of

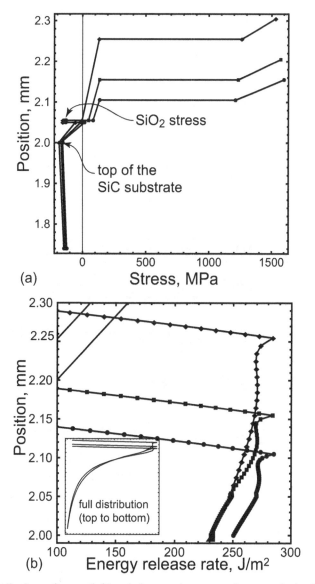

**Figure 14.5** Distributions of stress (left) and plane strain energy release rate in the five-layer system described in Table 14.2, for three different thicknesses of layer no. 4 (YbDS). The stress is computed using the temperature distribution in Figure 14.4 as the reference temperatures defining the stress free state; the inset shows the energy release rate over the full stack, illustrating the asymptote to zero for top and bottom locations.

system performance as follows:

$$S_{ik} = \frac{G[P_o^{ik}] - G[(1+\epsilon)P_o^{ik}]}{\epsilon P_o^{ik}} \cdot \frac{P_o^{ik}}{G[P_o^{ik}]} \tag{14.19}$$

where $P^{ik}$ is the property of interest, with $i$ being the layer number and $k$ being the property of interest. The first term is simply the derivative of the energy release rate with

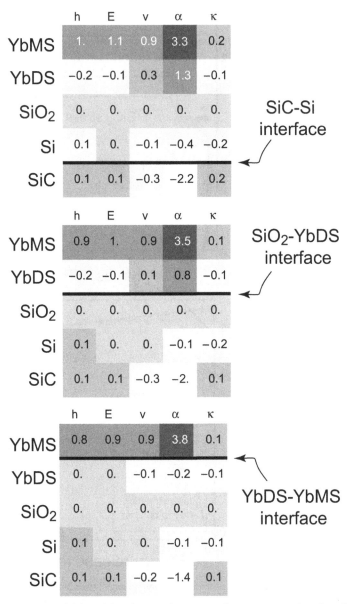

**Figure 14.6** Property sensitivity of the plane strain energy release rate at various interfaces in the five layer system described in Table 14.2 (with $h_{YbDS} = 100 \ \mu m$); the sensitivity is defined as the percentage change in $G$ per percentage change in $P$, where $P$ is the property being probed.

respect to the property of interest; multiplying by the second term produces a dimensionless measure of sensitivity, which reflects the fractional change in energy release rate in terms of the fractional change in property.

This calculation has been encoded in the module GSensitivity[*multilayer*, {*i, k*}, *iface*], where *multilayer* is the multilayer definition, *i* is the layer number and *k* indicates the property of interest in the list defined as (14.2), and *iface* is the interface number for which the energy release rate is computed. It should be noted that the response is generally nonlinear with respect to layer properties, such that this calculation computes a local derivative that may change if the properties are changed significantly from the base state.

Figure 14.6 illustrates the sensitivity of the current five-layer system to system properties, in its base state defined by the properties in Table 14.2, with $h_{YbDS} = 100$ µm. Results are shown for the sensitivities of the energy release rate corresponding to the three interfaces indicated in the figure. Results have been rounded to nearest decimal point, such that some sensitivities are not identically zero, but rather very small. Negative numbers imply that that the energy release rate will decrease if that property is increased; positive numbers indicate the energy release rate will increase if that property is increased. It is immediately apparent that decreasing the CTE mismatch between the top YbMS layer and the substrate has the greatest effect: note that the system is 2.5 times more sensitive to the CTE of the YbMS layer than to that of the substrate. Assuming the properties are fixed, we see that the greatest sensitivities to layer thickness are the top two layers, but only for delamination at the bottom interfaces.

It may be appreciated that the normalized sensitivities shown in Figure 14.6 reflect the scaling of the energy release rate with respect to properties. The values reflect, in an approximate manner, the exponent that is attached to that variable. For example, values close to unity reflect nearly linear scaling of the energy release rate with respect that variable (as in the case of the film thickness or modulus of the layers). These relationships are only approximate, as the true scaling with system properties is more complicated than a simple power law. The descriptions in Chapter 4 (for the bilayer) and Chapter 5 (for multilayers) clearly illustrate that the energy release rate will be a complicated function involving properties raised to different powers in the numerator and denominator. That is, the system is nonlinear with respect to properties, such that the scaling changes depending on the base set of properties. One can repeat the sensitivity study for other base states and explore how the system's sensitivity changes with more radical design changes.

# 15 Software for Semi-Infinite Multilayers: Transient Delamination

In this chapter, we describe software that analyzes the time-dependent thermal response of multilayers, called *LayerSlayer Transient (LST)*. As with its counterpart for steady-state analysis described in the previous chapter, the software is based on the one-dimensional heat conduction framework described in Section 2.6 and the general framework for multilayer mechanical analysis described in Chapter 5. The distinction is that *LST* solves for transient, time-dependent temperatures, whereas *LS* solves only for the steady-state response. *LST* is subject to the same geometry restrictions as *LS* and the same assumptions regarding deformation. Though each code can be used independently of the other, the reader is encouraged to become familiar with the conceptual framework of *LS* prior to exploring *LST*.

In *LST*, all of the governing equations described in Chapter 14 for *LS* now become functions of time. Whereas *LS* solved for temperatures at discrete points via linear algebra using $[K]\{T\} = [q]$, *LST* determines temperatures at discrete points as a function of time by solving the coupled first-order differential equations $[C]\{\dot{T}(t)\} + [K]\{T(t)\} = [q(t)]$. It is important for the user to keep this difference in mind, particularly in light of the fact that the inputs and outputs of both pieces of software are extremely similar. However, in *LST*, one must always specify the time at which an output variable, such as stress or energy release rate, is sought. A powerful feature of *Mathematica* is that it solves coupled systems of differential equations with ease and automatically produces interpolation functions for nodal temperatures that enable one to easily extract results for any point in time.

The second important distinction between *LS* and *LST* is that *LST* requires that the physical layers that make up a multilayer are subdivided into 'elements', such that nonlinear spatial distributions of temperature within a single layer can be captured. This means that temperatures are computed at discrete points within each layer (not just at the interfaces), leading to much larger systems of equations. Within each element, the temperature distribution is assumed to be linear; *LST* essentially determines a piecewise linear distribution of temperatures that converges to the true nonlinear distribution as the number of elements are increased. It may be recognized that since the temperature is in the elements is linear, the elements can be thought of as additional layers with identical mechanical properties as their parent layer, and *LS* can be used to perform the mechanical analysis. This is indeed the case, although the modules in *LST* perform this analysis and there is no need to invoke *LS*.

Finally, the reader is reminded once again that the analysis in Chapter 5, which forms the basis for both mechanical analysis in *LS* and *LST*, does *not* yield the mode mix, which requires full solution of the associated elasticity problem which is covered in Chapter 16.

## 15.1 Overview of *LayerSlayer Transients*'s Assumptions, Capabilities and Interface

*LayerSlayer Transient* (LST) can be used to analyze a multilayer comprising any number of semi-infinite layers, as shown in Figure 15.1. The layers are assumed to have much larger in-plane dimensions than their through thickness dimension, and plane strain deformation ($\epsilon_z = 0$) is enforced for the intact multilayer and any sublayers formed by a delamination crack. (*N.B.: LST* does not handle equi-biaxial deformation.) Temperatures are assumed to vary only in the thickness direction, or $y$-direction in Figure 15.1. *LST* predicts the following:

- the temperature distribution through the layers as a function of time; one must prescribe either both surface temperatures as a function of time, or one surface temperature and the ambient temperature on the other surface that dictates convective heat transfer.
- the stress distribution associated with a given temperature distribution and a specified time, subject to plane strain deformation.
- energy release rate for plane strain delamination ($\epsilon_z = 0$ both ahead of and behind the crack), assuming no external loads are present, predicted as a function of time

The temperature analysis follows the transient framework outlined in Sections 2.6, 13.3 and 13.4; the last shows examples of solutions generated with *LST*. Once the transient thermal analysis is completed, *LST* utilizes the framework described in Chapter 5 for plane strain deformation. As in *LS*, the key assumption is that the strain distribution in bonded layers is given by

$$\epsilon_x(y, t) = \epsilon_o^x(t) - \kappa_z(t)y + \alpha(T(y, t) - T_o(y)) + \epsilon_g, \tag{15.1}$$

where $\epsilon_o^x(t)$ is the elongation in the $x$-direction at the bottom of the stack, $\kappa_z(t)$ is the curvature of the stack about the $z$-axis, $\alpha$ is the coefficient of thermal expansion in the layer, $T_o(y)$ is a reference temperature and $T(y, t)$ is the temperature distribution. The specified parameter $\epsilon_g$ represents a spatially uniform (in each layer) and constant misfit strain that arises from sources other than thermal expansion.

The thermal analysis solves for the transient temperature distribution through the stack by dividing each layer into elements and yields a piecewise linear temperature profile that converges to the full nonlinear distribution in the limit of small elements. This discretization is automated, such that the input required by the user is virtually identical to *LS*. (The output will necessarily involve more information that is returned by *LS*.) The temperature is not continuous across the interface but rather experiences jumps proportional to the inverse of the interface conductance: hence, the temperature

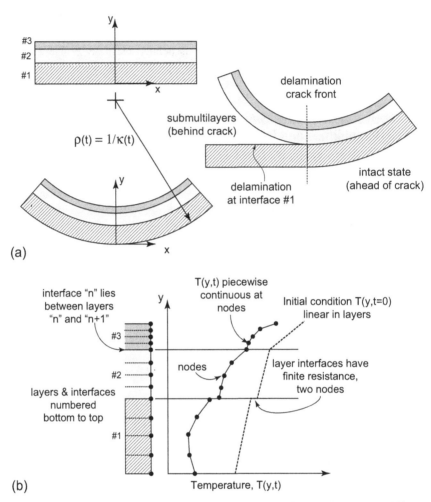

**Figure 15.1** Conventions for LayerSlayer Transient (LST): (a) the layers of the total multilayer are numbered from top to bottom, and LST treats only plane strain deformation with $\epsilon_z = 0$; the same sign conventions as *LS* are used. (b) *LST* divides the multilayer into 'elements' that each contain a linear variation in temperature; the temperatures are continuous at nodes internal to a layer, but two nodes per physical interface are used to capture discrete temperature jumps associated with the thermal resistance of the interface.

on the bottom of one layer and the top of the layer beneath it are retained as possibly independent variables. It should be noted that the interfaces are assumed to have zero heat capacity, such that they do not influence the time dependence of the solution.

When analyzing cracking, it is assumed that the crack has no influence on the temperature distribution, that is, the thermal response of the intact stack is analyzed. While cracking would naturally alter the heat flow in the cracked region, in this book it is assumed that cracking events are sufficiently fast so as to not allow for changes associated with debonding. That is, the scenario being analysis consists of a steady-state crack growing at speeds faster than thermal transport.

A given layer is defined by the list

$$Layer = \{h, E, v, \alpha, \{T_o^b, T_o^t\}, \epsilon_g, k, \rho c_p, N_{el}, \{T^b, T^t\}\} \tag{15.2}$$

where

- $h$ is the layer thickness
- $E$ is the Young's modulus of the layer
- $\alpha$ is the coefficient of thermal expansion of the layer
- $\{T_o^b, T_o^t\}$ are the reference temperatures at the bottom and top of the layer
- $\epsilon_g$ is a growth or deposition strain (defined below)
- $k$ is the thermal conductivity of the layer
- $\rho c_p$ is the heat capacitance of the layer
- $N_{el}$ is the number of elements in the layer used to determine the temperature profile in that layer
- $T^b$, $T^t$ are the current temperatures at the the bottom and top of the layer

Naturally, the order of the properties in the stack is fixed, and one must always be sure that the properties are properly loaded. Note that the code assumes that the initial reference temperature and current temperature distribution are linear through the layer, while the growth strain (if present) is constant through the layer. The temperature values in the layer definition function largely as placeholders: in almost all instances, either the current temperature distribution or the reference temperature distribution is replaced with the results of the thermal analysis. The mesh that is generated for each layer is uniform, that is, the element size in each layer is $h/N_{el}$.

A multilayer is defined by forming a list of layers

$$Multilayer = \{Layer1, Layer2, \ldots, LayerN\} \tag{15.3}$$

Once the multilayer is defined as a list of layers, with each layer represented as a list of properties, the geometry is completely defined and ready for subsequent analysis.

In the software implementation, modules are created to calculate the response of the multilayer, with each module producing different quantities of interest. The general use of a module in *Mathematica* is

$$output = \texttt{Command}[Multilayer, iflag1, iflag2, iflag3, iflag4]$$

where *output* is the name of a variable that is being set equal to the output of the module `Command[]`. Generally, the first argument in the commands is the user-defined name of the multilayer being analyzed. Additional parameters that are needed are described in detail below, but, for example, *iflag1*, *iflag2* and so forth might be a list thermal conditions imposed in the analysis, the stress component of interest, the interface at which the energy release rate is sought and so forth. In this chapter, anything written in `Courier font` is identifies the text as a *Mathematica* module (or command).

## 15.2    Transient Temperature Distributions

**Numerical Formulation for Transient Temperatures**

In this section, we briefly describe the numerical framework used to solve for the transient steady-state temperature profiles for a semi-infinite multilayer; these temperature distributions serve as inputs to the mechanical analysis described in subsequent sections. The reader is referred to Section 2.6 for a description of one-dimensional heat conduction that provides the theoretical foundation for the framework described here.

In *LST*, each multilayer is broken into small segments (referred to as elements), which describe the relationship between heat fluxes, temperatures and the time derivatives of temperature at its end points, called nodes. For each element, the heat flux through layer $i$ (with positive heat flux occurring in the positive $y$-direction) can be written as

$$\begin{pmatrix} q_{in} \\ q_{out} \end{pmatrix} = \frac{k_i}{\ell_i} \begin{bmatrix} -1 & 1 \\ 1 & -1 \end{bmatrix} \begin{pmatrix} T_{2i-1} \\ T_{2i} \end{pmatrix} + \rho c_p \ell_i \begin{bmatrix} -1 & 1 \\ 1 & -1 \end{bmatrix} \begin{pmatrix} \dot{T}_{2i-1} \\ \dot{T}_{2i} \end{pmatrix} \qquad (15.4)$$

where $\ell_i$ is the size of the elements in layer $i$, $T_{2i-1}$ is the temperature at the start of the element $i$, $T_{2i}$ is the temperature at the end of the element $i$, and dots denote differentiation with respect to time. It should be noted that all $T_i$ are functions of time. Using the same approach we can write the equation governing the the jump condition across the interfaces, which arises due to finite interface conductance. An element with zero length is included at the interface, with the parameter $k_i/\ell_i$ replaced with $k_{INT}$, the conductance of the interface that controls the temperature changes from one layer to the next as in (2.50). Note that interface elements (of zero length) have no heat capacity, and therefore the temperature drop across such interfaces is established instantaneously.

*LayerSlayer Transient* combines all the elemental equations including those associated with layer interfaces, while imposing conservation of heat flux at the internal nodes, to form a global system of equations described by

$$[C]_{MxM} \{\dot{T}\}_{Mx1} + [K]_{MxM} \{T\}_{Mx1} = \begin{bmatrix} -q_{in} \\ 0 \\ \vdots \\ 0 \\ 0 \\ q_{out} \end{bmatrix}_{Mx1} \qquad (15.5)$$

where $[C]$ is a global heat capacity matrix involving the properties $\rho c_p \ell_i$, and $[K]$ is a global thermal conductivity matrix involving properties for the layers $k_i/\ell_i$ and the interfaces $k_{INT}$. In (15.5), $q_{out}$ defines the condition governing heat flux coming out of the bottom surface, and $q_{in}$ defines the flux coming into the top surface. Conceptually, the heat fluxes can either be eliminated using prescribed temperatures or rewritten in terms of convective heat transfer conditions as in (2.49).

Together with the boundary conditions, (15.5) represents a system of coupled first-order differential equations, whose variables are the time-dependent nodal temperatures. Their solution also requires specification of the initial temperature distribution.

In *LST* this is either uniform or a piecewise linear distribution associated with a steady-state condition. The numerical solution to (15.5) is generated automatically within the module, using the packaged NDSolve[] algorithm in *Mathematica*. The algorithms for differential equation solving in *Mathematica* are numerous, and they allow full specification of numerical parameters controlling this solution, such as maximum time step, maximum number of iterations, precision tolerances and so forth. These have been set to defaults that have proven to work well for the problems described in this chapter, Chapter 13 and the paper by Jackson & Begley (2014).

The output of the temperature analysis in *LST* is quite similar in format to that of *LS*, that is, it consists a list of temperatures at the nodes. In *LST*, these temperatures are provided in the form of interpolation functions, which allow users to compute a list of numerical values of the temperatures of the nodes at any time. This provides an easy way to compute the relevant quantities of interest, such as the stresses or energy release rate for delamination, as a function of time.

**Thermal Analysis Modules**

As with its steady-state counterpart in the previous chapter, *LayerSlayer Transient (LST)* utilizes interface elements between physical layers that have zero thickness to allow for temperature drops between layers associated with finite interface conductance. As such, one must provide a list of interface conductances for the thermal analysis modules, with the interface numbering from bottom to top for example

$$ifaces = \left\{ k_{INT}^{12}, k_{INT}^{23}, \ldots, k_{INT}^{(N-1)(N)} \right\}, \tag{15.6}$$

where *ifaces* is a user-defined name of the list containing interface properties, $k_{INT}^{12}$ is the interface conductance between layers 1 and 2, $k_{INT}^{23}$ is the interface conductance between layers 2 and 3 and so forth. Setting the interface conductances to a large number (in comparison to the effect layer conductances, characterized by $k_i/h_i$) results in nearly perfect conduction, with (piecewise) continuous temperatures throughout the stack.

Two modules for transient thermal analysis are included, distinguished by different types of boundary conditions. The module GetThermalSolution[] requires the user to prescribe the temperature on the top and bottom surfaces of the multilayer as a function of time. As will be illustrated below, the user is allowed to define any time-dependent function they wish for either temperature. For this option, no other boundary conditions are needed (or even possible). The format of the module input is

GetThermalSolution[*multilayer*, *ifaces*, $\{T_b, T_t\}$, -1, *tmax*]

where *multilayer* is the user-defined list of layers (each layer being defined as a list of properties), *ifaces* is the user-defined list of interfaced conductivities, $T_b$ is the user-defined function of time that defines the temperature on the bottom surface of the multilayer, $T_t$ is the user-defined function of time that defines the temperature on the top surface of the multilayer, and *tmax* is the user-defined maximum time for which a solution is sought. (The '−1' in the function call is a placeholder for a future capability and should not be changed.)

The initial conditions for `GetThermalSolution[]` are taken to be the *steady-state* temperature distribution that arises from evaluating the user-prescribed functions $T_b$ and $T_t$ at $t = 0$. That is, the initial conditions are a collection of piecewise linear temperature distributions that are identical to those that one would obtain from the two-temperature option of the steady-state code *LS*. Upon executing `GetThermalSolution[]`, *LST* returns a list of interpolation functions which define the nodal temperatures as a function of time that is valid over the range $0 < t < tmax$. Note that the list will have two entries for physical interfaces, each with the same location: for example, it will include both $\{y_i, T_t^j\}$ and $\{y_i, T_b^{j+1}\}$, where $y_i$ is the position of the interface between the $j$ and $j + 1$ layers, with the first temperature associated with the top of the bottom layer (layer $j$), and the second temperature associated with the bottom of the top layer (layer $j + 1$).

The second type of problem that can be addressed is one with convective heat transfer on one of the surfaces. For such cases, the user prescribes the bottom surface temperature (fixed in time) and convective heat transfer on the top surface. The command/module is named `GetThermalTQ[]`. The user is allowed to define any time-dependent function they wish for the ambient temperature that drives heating/cooling of the top surface, according to $q_{top} = h\left[T_{top}(t) - T_o(t)\right] = kT'(top, t)$ where $T'(top)$ is the temperature gradient at the top of the multilayer. No other boundary conditions are needed (or even possible). Again, the initial conditions in this module are taken to be the steady-state temperature profile that arises from prescribed temperatures at the top and bottom of the stack. The format of the module input is

$$\texttt{GetThermalTQ}[\textit{multilayer}, \textit{iface}, \{T_b, T_t^o, h_T\}, \textit{tmax}, T_t^a]$$

where *multilayer* is the user-defined list of layers (each layer being defined as a list of properties), *iface* is the user-defined list of interfaced conductivities, $T_b$ is the fixed temperature on the bottom surface of the multilayer (also used to determine the initial temperature distribution), $T_t^o$ is the temperature on the top surface of the multilayer used to calculate the initial conditions, *tmax* is the user-defined maximum time for which a solution is sought, and $T_t^o$ is a user-defined function of time that defines the ambient temperature that drives heat or cooling. Once again, the code returns a list of interpolation functions that define the nodal temperatures as a function of time that is valid over the range $0 < t < tmax$.

For convenience, a simple module has been written to plot temperature distributions at various times, with coloring of the distribution of each time scaled to make it easier to track the evolution of temperature profiles. The module is

$$\texttt{PlotTemperature}[\textit{out}, \{t1, t2, t3, t4 \ldots\}]$$

where *out* is the output from `GetThermal` and $\{t1, t2, t3, t4, \ldots\}$ is a user-defined list of times at which the temperature distribution should be plotted. The user can choose any time that falls between $t = 0$ and the user-defined *tmax*. The curves are automatically colored by scaling their hue in accordance with the minimum and maximum time in the list, such that the curves span from red (minimum time) to blue (maximum time). Curves for intermediate times transition between these two extremes with orange, yellows and greens.

## 15.3    Deformation and Stresses

Once the temperatures throughout the stack have been computed, the deformation and stress in the layers are computed following the framework described in Chapter 5. It should be noted that the deformation and stress do *not* need to be computed a priori to compute the energy release rate, which has its own self-contained module. The provided version of *LST* computes only plane strain results with zero applied forces and moments. The relevant module for retrieving deformation variables, $\epsilon_o(t)$ and $\kappa(t)$ in (5.6), is

$$\mathtt{GetDeformation}[multilayer,\ results,\ treq,\ tref]$$

where *multilayer* is the user-defined multilayer (list of layers, each of which is a list of properties), *results* are the output from either $\mathtt{GetThermalSolution[]}$ or $\mathtt{GetThermalTQ[]}$, and *treq* is the time at which the deformation is sought. One can easily tabulate deformation as a function of time by repeatedly calling the module $\mathtt{GetDeformation}$ with different *treq*.

The parameter *tref* defines the reference temperatures that define the stress-free state; setting $tref = -1$ dictates that the original reference temperatures provided with the layer definition should be used. Alternatively, setting $tref = 0$ dictates that the steady-state temperature profile determined by the initial surface temperatures defined in the thermal analysis modulus should be used, such that the initial condition also corresponds to a zero stress condition. If the reference state that defines zero stress is reached during the temperature cycle, $0 < t < t_{max}$, one can set *tref* equal to the simulation time associated with the stress-free condition. This is a useful feature that enables one to compute mechanical behavior assuming that all stresses relax at elevated temperature, as will be illustrated below.

Similarly, the stress distribution in the multilayer can be computed once the temperatures throughout the stack have been computed, following the framework described in Chapter 5. It should be noted that the deformation does not need to be computed a priori to compute the stresses. (This is different from *LS*, which requires deformation to be computed first.) At this time, *LST* provides only the option of computing plane strain results with zero applied forces and moments:

$$\mathtt{GetStressDistribution}[multilayer,\ results,\ treq,\ iflag]$$

where *multilayer* is the user-defined multilayer (list of layers, each of which is a list of properties), *results* are the output from either $\mathtt{GetThermalSolution[]}$ or $\mathtt{GetThermalTQ[]}$, and *treq* is the time at which the deformation is sought. Once again, the parameter *tref* defines the reference temperatures that define the stress-free state; setting $tref = -1$ dictates that the original reference temperatures be used, while any $0 < tref < t_{max}$ refers to a distribution computed from the solution.

$\mathtt{GetStressDistribution[]}$ returns a list of $\{y_i, \sigma_{11}^i\}$ pairs that define the stress distribution in the in the multilayer. Typically, the evolution of stress with time is of central interest: to facilitate viewing the results, a plotting module is provided:

$$\mathtt{PlotStress}[multilayer,\ results,\ tlist,\ iflag]$$

with the same variable definitions as above, only with *tlist* providing a list of times for which stress distributions are sought. The plot automatically scales the colors of the stress distributions, such that the color of the curves transitions from red to blue: the curve associated with the first time in *tlist* is pure red, and the curve associated with the final time in *tlist* is plotted in blue.

## 15.4     Energy Release Rates

After the transient temperature distribution has been found, the energy release rate can be computed assuming plane strain deformation ahead of and behind the crack. The framework for the computation is given in Chapter 5. The module for the energy release rate is self-contained, such that neither the deformation nor the stress needs to be computed a priori. The module format is

GetERRinterface[*multilayer, results, iface, treq, tref*]

where *multilayer* is the user-defined list of layers (each a list of properties), *results* are the output from either GetThermalSolution or GetThermalTQ, *iface* is the interface number defining the location of the interface crack (where *iface* = $N$ for a crack just above the $N^{th}$ layer), and *treq* is the time of interest. Once again, the parameter *tref* defines the reference temperatures that define the stress-free state, as described above for the deformation and stress modules.

The discretization required for the thermal analysis implies that one can easily compute the energy release rate associated with a delamination crack at any node, by performing the relevant integrations described in Chapter 5. As such, a second module is provided that returns the energy release rate for a delamination crack located at each nodal position. This module is

GetERRDistribution[*multilayer, results, treq, tref*]

where *multilayer* is the user-defined list of layers (each a list of properties), *results* are the output from either GetThermalSolution or GetThermalTQ, *treq* is the time of interest, and *tref* as the same definition as above. The module returns a list of pairs, $\{y_i, G_i\}$, where $y_i$ is the location of the delamination crack and $G_i$ is the plane strain energy release rate associated with that location. If one wishes to compute the energy release rate at a specific location (e.g., that is associated with the midpoint in a layer as a function of time), one must calculate the closest node to the desired location and extract the relevant pair.

An interesting feature of these modules is that one can plot the energy release rate as a function of time for all locations: to facilitate viewing the results, a plotting module is provided:

PlotERR[*multilayer, results, tlist, iflag*]

with the same variable definitions as above, only with *tlist* providing a list of times for which energy release rate distributions are sought. As with the stress-plotting module, the plot automatically scales the colors of the energy release rate curves, such that the

**Table 15.1** List of Properties Used in the Transient Case Study. The Initial Temperature Distribution is Assumed to be Uniform at 20°C; the Study Illustrates the Impact of Choosing Different Reference Temperatures that Define the Stress-Free State

| Layer | Thickness, m | $E$, Pa | $\upsilon$ | $\alpha$, 1/°C | $\kappa$, W/(mC) | $\rho c_p$, |
|---|---|---|---|---|---|---|
| Layer1 | 0.003 | $200 \times 10^9$ | 0.3 | $18 \times 10^{-6}$ | 50 | $4 \times 10^6$ |
| Layer2 | 0.002 | $40 \times 10^9$ | 0.2 | $11 \times 10^{-6}$ | 25 | $2.5 \times 10^6$ |

colors of the curves transition from red to blue: the curve associated with the first time in *tlist* is pure red, while the curve associated with the final time in *tlist* is plotted in blue.

## 15.5 A Case Study of a Bilayer Subjected to a Sudden Rise and Fall in Ambient Temperature

In this example, the basic analysis steps and capabilities of *LayerSlayer Transient* are illustrated for bilayer system shown in Figure 15.2 with the properties listed in Table 15.1. The properties loosely correspond to a ceramic coating with relative high thermal conductivity (for a ceramic) on top of a steel substrate; they have been adjusted somewhat to produce results that highlight certain behaviors and illustrate the uses of the software. It should be emphasized at the outset that the steps taken while analyzing more sophisticated multilayers with more layers are identical, although computation times may increase considerably as more layers are added.

The bilayer is subjected to a heating cycle, wherein hot gas flows over the top surface for a brief (2 sec) interval; Figure 15.2 illustrates the bilayer and the time history of the ambient gas that drives heating. The study there explores the impact of the rate at which the ambient temperature rises and falls: a convective heat transfer coefficient of 8000 $W/°C$ was chosen to drive rapid temperature changes and highlight the influence of temperature transients. In this example, the interface conductivity is set to a high number, such that there is essentially perfect conductance between the coating the substrate. The initial temperature distribution is assumed to be uniform and equal to 20°C; in what follows, different reference temperatures that define the stress-free state are considered. The reader is referred to the example file accompanying this chapter for complete details.

### Layer Definition
The properties in Table 15.1 are used to define the layers, with zero growth strain through the bilayer and uniform reference temperatures set for each layer. The multilayer is defined as follows, using the property order given as (15.2)

Layer1 = {0.002, 2 · 10$^{11}$, 0.35, 18 · 10$^{-6}$, {20, 20}, {0, 0}, 50, 4 · 10$^6$, 20, {20., 20.}};

Layer2 = {0.0005, 4 · 10$^{11}$, 0.20, 11 · 10$^{-6}$, {20, 20}, {0, 0}, 25, 2.5 · 10$^6$, 20,

{20., 20.}};

Bilayer = {Layer1,Layer2}

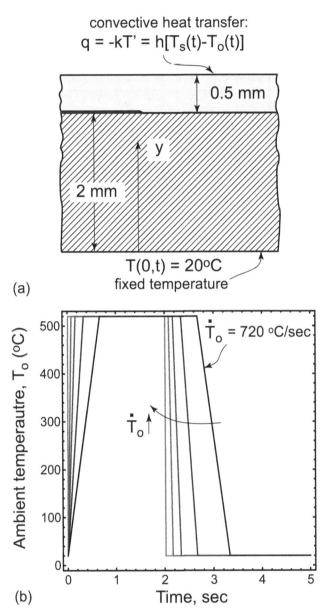

(a)

(b)

**Figure 15.2** The bilayer problem considered in the case study, assuming the properties in Table 15.1; the bottom of the substrate is fixed at room temperature, while the top is exposed to convective heat transfer with a hot gas whose temperature is ramped, held constant and then decreased at the same rate. The results that follow are for the five cycles in ambient temperature shown above.

Thus, the layers are named appropriately from bottom to top, that is, `Layer1` is on the bottom, and `Layer2` is on the top. We define the list of interface conductivities as `ifacek=`$\{1 \cdot 10^8\}$, with only one entry because there is only one interface. Note that the above layer definition uses 20 elements in each layer; this produces reasonably smooth temperature and stress distributions and energy release rates that are convergent (in the sense that more elements do not alter the results). One should use as a few elements as possible; increasing the number of elements requires decreasing time steps. Models with a large number of elements may require additional attention on the numerical parameters used to solve the associated ordinary differential equations.

**Thermal Analyses**

In this example, the module `GetThermalTQ` is used, such that one must prescribe a heat transfer coefficient and function of time that defines the time history of the ambient temperature. The histories of ambient temperature considered in this example are shown in Figure 15.2. They are defined by the function

$$T[a\_]:=20 + 500\text{Piecewise}[\{\{at, t \leq 1/a\}, \{1.0, 2 + 1/a \geq t \geq 1/a\},$$
$$\{1 - a(t - (2 + 1/a)), 2 + 1/a \leq t \leq 2 + 2/a\}\}]$$

where 'a' defines the rate at which the ambient temperature is ramped upwards and downwards at the beginning and end of the hold cycle, as shown in Figure 15.2. For example, we use pass 'T[1]' for a 1 *sec* ramp period. We define the heat transfer coefficient to be 8000 $W/°C$, and initial temperatures on the top and bottom surfaces to be 20°C, and load these into the list `Props=`$\{20., 20., 8000.\}$. Hence, the transient temperature profiles from $t = 0$ to $t = 4$ (*sec*) are obtained by

$$\text{out=GetThermalTQ[Bilayer, ifacek, Props, 4.0, T[1]]} \qquad (15.7)$$

This returns a list where each entry has a paired list of the position of the node and the interpolation function that defines the temperature of that node (as a function of time).

Figure 15.3 shows the temperature distributions at various times for the second thermal cycle, that is, the ramp rate $a = 12$. Results are shown both for the the ramp to elevated temperature and for cooling back to room temperature; this plot was created using the `PlotTemperature` module. Figure 15.4 shows the time history of the surface temperature and the time history of the interface temperature for the five thermal cycles shown in Figure 15.2. The curves are found by making a list of (time, temperature) for the appropriate nodes, as provided in the example file accompanying this chapter. The results for the fastest temperature rise, $a = 100$, are essentially identical to a step function in temperature; one does not observe changes in the temperature profile for faster loading rates. This simply means that the ambient temperature is ramped much faster than the system can respond. Note that for the slowest loading rate, $a = 1.5$, the temperature almost reaches a quasi-steady state, in which the temperature rise of the structure rises linearly in time in concert with the rise in ambient temperature.

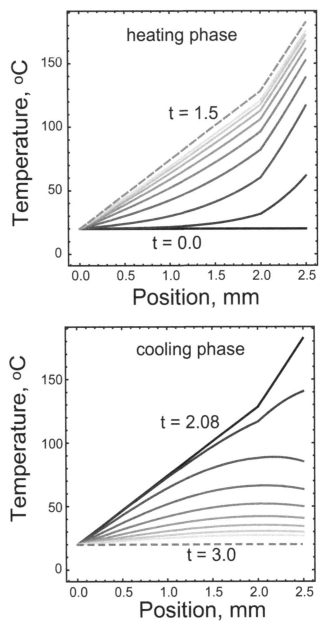

**Figure 15.3** Temperature profiles for the bilayer case study at various times, during the heating phase and cooling phase of the cycles shown in Figure 15.2 for the second fastest heating/cooling rate. The curves span from the initial time (black) to the final time (light) in increments of 0.05 *sec*, except for the final time shown.

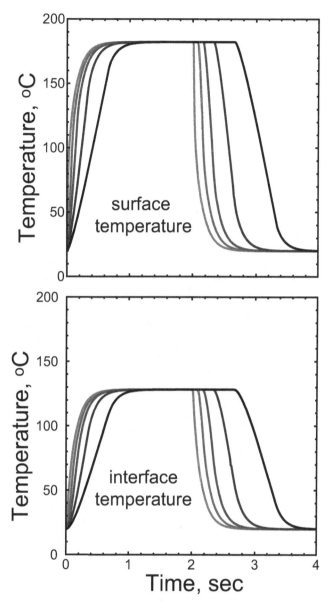

**Figure 15.4** The surface temperature and interface temperature as a function of time, for the bilayer described in Table 15.1 and the heating cycles in Figure 15.2.

## Mechanical Analysis

The deformation response of the bilayer for each of the loading cases shown in Figure 15.2 is shown in Figure 15.5. These results are generated with the command GetDeformation[Bilayer, out, *treq*, 0.], which provides the multilayer definition, the results generated via the thermal analysis module, and the requested time. To make the curves, a list is compiled by looping through time. The last entry in the

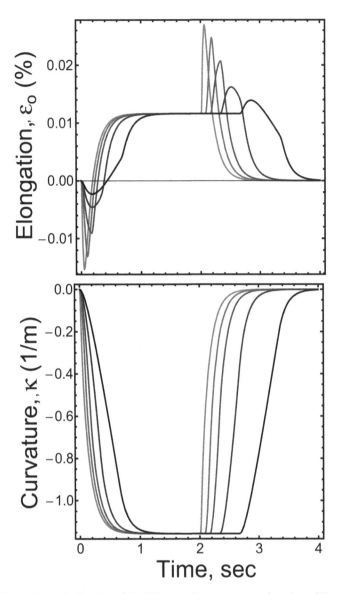

**Figure 15.5** Elongation at the bottom of the bilayer and curvature as a function of time, for the bilayer described in Table 15.1 and the heating cycles shown in Figure 15.2. One can associate each curve by noting the rate of change: faster heating and cooling rates correspond to steep slopes in the above figures. Note the strong transients in the elongation during the ramp-up and ramp-down portions of the heating cycle.

GetDeformation module is set to zero, which implies that the results correspond to the assumption that the temperatures at $t = 0$ define the stress-free state. For this problem, this corresponds to a uniform temperature of 20°C. The effect of transients on the elongation at the bottom of the stack is clearly observed in Figure 15.5; the bottom of the bilayer initially contracts, as localized heating in the coating causes the top surface to elongate and the bilayer to curve downward. Eventually, even though the bilayer remains curved downward, the temperature rise in the substrate is sufficient to cause the elongation along its bottom to become positive. When heating is turned off, the moment created by surface heating is lessened, and the elongation increases briefly during the transient.

In Figure 15.6, the stress distribution in the bilayer is shown for one case (the second fastest heating rate in Figure 15.2) at several times. Again, the $t = 0$ results are used to define the stress-free state; as such, the curves were generated with `PlotStress[Bilayer, out, {0, 0.05, 0.01.... 1.5}, 0]`. Results during the heating phase (ramp up) and cooling phase (ramp down) are shown in separate plots. The time increment between curves is 0.05 sec; on the left, the black curve corresponds to the steady-state result after the transient associated with the ramp-up is finished. On the bottom, the dashed gray curve corresponds to the steady state result just prior to the start of the ramp-down phase. One can see that, at steady state, the coating is in tension due to the larger CTE of the substrate; bending effects, induced by the temperature gradient, lead to the stress being highest at the interface.

The results for the cooling phase (on the bottom in Figure 15.6) illustrate that the stress at the very top of the film increases as the cooling is initiated. This is a result of the fact that the temperature decrease near the top surface decreases the curvature of the stack, such that the influence of substrate elongation on the coating is stronger.

The energy release rate for delamination at the interface is shown in Figure 15.7 as a function of time, for the five heating cycles shown in Figure 15.2. On the top, the reference temperature defining the stress-free state is set to $t = 0$; to generate the curve, a table of $\{t_i, G_i\}$ as created by repeated calling the module `GetERRInterface[Bilayer, out, 1, ` $t_i$ `, 0.]` for different times $t_i$, where the '1' indicates that the results are sought at the first (and in this example, only) interface in the multilayer. One observes that transients increase the energy release rate, as seen in the examples in Chapter 13. In this instance, the transient spike is much larger during cooling than during heating. This is purely a result of the fact that the stress-free state was defined to be the initial condition in the simulation.

On the bottom in Figure 15.7, the reference temperature is defined as $t = 1.99$, which for all cases corresponds to the steady-state profile reached during the hold segment of the heating cycle. (That is, the command to recover the results on the bottom of Figure 15.7 is `GetERRInterface[Bilayer, out, 1, ` $t_i$ `, 1.99]`). While 2 sec is obviously not enough to cause stress relaxation, the results clearly indicate that the steady state is reached during the the hold portion of the heating cycle: as such, the results would not change if the hold phase was lengthened. This change in reference temperature (defining zero stress) is therefore meaningful for materials subject to creep and heating cycles with long hold periods. In this scenario, the spike in energy release rate occurs during heating, not cooling: note the the energy release rate at the start of the cycle is nonzero. At steady state during the hold phase of the heating cycle, the energy release rate is zero because this was defined to be the stress-free state. In this instance, it is the transient during cooling that is inconsequential.

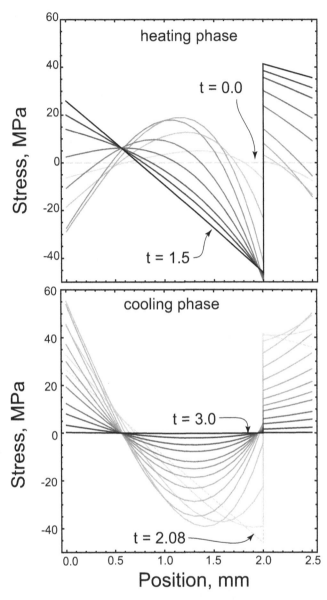

**Figure 15.6** Stress distributions at various times during the heating phase and cooling phase of the thermal cycle with $a = 12$ shown in Figure 15.2, for a bilayer with properties listed in Table 15.1. One can associate each curve by noting the rate of change: faster heating and cooling rates correspond to steep slopes in the above figures. The stress-free state is taken to be room temperature.

Finally, we can use *LST* to confirm that the maximum energy release rate for a delamination crack occurs at the interface for this problem. Figure 15.8 plots the energy release rate for delamination cracks located at the nodal positions, for several instances in time. The initial condition of uniform temperature is used to define the stress-free state. For a

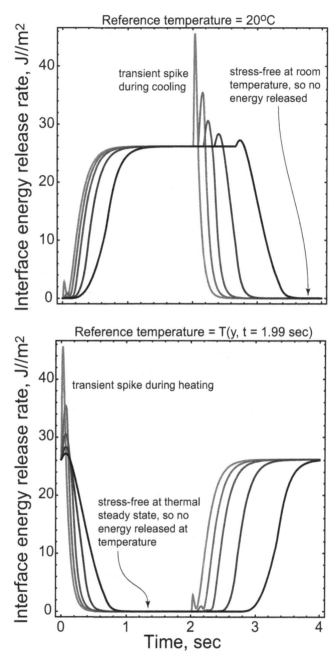

**Figure 15.7** The energy release rate at the interface of a bilayer, with the properties listed in Table 15.1 and heating cycles shown in Figure 15.2. On the top, *tref* = 0, implying the initial condition is the stress-free state; on the bottom, *tref* = 1.99, implying that the elevated, steady-state temperature state is stress-free.

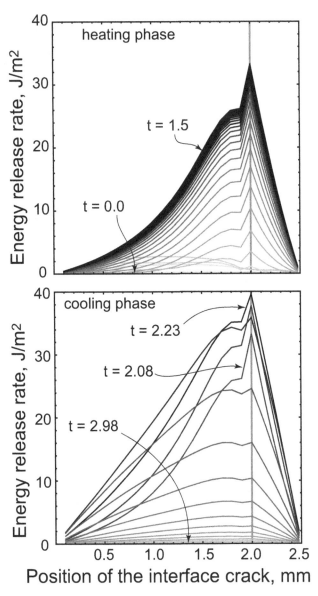

**Figure 15.8** The energy release rate in a bilayer as a function of the position of the delamination crack, for a bilayer with the properties listed in Table 15.1 and the second fastest heating cycle shown in Figure 15.2. Times span from 0 to 1.5 sec (on the top) and from 2.18 to 2.98 (on the bottom). The time increment between curves is 0.05 sec.

brittle film on a ductile substrate, the fact that the peak energy release rate in the initial heating transient (or during the final stages of cooling) occurs in the substrate is of no consequence. Typically, the interfaces represent the locations with the lowest toughness and hence are the location of interest.

# 16 Finite Element Software for Multilayers: *LayerSlayer FEA*

Previous chapters have frequently pointed out that certain aspects of fracture in multilayers require complete solutions for stress distributions close to the crack tip. Important examples include mode mix, channeling/tunneling cracks, crack kinking and features with finite length (such as edge cracks or ligaments). While analytical solutions have been presented for a few important cases, a numerical approach is often needed even for relatively simple geometries.

Finite element analysis (FEA) is a well-suited approach, as it is capable of handling virtually any geometry, and all relevant crack tip parameters can be computed using this approach. The fact that both commercial and freeware codes are widely available begs the question as to why we have even bothered to provide an FEA code. Surely, many analyses – especially those involving complicated effects such as plasticity or crack face contact – can be conducted with existing codes. While this is true, there are two important considerations that make a stand-alone code worthwhile.

First, and most importantly, an accurate analysis of cracking in thin films often requires specialized meshes that can be cumbersome to create with other tools. Here, meshing software is provided that is specifically tailored to the analysis of thin film systems with cracks. We provide complete 'model generators' that define the geometry, create appropriate meshes and specify all boundary conditions and loading with an absolute minimum of user input. Second, the software also provides efficient, accurate and reliable subroutines to postprocess the results to calculate crack tip parameters. Such computations are by no means universally available in commercial codes. In these two important aspects, there is considerable computational expertise is embedded in the codes, which makes multilayer analysis accessible to those without significant experience with other codes.

Although the provided code lacks the ability to analyze complex geometries and material response, it is highly efficient, both with regard to the effort required for comprehension and for use. Specifically, the software is designed for automation, such that one can set up and analyze hundreds of possible multilayer/crack configurations in minutes. The prepackaged analysis modules (largely) focus on rectangular layers with straight cracks that run along interfaces or perpendicular to the layers. A single module can handle a broad range of geometry, e.g., edge cracks in multilayers – with any crack length, in a component of any width, and with any number of layers. The entire analysis can be set up with just a few input parameters.

At the outset, is important to keep in mind the limiting assumptions that are built into the *LayerSlayer* FEA code and capabilities provided by existing subroutines:

- linear, isotropic elastic materials with uniform thermal strains (and uniform growth strains), subject to *plane strain* $(\epsilon_z = 0)$, small deformation (linear strain-displacement relationships), with no treatment of contact.
- eight-noded quadrilateral elements with quadratic displacement distributions, and linear stress/strain distributions, with limited postprocessing (e.g., deformed plots, contour plots of stresses).
- simple mechanical loading scenarios, e.g., point loading and uniform pressure loading on faces; uniform misfit strains in each layer without temperature gradients.
- $G$, $K_1$ and $K_2$ computation (and hence phase angle $\psi$)
- prepackaged modules for edge delamination cracks, penetrating cracks and channeling cracks

This chapter describes the basics of the finite element approach and how it is used to compute relevant fracture parameters; the chapter concludes with a description of the provided software *LayerSlayer FEA* (*LS-FEA*) and its capabilities. The next chapter utilizes *LS-FEA* to illustrate the impact of numerical parameters on solution accuracy. For many users and many cases, this chapter will be all that is needed, as the key conclusions from the final chapter are summarized in this chapter.

## 16.1    Basics of the Finite Element Approach

Here, we review only the essential elements of a finite element analysis (FEA) for linear, isotropic thermoelastic materials, and refer the reader to a large number of excellent texts on the subject (e.g., Bathe 1997; Bower 2010; Buchanan 1994). Those familiar with FEA may choose to skip to sections that cover the computation of fracture parameters (Section 16.2) and/or the user interface and meshing strategy (Section 16.3).

The central concept of finite elements is the discretization of a given domain into elements used to define approximate spatial variations in displacements, which in turn define approximate spatial variations in strains and stresses. The elements form a contiguous patch work of quadrilaterals that cover the domain of interest. The element sides are defined by material points called nodes. Boundary conditions and external forces are applied at the nodes, and the solution computes a list of nodal displacements.

The provided code *LS-FEA* uses conventional eight-noded quadrilateral elements; the displacements vary quadratically with the element. Hence, the strains (and stress) vary linearly within the element. It is worth mentioning here that the limited postprocessing capabilities of *LS-FEA* produce elemental averages for stress/strain, such that a contour plot appears a patchwork quilt. However, the computations in *LS-FEA* produce enough information to produce far more aesthetic plots showing actual spatial distributions that are much smoother.

Locations in the domain that experience significant strain gradients require small elements to accurately capture the variations inherent to the solution. The severe (singular)

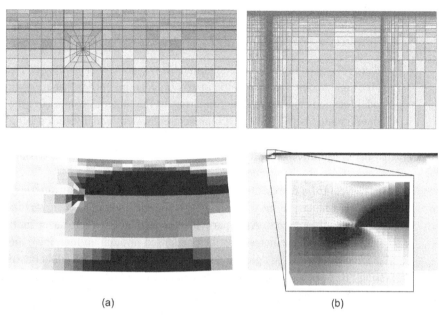

(a)                                                    (b)

**Figure 16.1** Examples from the multilayer module `DelamFEA`, which creates models of delamination edge cracks, with the top figures showing the undeformed mesh and the bottom figure showing patchwork contours of $\sigma_{xx}$ on the deformed mesh: (a) A coarse mesh illustrating the combination of fan blocks and rectangular blocks that comprise a bilayer; black lines indicate the boundaries of the modular blocks used to define the model. All block interfaces except for the edge crack are tied together by eliminating redundant nodes. (b) A fine mesh for a very thin film on a substrate that is representative of that required for accurate results: the mesh is created with the same code as in (a), only with different dimensions and different meshing parameters.

gradients surrounding crack tips – and especially those at an interface (see Chapter 3) – therefore require highly focused meshes near the crack tip with small elements in all directions. Indeed, the ideal mesh to compute near tip fields is a focused, radial fan mesh with nodes along rays that emanate from the crack tip. To make calculation times tractable (which scale linearly with the number of nodes), this high-focused region must be smoothly transitioned to regions with larger elements. Those inexperienced with FEA may wish to look ahead to Figures 16.1 and 16.2 to see illustrations of these mesh characteristics.

Meshing can be quite a challenge for off-the-shelf meshing software when modeling multilayers that involve domains with large aspect ratios. *LS-FEA* is specifically designed to address this challenge by embedding a knowledge of 'acceptable' meshes within the source code, such that users do not require prior experience or expertise. Section 16.3 provides details of the underlying meshing strategies used in *LS-FEA*.

To set the stage for the explanation of how crack tip parameters are calculated, and provide context for the *LS-FEA* interface, we now review the basic mathematics of the finite element approach. In two dimensions, for a single element with eight nodes, the total potential energy of the element can be written as

$$\Pi_e = \frac{1}{2}\{q\}^T_{1x16} [k]_{16x16} \{q\}_{16x1} - [F_{th}]^T_{1x16} \{q\}_{16x1} - [F_a]^T_{1x16} \{q\}_{16x1} \tag{16.1}$$

where $\{q\}$ is a column matrix containing the two components of the displacement vector at the eight nodes. The symmetric matrix $[k]$ contains entries that are computed from the nodal locations and the elastic properties of the material comprising the element. The column matrices $[F_{th}]$ and $[F_a]$ represent equivalent nodal forces arising from thermal terms and applied tractions (or point forces), respectively. Note that the thermal terms in the constitutive law produce contributions identical to those arising from external forces; these are the source of thermal stresses.

Above, (16.1) represents the sum of strain energy (the first two terms) and the external work (the last term) in one element; the total potential energy of the model is the sum over all elements. As a particular node might participate in multiple elements, this sum over all elements must be condensed through an inordinate amount of bookkeeping referred to as 'global assembly'. *LS-FEA* performs global assembly by summing (16.1) over all elements, such that the total potential energy of the model is computed by

$$\Pi = \frac{1}{2}\{Q\}^T_{1x2N} [K]_{2Nx2N} \{Q\}_{16x1} - [F]^T_{1x2N} \{Q\}_{2Nx1} \tag{16.2}$$

where $\{Q\}$ is the list of all nodal displacements (with $N$ nodes in the system, in two dimensions there are $2N$ nodal displacement), $[K]$ is the global (symmetric) stiffness matrix obtained by summation of stiffness contributions from each element, and $[F]$ is a global force vector that includes contributions from both external loads and thermal terms.

For an elastic body, the solution minimizes the potential energy: setting $\partial \Pi / \partial Q_j = 0$ implies[1]

$$[K]_{2Nx2N} \{Q\}_{2Nx1} = [F]_{2Nx1} \tag{16.3}$$

Prior to solving (16.3) for the nodal displacements, we must impose boundary conditions; otherwise, $[K]$ is singular. The boundary conditions manifest themselves as a subset of $\{Q\}$ where displacements are specified by the problem statement. For example, the FEA model of a cantilever beam glued to a rigid wall implies that all the nodes that lie along the wall have zero displacement.

Rather than eliminate variables corresponding to prescribed nodal displacements in (16.3), *LS-FEA* uses Lagrange multipliers to impose all boundary conditions. In this approach, (16.2) is modified by subtracting the term $\sum_{k=1}^{M} \lambda_k (Q_j - Q_j^*)$, where $Q_j^*$ is the value of the variable $Q_j$ that is being prescribed, $\lambda_k$ is a new unknown (the Lagrange multiplier) and $M$ is the number of prescribed displacements in the model. The minimization procedure then leads to

$$[K_g]_{(2N+M)x(2N+M)} \{\tilde{Q}\}_{(2N+M)x1} = [F]_{(2N+M)x1} \tag{16.4}$$

---

[1] There are numerous ways to demonstrate that the extremum in this quadratic potential is a minimum, and that there is a unique solution $\{Q\}$ that minimizes this energy.

where $\{\tilde{Q}\}$ now includes both unspecified nodal displacements and unknown Lagrange multipliers. Though this approach involves a larger set of equations than those obtained by eliminating variables, linear solutions are quite inexpensive and the difference in computation times is negligible.

With the solution of (16.4), one obtains the FEA approximation to the full elasticity solution: strains can be computed at any position in any element through simple matrix operations. The details of this can be found in any introductory text on finite element methods. Stresses are then computed using the constitutive law described in Chapter 2.

A very important consideration when applying FEA to fracture problems is that the approach provides highly accurate estimates of the strain energy, even for relative coarse meshes. Any FEA-based quantity – displacements, strains, stresses and so forth – will converge to the exact elasticity solution (defined by pointwise satisfaction of the underlying PDEs) as the element size is decreased. However, the computed strain energy of the system will converge much 'faster', in the sense that continued mesh refinements do not affect the computed strain energy as much as the strain at a particular point. Simply put, accurate total strain energy in the system can be obtained with coarse meshes, while accurate strain distributions typically require much finer meshes. In the present context, this implies that accurate estimates of the energy release rate can be obtained with relatively coarse meshes. The meshes required to get accurate values of crack tip stress intensity factors are much finer. Following the section on computing crack tip parameters in Section 16.2, appropriate meshing strategies are discussed in Section 16.3.

## 16.2    Computation of Fracture Parameters

*LayerSlayer FEA* uses the stiffness derivative method to compute crack tip fracture parameters, as first pioneered via Parks for computing the energy release rate using finite element models (Parks 1974). This method was extended by Matos et al. (1989) to compute stress intensity factors at bimaterial interfaces. It is worth nothing that there are other methods to computing such parameters; e.g., see the review in Matos et al. (1989). However, the following stiffness derivative method has proven effective for interface crack problems and is particularly simple to implement.

From Chapter 3, the strain energy release rate for a crack in a two-dimensional problem is defined as

$$G = -\frac{\partial \Pi}{\partial a} \tag{16.5}$$

where $\Pi$ is the total potential energy per unit thickness and $a$ is the crack length. It should be kept in mind that external forces and loads imposed in the model should be expressed on a per unit depth basis as well. Equation (16.5) implies the energy release rate can be computed by taking the numerical derivative of the potential energy from FEA with respect to crack length.

Instead of computing (16.5) through direct computation using multiple simulations with different crack lengths, one may note that differentiation of (16.2) implies the following:

$$G = -\frac{\partial \Pi}{\partial a} \cong -\frac{1}{2}\{Q\}^T_{1x2N}\left(\frac{\partial [K]_{2Nx2N}}{\partial a}\right)\{Q\}_{2Nx1} + \left(\frac{\partial [F]^T_{1x2N}}{\partial a}\right)\{Q\}_{2Nx1}$$

$$\cong -\frac{1}{2}\{Q\}^T_{1x2N}[dK]_{2Nx2N}\{Q\}_{2Nx1} + [dF]^T_{1x2N}\{Q\}_{2Nx1}$$

(16.6)

where $[dK]$ is the derivative of the stiffness matrix with respect to crack tip position, and $[dF]$ is the derivative of the force vector with respect to crack tip position. The derivatives of the stiffness matrix and nodal force vector (arising from thermal terms) can be approximated in an efficient manner as follows.

Since $[dK]$ and $[dF]$ are sought for infinitessimal changes in crack length, it is not necessary to construct entirely new meshes. Instead, one can use the original mesh and simply shift the nodal positions of nodes surrounding the crack tip to generate the new mesh. The mesh generation step does not need to be repeated, which saves considerable time. With the vector $\underline{\Delta a}$ defining the direction of crack advance, we define the new mesh needed to compute the derivatives by

$$\underline{x}^S_i = \underline{x}_i + \underline{\Delta a} \text{ for } |\underline{x}_i - \underline{x}_t| \le r_S$$

$$\underline{x}^S_i = \underline{x}_i \text{ for } |\underline{x}_i - \underline{x}_t| > r_S$$

(16.7)

where $\underline{x}^S_i$ is the shifted position of node $i$, $\underline{x}_i$ is the original global nodal position of node $i$, $\underline{\Delta a}$ is the shift in the position of the crack tip, $\underline{x}_t$ is the position of the crack tip, and $r_S$ is the size of a circular region over which the nodes will be moved. That is, all nodes falling within one distance $r_S$ from the crack tip are shifted by $\Delta a$ (the magnitude of the vector defining the direction of crack advance), while all others remain unchanged.

The first-order forward finite difference estimate for the stiffness matrix derivative is

$$[dK] = \frac{\partial [K]}{\partial a} \approx \frac{1}{\Delta a}\left([K(\underline{x}^S_i)] - [K(\underline{x}_i)]\right) + O[\Delta a],$$

(16.8)

with a similar expression for the first-order estimate $[dF]$.

There are two sources of error in the derivatives (and hence, two contributions to error in the estimate for $G$): formula error (i.e., $O[\Delta a]$ term in 16.8) and round-off error. For small $\Delta a$, the difference between the stiffness entries computed for the shifted mesh and the original mesh is vanishingly small, and errors arise with the magnitude $\epsilon/\Delta a$, where $\epsilon$ is the inherent round-off error arising from the computation of the relevant $[K]$ and $[F]$ matrices. Thus, the error initially decreases with increasing $\Delta a$ (when round-off error dominates) and then increases with increasing $\Delta a$ (when formula error kicks in).

The formula error can be reduced with a higher-order estimate of the derivative, for example:

$$[dK] = \frac{1}{\Delta a}\left(-\frac{11}{6}[K(\underline{x}_i)] + 3[K(\underline{x}^S_i)] - \frac{3}{2}[K(\underline{x}^{2S}_i)] + \frac{1}{3}[K(\underline{x}^{3S}_i)]\right) + O[\Delta a^3]$$

(16.9)

where $\underline{x}^{jS}_i = \underline{x}_i + j\underline{\Delta a}$. This approximation allows for larger values of $\Delta a$ to be used, conceptually avoiding round-off errors at small values of $\Delta a$.

To compute the stress intensity factors, the approach involves the superposition of the problem at hand with asymptotic bimaterial crack tip fields. Consider two different loading states that are imposed on the same geometry, and denote them as problem $A$ and problem $B$. Let problem $A$ be the problem of interest, and problem $B$ is a problem for which the stress intensity factors are known. Specifically, the second problem is one in which $K_1^B \neq 0$ and $K_2^B = 0$. The stress intensity factors for the case where both loading states are imposed simultaneously are given by $K^{A+B} = K^A + K^B$. However, the energy release rates for the case where both $A$ and $B$ are imposed simultaneously do not superpose directly, but instead are given by

$$G_{A+B} = -\frac{1}{2}\{Q_A + Q_B\}_{1x2n}^T [dK]_{2nx2n} \{Q_A + Q_B\}_{2nx1} + [dF]_{1x2n}^T \{Q_A + Q_B\}_{nx1}$$
(16.10)

where $\{Q_A\}$ are the nodal displacements associated with problem $A$ and likewise for problem $B$. Irwin's relationship for mixed-mode fracture and superposition then dictate that

$$G_{A+B} = \frac{1 - \beta_D^2}{E_*}\left[\left(K_1^A + K_1^B\right)^2 + \left(K_2^A\right)^2\right]$$

$$= \frac{1 - \beta_D^2}{E_*}\left[\left(K_1^A\right)^2 + \left(K_1^B\right)^2 + \left(K_2^A\right)^2 + 2K_1^A K_1^B\right]$$
(16.11)

$$= G_A + \frac{1 - \beta_D^2}{E_*}\left[\left(K_1^B\right)^2 + 2K_1^A K_1^B\right]$$

Recall that $K_1^B$ is known and has with it the displacements $Q_B$. This implies that the stress intensity factor to the problem of interest can be computed from

$$K_1^A = \frac{1}{2}\left[\frac{E_*}{1 - \beta_D^2}\frac{(G_{A+B} - G_A)}{K_1^B} - K_1^B\right]$$
(16.12)

Thus, one can compute the stress intensity factor for the problem of interest, $K_1^A$, provided a known solution exists for the same geometry that is defined by $Q_B$ and $K_1^B$.

For case $B$, one can use the asymptotic crack tip displacements for a bimaterial crack, which can be expressed as

$$u_i = \frac{8K_1^B}{(1 + 2i\epsilon)\cosh(\pi\epsilon)E_*}\sqrt{\frac{r}{2\pi}}r^{i\epsilon}f_i(\theta, \epsilon)$$
(16.13)

where $f_i(\theta, \epsilon)$ are tabulated in the paper by Matos et al. (1989). These can be used to compute the nodal displacements $Q_B$, which are valid nodal displacements because the asymptotic solution satisfies all governing equations (such as equilibrium). That is, $Q_B$ determined from (16.13) are the proper displacements for a problem with the loading defined by $K_1^B$.

Therefore, the steps to compute $K_1^A$ are as follows: first, one computers $Q_A$ for the problem of interest and $G_A$ using (16.6) and (16.8). Then one computes $Q_B$ from the asymptotic fields for a given value of $K_1^B$ using the nodal positions with (16.13), and subsequently computes $G_{A+B}$ using (16.10). Finally, with these results in hand, one computes $K_1^A$ using (16.12). Naturally, we can repeat the process with $K_1^B = 0$ and $K_2^B \neq 0$

to get mode 2, or, use Irwin's relationship to solve for $K_2^A$ once $G_A$ and $K_1^A$ have been computed.

In *LS-FEA*, one must prescribe three parameters to compute the stress intensity factors: $\Delta a$, $r_S$ and $K_{1,2}^B$. An exhaustive numerical parameter study has led to the selection of default values listed later in this chapter, which have proven effective for the myriad of cracking problems discussed in this book. These numerical studies are described in Chapter 17.

## 16.3    An Overview of the *LayerSlayer FEA* Framework

The *LayerSlayer FEA* framework is a collection of subroutines (written in *Mathematica*) that perform the basic steps of finite element analysis. While the user is given access to individual subroutines, most users will find the 'modules' described in this section to be the most convenient and, in many instances, all that is needed. These modules are designed to (1) perform all the necessary steps with the absolute bare minimum of user input, (2) enable broad parameter studies by handling arbitrary dimensions and number of layers and (3) empower novice users by embedding the required expertise in computational mechanics.

This section provides an overview of the modules' internal analysis steps and contextualizes the required user-defined inputs. User-defined modules for other types of analysis can be easily developed using the existing modules as a template. While only two modules are described in detail in this book, there are many many more examples associated with the multilayer problems described throughout this book; all of them follow the basic analysis steps outlined below.

Consider the analysis of the the bilayer delamination problem shown in Figure 16.1, which subjects the multilayer to a uniform temperature change while allowing for bending. The entire analysis – including geometry definition, meshing, solution and postprocessing – is conducted by calling the module DelamFEA, as follows:

$$\text{DelamFEA[layers, geometry, mesh, load, jkparams]} \qquad (16.14)$$

where the input arguments contain information about the multilayer problem; complete details of these inputs and their formatting are provided in the next section. Both cases shown in Figure 16.1 – a coarse mesh with thick layers (Figure 16.1(a)) and a fine mesh of a very thin film on a thick substrate (Figure 16.1(b)) – simply have different inputs to DelamFEA. A package is also provided for analyzing penetrating cracks and has similar inputs.

Both provided modules utilize the following information and perform the following steps internally:

- **Layer and geometry definition:** the number of layers and their physical properties (such as thickness, modulus, CTE etc.) are provided in lists that define the multilayer stack (layers in 16.14 above). A very simple list of specimen characteristics are also defined: e.g., the width of the multilayer specimen, the crack length and position of the delamination crack (geometry in 16.14 above).

- **Mesh generation:** using the geometry and mesh information passed to the subroutine (contained in mesh in 16.14 above), a mesh is automatically generated using a variety of internal subroutines. The mesh information is a very brief list of the number of elements to use near the crack tip and outer regions that connect the crack tip mesh to the edges of the specimen. This step generates the node and element information, which in turn defines the node numbers used to impose boundary conditions in the problem (such as those at the extreme corners of the multilayer, where forces might applied to impose three-point bending). The mesh generators all automatically perform any required mesh (block) consolidation described in later in this section.

- **Boundary conditions and loading:** the user passes a very limited amount of information regarding the type of loading and boundary conditions to be imposed (load in 16.14). This information is used to specify prescribed nodal displacements using the node numbers and locations generating by the meshing algorithm. For the bending problem in Figure 16.1, the left bottom corner node is pinned while the right bottom corner node is constrained against motion in the $y$-direction.

- **Displacement solution:** internally, the solution is obtained by passing the information defined in the above steps to the solver, as in SolveFEA[job, bcs, forces], where job contains node and element information, bcs is the list of imposed displacements and forces is the list of applied forces, all of which are generated internally based on user inputs and the results of the embedded mesh generator.

- **Postprocessing:** in some modules, a contour plot of stress is generated for the deformed shape; this is useful to provide an easy qualitative check to ensure the model has been generated properly (without any erroneous gaps between meshing blocks).

- **Crack tip parameters:** the energy release rate and stress intensity factors are obtained by passing the finite element model and solution, interface properties, and numerical parameters to GetJK[job, disp, jkparam], where job contains the mesh information generated in the earlier step, disp is the output displacement from the analysis and jkparam is a list of the numerical parameters used to compute the crack tip parameters, as described in Section 16.2.

### *LayerSlayer FEA*'s Meshing Strategy

Many, if not most, users will find the automatic meshing algorithms that are embedded in the provided modules described later to be perfectly sufficient. That is, most users will not need to perform the steps described in this section, as these will be performed internally. However, a review of the core meshing strategy of *LS-FEA* provides some insight into the potential source of numerical difficulties, and enables advanced users to create their own modules for other geometries.

The core meshing strategy in *LayerSlayer FEA* is based on the following objectives: (1) to use a radial fan mesh surrounding any crack tip, with elements focused along rays emanating from the crack tip, and (2) to use the simplest possible strategy for a mesh that connects the the focused region near the crack tip to the rest of the multilayer. These

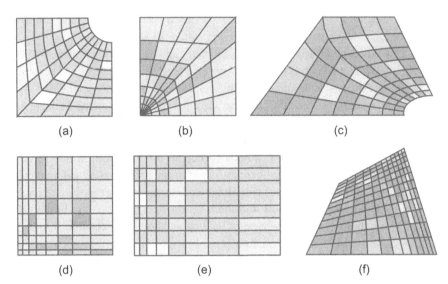

(a)                    (b)                    (c)

(d)                    (e)                    (f)

**Figure 16.2** Examples of the two meshing modules in LS-FEA: (a–c) the fan mesh module, and (d–f) the rectangular block module. The radial fan mesh with collapsed nodes (b) are used for all crack tips; the rectangular meshes are used for 'fill' regions that define the outer regions of the multilayer.

two objectives are met with two different meshing subroutines; examples of the output from these subroutines are shown in Figure 16.1.

In the first subroutine, a radial fan mesh is created by defining a square block with a semicircular cut in one corner, as shown in Figure 16.2(a). The user specifies the number of elements along the arc of the cut and the two opposing sides of the block. Radial lines are drawn from the arc to the opposite faces. The user also specifies the number of elements along these radial lines. One can prescribe a zero radius for the arc, as shown in Figure 16.2(b); this leads to the definition of redundant nodes at the focal point of the arrays, which are removed by subsequent postprocessing. In all *LS-FEA* prepackaged modules, four radial fan blocks are defined such that the crack tip is surrounded by a complete radial fan, as shown in Figure 16.1(a).

The subroutines for meshing *LS-FEA* allow the user to prescribe arbitrary locations for the four corners of the parent block, including the corner that serves as the focal point of the fan. The final nodal positions are determined by mapping a unit square into the four-sided block defined by the corner locations. This procedure generally produces acceptable meshes, even when the corner locations form uneven global blocks, as in Figures 16.2(c, f). One should also note that if the global block has uneven sides, the cut corner maps into an ellipse if the radius of the cut is finite (as shown in Figure 16.2(c)). Again, it should be noted that users these steps are handled automatically by the prepackaged delamination or penetrating modules described in Sections 16.4 and 16.5.

To complete the meshing strategy, a technique is required to connect the radial fan regions to the rest of the multilayer. The simplest possible approach is adopted, that is, four-sided regions with straight sides and a regular grid, as shown in Figures 16.2(d–f). Here, the user prescribes the number of elements on each side of the block and a bias parameter that dictates how the nodes should be distributed along each edge. Elements can be shaded towards the same corner, as in Figures 16.2(d, f), or only one side, as in Figure 16.2(e). The same mapping procedure outlined above for radial fan meshes is used; specifying unequal sides allows one to mesh trapezoids, as shown in Figure 16.2(f), such as would be needed for films with tapered thickness.

### Creating a Coherent Mesh from Multiple Mesh Blocks

The *LS-FEA* modules for delamination and penetration create a multilayer geometry using combinations of coherent mesh blocks; they do this internally without required user actions. Figure 16.1(a) provides an example of how the above subroutines are used by these modules. To create the mesh for delamination shown in Figure 16.1(a), four radial fan mesh blocks are placed surrounding the interface crack tip. In each layer, five rectangular mesh blocks are used to fill out the remainder of the layer. The meshing subroutines internal to the modules ensure equivalency between the nodal positions along the edges of one block (e.g., the radial fan mesh near a crack tip) and an adjacent block (e.g., a rectangular mesh that 'fills in' the layer region spanning from the crack tip region to the outer edge of the layer). When two blocks are meshed properly, redundant nodes are created along the interface between the adjacent blocks. These redundant nodes are removed via postprocessing, wherein one node is kept and used to replace all redundant nodes in the element definitions.

In this strategy, one must ensure that the proper number of elements along each adjacent side of the mesh blocks are specified. The provided modules solve this bookkeeping headache internally; one merely specifies the number of elements surrounding the crack tip and the remaining number of elements in the 'fill' regions. The model with the fine mesh shown in Figure 16.1(b) is made with the same module/algorithm as used to create the coarse mesh Figure 16.1(a); the only distinction is that different specifications for layer dimensions and the number of elements are used as inputs. In *LS-FEA*'s modules, one specifies only a very limited number of mesh parameters; typically, this includes (1) the number of elements around the crack tip, (2) the number of elements along a ray, (3) the number of elements to fill in the vertical direction and (4) the number of elements to fill in the horizontal direction.

To define a crack, the nodes that fall along the crack face may be redundant (i.e., have the same positions), but should not be eliminated. This is handled internally in the provided modules by artificially separating the nodes along a crack face such that they lie just beyond the separation tolerance used to identify redundant nodes.

With meshing tools developed for the geometries addressed in this book (and described in Sections 16.4 and and 16.5), the user needs only to specify dimensions and a very limited set of parameters. Details of the meshing strategy are, by and large, irrelevant to the user. However, for geometries not considered in the provided modules, the advanced user will have to write a new module utilizing the base subroutines provided with this book.

## Troubleshooting Mesh Problems and the Critical Role of Units

Aside from numerical performance issues associated with undesirable meshes (which are general to any FEA code), most problems that arise with *LS-FEA* are related to the step that consolidates the mesh blocks described earlier. The root of these problems lies with the method used to identify redundant nodes along the faces of adjacent mesh blocks. *LS-FEA* identifies redundant nodes using a brute force method that checks the distance between all nodes and eliminates those that fall with a very small distance of one another. This procedure implies the mesh consolidation algorithm can fail in one of two ways: either redundant nodes are not eliminated because the mesh is not coherent (to within the resolution of the consolidation tolerance), or conversely, distinct nodes are improperly eliminated because the element size is on the order of the consolidation tolerance.

*It is critical to note that when this happens, the analysis may still (and usually does) execute: untied faces between mesh blocks introduces spurious cracks into the model, while improperly tied faces lead to spurious deformation.* Simply put, the analysis may not fail to execute, but rather return erroneous results. Plotting the deformed shape of the model with a significant amplification factor applied to the nodal displacements (i.e., plotting an exaggerated deformed shape) almost always reveals such difficulties immediately.

To avoid improper elimination of nodes, the minimum layer thickness should be kept close to unity. That is, one should choose thickness units such that the numerical value of the minimum layer thickness is on the order of 1, and not a very small value such as $10^{-6}$. This is simply a distinction of the units chosen to express the properties of the layers. For example, one can choose to express the layer thickness in terms of micrometers and the moduli in terms of $N/\mu m^2$; the inferred units on the computed energy release rate will then be $N/\mu m$.

The rationale for choosing units that lead to layer thickness close to one is that effective meshes rarely require elements less than about 1% of the minimum layer thickness. With the scaling of the thinnest layer set to something on the order of unity, the minimum element sizes will on the order of 0.01; this is orders of magnitudes larger than the default cutoff distance used to identify whether or not two nodes are redundant, $10^{-8}$. On the other hand, choosing very large dimensions could lead to round-off errors introduced during mesh generation that place truly redundant nodes too far apart (outside the cutoff threshold). However, this error has not been encountered when the minimum layer thickness is $O[1 - 10^3]$, even when other dimensions (such as specimen widths) are 2 orders of magnitude larger. This issue is worth keeping in mind, however, should mesh consolidation fail.

Of course, the mesh consolidation step will fail if the mesh blocks are not defined properly; that is, if they do not have the same number of nodes and the same spatial distributions along matching interfaces. The provided modules involve careful 'mesh block' constructions that have been vetted exhaustively. Nevertheless, given that logic commands are embedded in the provided modules to handle different geometries, it is possible that the user can find rare combinations of dimensions that lead to improper block definitions. A plot of the deformed shape typically clearly reveals such occurrences.

## 16.4    Multilayer Delamination Module

This section describes the *LS-FEA* module named `DelamFEA`, which analyzes delamination cracks in rectangular multilayers, shown in Figure 16.3. This module provides the basis to conduct a large number of the delamination analyses discussed in previous chapters – e.g., those for interior edge cracks discussed in Chapter 9.

The module utilizes the block meshing strategy described in Section 16.3, which has been generalized for delamination edge cracks in a multilayer with an arbitrary number of layers. The module calls a core meshing subroutine, `MeshMultilayerD`, which generates a complete finite element mesh, including a list of the control nodes labeled in Figure 16.3(a), and their spatial locations. This information is used to impose the boundary conditions and nodal forces corresponding to the cases shown in Figures 16.3(b–d).

The preprogrammed boundary conditions in `DelamFEA` are free bending driven by misfit strains (Figure 16.3(b)), four-point bending with or without misfit strains (Figure 16.3(c)) and three-point bending (Figure 16.3(d)). It is relatively straightforward to modify the content of `DelamFEA` to handle other boundary conditions and loading scenarios: significant modifications are needed only if the geometry is different from Figure 16.3(a).

The module `DelamFEA` returns the energy release and stress intensity factors for the delamination crack. Additional results, such as a contour plot of the stresses in the multilayer, can be obtained using various postprocessing modules; for details, the reader should consult the example files accompanying this text.

A summary of the information and its format that must be provided to `DelamFEA` are as follows:

**Input format:**
The module format is

```
DelamFEA[multilayer, geometry, mesh, load, jkparams]
```

where the input arguments are explained in detail below. A complete example, that is, the list of inputs needed to analyze the case shown in Figure 16.1, is provided at the end of this section.

**Layer definition:**
The multilayer is defined as a list of layers, each consisting of a list of properties. That is, individual layers and the multilayer are defined as

$$\text{Layer1}=\{h, E, v, CTE, \{T_{ref}, T_{ref}\}, \epsilon_g, k, \{T_{cur}, T_{cur}\}\}$$
$$\text{Layer2}=\{h, E, v, CTE, \{T_{ref}, T_{ref}\}, \epsilon_g, k, \{T_{cur}, T_{cur}\}\}$$
$$\text{multilayer}=\{\text{Layer1, Layer2, ..., LayerN}\}$$

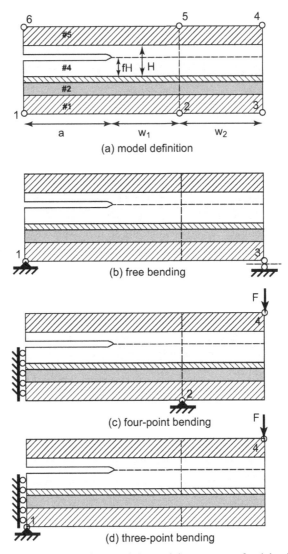

**Figure 16.3** Geometry and control points used the module DelamFEA for delamination edge cracks: (a) the crack can be placed within any layer or at any interface, (b) the free bending case simply prevents rigid body translation and rotation, (c) the four-point bending case assumes symmetry across the left edge and (d) the three-point bending condition imposes symmetry across the left edge and prevents rigid body translation in the vertical direction. All cases use the same meshing module, which returns the node sets needed to impose different boundary conditions.

where $h$ is the layer thickness, $E$ is the elastic modulus, $v$ is the Poisson's ratio, $CTE$ is the coefficient of thermal expansion and $T_{ref}$ is the reference temperature for that layer. The reference temperature is repeated so that multilayer definitions from *LS* can be copied and used with *LS-FEA*; only the first value listed is actually used in the code. *LS-FEA* does not allow for temperature gradients to be imposed. $T_{cur}$ is the current temperature at which the analysis is conducted. $\epsilon_g$ is a growth strain (uniform in the layer), while $k$ is not used but represents the thermal conductivity (see Chapter 14). The layers are numbered from bottom to top, as shown in Figure 16.3(a).

### Geometry Definition:

The geometry shown in Figure 16.3 requires the following input:

$$\texttt{geometry} = \{\{l_c, f, a\}, w_1, w_2\}$$

where $l_c$ is the layer number containing the crack, $0 \le f \le 1$ is the fractional distance from the bottom of that layer to the crack plane, and $a$ is the length of the edge crack. The parameter $w_1$ defines the distance from the crack tip to the internal rollers. (The parameter $w_2$ defines the distance from the internal rollers to the outer right edge of the specimen). Hence, the total specimen width is $W = a + w_1 + w_2$.

For some loading conditions, the control points no. 2 and no. 5 in Figure 16.3 may not be relevant: however, the user must still prescribe values for both $w_1$ and $w_2$. A convenient choice in such situations is simply to divide the distance between the crack tip and the specimen edge in half and set $w_1$ and $w_2$ to this value.

If the fractional distance, $f$, is equal to unity, the crack is placed at the interface on top of the layer with the specified crack; if it is set to zero, the crack is placed at the interface on the bottom the specified layer. Hence, one obtains identical results with $\{\{l_c, f, a\}, w_1, w_2\} = \{\{1, 1.0, a\}, w_1, w_2\}$ and $\{\{l_c, f, a\}, w_1, w_2\} = \{\{2, 0.0, a\}, w_1, w_2\}$; both place the crack at the interface between the first and second layer.

### Meshing Parameters:

The meshing strategy follows the scheme illustrated at the top of Figure 16.1(a), with additional layers meshed similarly to the top half of the top layer. The mesh is automatically generated in `DelamFEA` using the following input:

$$\texttt{mesh} = \{N_{arc}, N_{ray}, \{N_v, N_h\}\}$$

where $N_{arc}$ defines the number of elements along a circumference surrounding the crack tip. The crack tip mesh is a radial fan mesh that is broken into eight arc segments (i.e., a pie slice every $45°$), each with $N_{arc}$ in around the circumference; since the total number of elements surrounding a crack tip is $8N_{arc}$, this parameter typically can be small (8 or less). Similarly, $N_{ray}$ is the number elements in the radial direction of the fan mesh; since the fan mesh region is localized, accurate results are obtained with $8 < N_{ray} < 12$. The parameter $N_v$ defines the number of elements used to fill in the vertical direction; in the layer with the crack, this is the number of elements spanning from the radial fan mesh box to the outer edge of that layer. For layers without cracks, $N_v$ specifies the total number of elements through the thickness of the layer. $N_h$ specifies the number of elements used to fill out the mesh in the horizontal direction.

Table 16.1 provides recommendations for meshing parameters, based on exhaustive convergence studies described in Chapter 17.

**Table 16.1** Recommendations for Numerical Parameters for Delamination and Penetrating Cracks. In the Prepackaged Modules, $L_e = \ell/N_{ray}$, Where $\ell$ is the Minimum of Either the Crack Length or the Minimum Thickness of the Layers Participating in the Crack

| Feature | Parameter | Suggested Minimum | Suggested Default | Suggested Maximum |
|---|---|---|---|---|
| Thickness | $h$ | $O[0.1]$ | $O[1]$ | $O[10^3]$ |
| Circumferential elements | $N_{arc}$ | 2 | 4* | 8 |
| Ray elements | $N_{ray}$ | 4 | 12 | 16 |
| Vertical fill elements | $N_v$ | 5 | 10 | $20^+$ |
| Horizontal fill elements | $N_h$ | 5 | 10 | $20^+$ |
| Crack extension | $\Delta a$ | $10^{-6}L_e$ | $10^{-5}L_e$ | $10^{-3}L_e$ |
| Shift zone size | $r_S$ | $2L_e$ | $4L_e$ | $(N_{ray} - 1)L_e$ |

* Total number of elements around circumference of delamination crack is $8N_{arc}$, while that around the penetrating crack is $4N_{arc}$.
$^+$ These can and should be adjusted based on specimen aspect ratio; to date, the only aspect strained by using large numbers is one's patience with computation times. The basis for these recommendations is discussed in detail in Chapter 17.

The coherent meshing procedure implies the total number of elements in the horizontal direction (outside the fan mesh region) is given by $N_{h,tot} = 3N_h + 2N_{ray}$. Referencing Figure 16.1(a) and starting from the left edge, this total reflects the elements in the fill from the left edge to the focused region (the first $N_h$), the elements transversing the outer boundary of the fan region (the two times $N_{ray}$), the number of elements from the focused region to the inner rollers (the second $N_h$), and the number of elements from the inner rollers to the outer right edge of the specimen (the third $N_h$).

The routine MeshMultilayerD (which is used internally in the module DelamFEA) produces a list of model information used to impose boundary conditions and loading. The output of the MeshMultilayerD module is the following list:

```
{finaljob, {lefttop, leftbot, bot, right, top}, iprops, bendpts}
```

where finaljob is a compound list of the final nodal locations and the element definitions. The second entry in the above list provides a set of coordinates that can be used to impose boundary conditions on various edges of the model. For instance, lefttop is a pair of $(x, y)$ points that define the left edge of the specimen above the delamination crack, leftbot is a list of end points on the left edge of the geometry below the delamination crack, and so forth. The list *iprops* contains the elastic properties above and below the crack, which are used to compute crack tip parameters.

The list bendpts contains the six node numbers corresponding to the 'control' locations shown in Figure 16.3(a). The settings for various loading conditions are described next; one can easily adapt them, using the output of the meshing routine MeshMultilayerD to create additional modules (such as tensile forces applied to the multilayer).

## Loading and Boundary Conditions:

The delamination case to be analyzed are selected from those shown in Figure 16.3 using the input

$$\text{load} = \{case, value\}$$

where *case* indicates which of the three loading conditions is analyzed, and *value* indicates the value of the applied forces.

For the free bending case shown in Figure 16.3(b), *case* = 0: the bottom left corner of the specimen is fixed in place, while the bottom right corner is free to expand or contract in the horizontal direction. For *case* = 0, *value* is irrelevant but still must be specified (it can be set to anything).

For four-point bending shown in Figure 16.3(c), *case* = 4: the left edge of the specimen is constrained against horizontal motion, corresponding to a symmetry condition along that plane. The inner roller on the bottom is fixed to have zero displacement, while an applied force, with magnitude equal to *value*, is specified on the upper right corner of the multilayer.

For three-point bending shown in Figure 16.3, *case* = 3: the left edge of the specimen is constrained against horizontal motion, corresponding to a symmetry condition along that plane. The bottom of the left edge is pinned to have zero displacement, while an applied force, with magnitude equal to *value*, is specified on the upper right corner of the multilayer.

## Crack Tip Numerical Parameters:

The numerical parameters used to compute the energy release rate and stress intensity factors are specified in the input as

$$\text{jkparam} = \{\tilde{r}_S, \Delta\tilde{a}, \tilde{K}^B\}$$

where $\tilde{r}_S$ is the normalized size of the shift zone given by $r_S/L_e$, where $L_e$ is the element size in the focused radial fan mesh surrounding the crack tip. $L_e$ is computed within the module, such that the user in effect specifies the number of elements included in the shifted zone. Similarly, $\Delta\tilde{a}$ is the normalized shift in crack length given by $\Delta a/L_e$. The crack advance cannot be larger than the elements in the focused region, or the elements will invert; hence, $\Delta\tilde{a}$ must be less than unity. $\tilde{K}^B$ is the normalized size of the superimposed $K$ field, as described in Section 16.2. $\tilde{K}^B = K^B/\sqrt{E_* G_A/(1 - \beta_D^2)}$.

Appropriate values for these parameters are listed in Table 16.1 and discussed in detail in Chapter 17.

## Summary of Complete Input:

As an example, the bilayer shown in Figure 16.1 (subject to only thermal stressing, as shown in Figure 16.3(b)) is analyzed by providing the following list of input lines:

```
BottomLayer= {2,200.,0.33,0.x10⁻⁶,{1020., 1020.}, 0, 25.,{20.,20.}}
TopLayer= {1,200.,0.33,40.x10⁻⁶,{1020., 1020.}, 0, 2.,{20.,20.}}
BiLayerEx={BottomLayer,TopLayer}
```

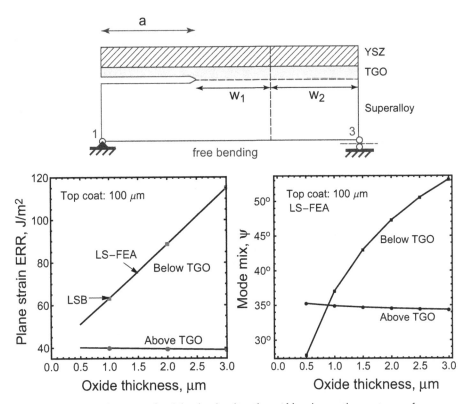

**Figure 16.4** Energy release rate for delamination in a thermal barrier coating system and associated mode mix: results are shown for debonding above and below the thermally grown oxide (TGO), computed using the module `DelamFEA`. The results from LS-FEA are in excellent agreement with those from the multilayer code LS described in Chapter 14, though the latter is not capable of predicting mode mix.

```
GeometryEx={{1.,1.0,2.0},2.,2.}
MeshparamEx = {2, 3, {5, 5}}
LoadEx={0,0}
JKparamEx = {0.5, 0.001, 0.1}
DelamFEA[BilayerEx, GeometryEx, MeshparamEx, LoadEx, JKparamsEx]
```

## Example No. 1: Delamination in a Thermal Barrier Coating System

To illustrate the application of *LS-FEA*, consider a thermal barrier coating system shown in Figure 16.4, consisting of a superalloy substrate, thermally grown oxide (TGO) and ceramic top coat. The specimen specifications are provided in Table 16.2. Isothermal conditions are imposed with the entire specimen subjected to the same temperature change. Figure 16.4 plots the energy release rate and mode mix for delamination cracks that lie above and below the TGO as a function of oxide thickness.

The energy release rates computed via the finite element-based *LS-FEA* are in excellent agreement with results generated with the software *LS* that is described in Chapter 14. This agreement indicates that the specimen dimensions chosen for the finite

**Table 16.2** Example No. 1—List of Properties Used in the Study of Delamination in a Thermal Barrier Coating System, Analyzed with the Module `DelamFEA`. The Current Temperature that is Imposed Throughout the Multilayer is $T = 20°C$.

| Layer | Thickness | $E$, GPa | $v$ | $\alpha$, ppm/°C | Ref. temp, °C |
|---|---|---|---|---|---|
| Top coat, YSZ | 100 μm | 25 | 0.2 | 13.0 | 1100 |
| TGO, $Al_2O_3$ | 0.5–3 μm | 380 | 0.2 | 8.1 | 1100 |
| Superalloy | 3000 μm | 200 | 0.3 | 16.0 | 1100 |

| Specimen dimensions | |
|---|---|
| Crack length, $a$ | 1000 μm |
| Inner span, $w_1$ (crack tip to roller point) | 2000 μm |
| Outer span, $w_2$ (roller point to outer edge) | 1000 μm |

element analysis (given in Table 16.2) are essentially equivalent to the semi-infinite (blanket) layer assumptions inherent to *LS*. The data points in Figure 16.4 are identical to those in Figure 5.6, which plots various *LS* results. It is worth reminding the reader that *LS-FEA* is needed to compute the mode mix, since this is not possible with *LS*.

For delamination at the TGO-top coat interface (i.e., above the TGO), the oxide thickness plays little role, as the stored elastic energy of the TGO is not released. When delamination occurs below the TGO, the TGO-top coat bilayer created by the delamination is strongly influenced by TGO thickness. While the energy release rate increases with oxide thickness, so does the mode mix (indicating an increase in mode 2 contributions), implying that the interfacial toughness is likely to increase as well. This example illustrates that a complete set of results, accessible using *LS-FEA*, is often needed to assess the likelihood of system failure. On the one hand, the driving force for failure at the TGO-top coat interface is smaller; on the other, so is the likely interface toughness, particularly in light of the fact that there is a significant mode 1 contribution. While the superalloy-TGO interface is subject to higher driving forces, the toughness of that interface is undoubtedly higher, and greater mode 2 contributions further mitigate the likelihood of failure.

Of course, isothermal conditions are likely only relevant for lab tests, as the purpose of the TBC is to create a temperature gradient through the stack in the vertical direction. With the current version of *LS-FEA*, one can approximate only by the temperature gradients, by imposing different uniform temperature changes in each layer. That is, the user can prescribe a different uniform reference temperature for each layer, and a different uniform current temperature for each layer, but not a gradient of reference or current temperatures within a layer. This is shown in the next example.

### Example No. 2: Influence of Molten Deposit on a Thermal Barrier Coating System

*LS-FEA* makes adding additional layers trivial. Many TBC systems are prone to attack from contaminants, such as calcium-magnesium-aluminum silicates (CMASs), which

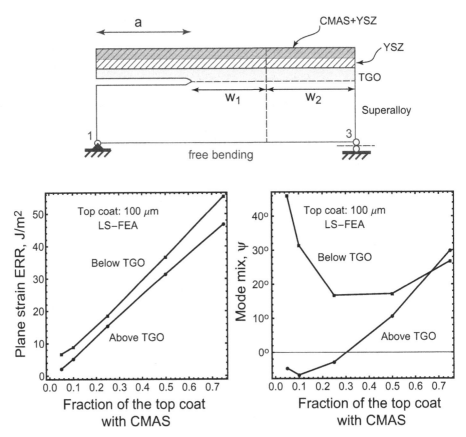

**Figure 16.5** Energy release rate and mode mix for delamination in a thermal barrier coating system that has experienced penetration by CMAS, as a function of the penetration distance into the top coat (expressed in terms of the fraction of top coat that has been infiltrated).

melt and penetrate the top coat. The portion of the top coat with CMAS penetration has a higher modulus than the pristine portion and stores more elastic energy during cooling, thus increasing the likelihood of failure. A four-layer system is shown in Figure 16.5, which represents a modified version of three-layer system of the previous example; the topmost layer reflecting the properties of a CMAS-infiltrated top coat. The total thickness of the top coat, including both the pristine and infiltrated regions, is kept constant. The properties of the system are provided in Table 16.3; note that different reference temperatures are prescribed for each layer.

As seen in Figure 16.5, the driving force for delamination is much smaller than in the previous example, even for low levels of CMAS penetration. This is because the substrate has been subjected to a smaller temperature change than the case shown in Figure 16.4. The driving force for delamination increases with the percentage of the top coat that experiences infiltration (i.e., the thickness of CMAS-infiltrated layer), due to the high modulus of that layer. Note that the infiltrated layer changes the mode mix; the presence of the CMAS-infiltrated layer with its high modulus increases the mode 1 component for failure beneath the TGO by increasing the bending experienced by

**Table 16.3** Example No. 2—List of Properties Used in the Study of Delamination in a Thermal Barrier Coating System, with an Added Layer Representing Infiltration of the Top by a Molten CMAS Deposit. Analysis is Conducted Using the Module `DelamFEA`. Note that the Reference Temperatures are Slightly Different than those Used in Example No. 1

| Layer | Thickness | $E$, GPa | $v$ | $\alpha$, ppm/$^\circ$C | Ref. T, $^\circ$C |
|---|---|---|---|---|---|
| CMAS & YSZ | $100 \cdot f$ μm | 100 | 0.2 | 10.5 | 1100 |
| Top coat, YSZ | $100 \cdot (1-f)$ μm | 25 | 0.2 | 13.0 | 1100 |
| TGO, $Al_2O_3$ | 0.5–3 μm | 380 | 0.2 | 8.1 | 900 |
| Superalloy | 3000 μm | 200 | 0.3 | 16.0 | 850 |

| Specimen dimensions | |
|---|---|
| Crack length, $a$ | 1000 μm |
| Inner span, $w_1$ (crack tip to roller point) | 2000 μm |
| Outer span, $w_2$ (roller point to outer edge) | 1000 μm |

the multilayer (TGO-top-coat-infiltrated region) that is delaminating. For delamination above the TGO, the residual stress in the infiltrated layer increases in the mode 2 component. This switch is due to a change in the bending behavior resulting from the presence (or absence) of the TGO layer in the sublayer formed by delamination.

## 16.5    Multilayer Penetration (Channeling) Module

The module `PenetrateFEA` is provided to analyze the multilayers with penetrating cracks shown in Figure 16.6(a); the geometry is assumed to be symmetric about the vertical plane containing the penetrating crack. The same model can be used to compute the release rates for both penetration and channeling out of the plane of the paper. Due to symmetry, $K_2 = 0$ and as such, only the energy release rates are computed. The energy release rate for penetration is calculated using the stiffness derivative method described in Section 16.2. The energy release rate for channeling is computed from the difference in potential energy in the intact and cracked states, as described in Chapter 6. It is worth noting that the inputs to the channeling module are virtually identical to the delamination module described in the previous section.

Three different loading scenarios are possible with `PenetrateFEA`: (1) free bending driven by misfit strains as in Figure 16.4(b), (4) four-point bending with misfit strains as in Figure 16.4(c) and (3) three-point bending with misfit strains as in Figure 16.4(d). Each case uses the same meshing algorithm and applies different boundary conditions and loads for each case. The embedded meshing subroutine called `MeshMultilayerP` uses the block strategy described earlier, with a half-fan (due to symmetry) surrounding the crack tip.

Note that complications arise when the crack tip lies exactly at a material interface with elastic mismatch, as described in Chapter 8. These cases should be avoided, as they lead to ambiguity with regard to the meaning of the energy release rate. Instead, the penetrating crack tip should be placed a small distance above or below the interface.

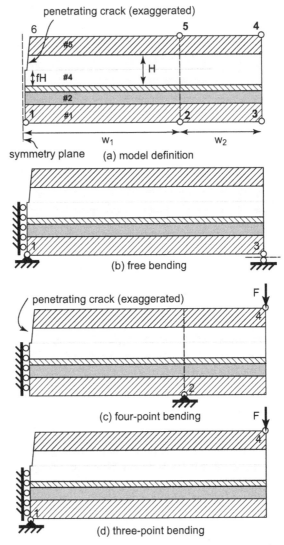

**Figure 16.6** Geometry and control points for the module to analyze penetrating/channeling cracks with the module `PenetrateFEA`: (a) the crack tip can be placed within any layer or at any interface and symmetry is enforced across the crack plane, (b) the free bending case simply prevents rigid body translation and rotation, (c) the four-point bending case assumes symmetry across the left edge and (d) the three-point bending condition imposes symmetry across the left edge and prevents rigid body translation in the vertical direction. All cases use the same meshing module, which returns the node sets needed to impose different boundary conditions.

### Input Format:

The module format is

```
PenetrateFEA[multilayer, geometry, mesh, load, jkparam]    (16.15)
```

with the input arguments explained in detail below. A complete set of input commands is provided at the end of this section.

### Layer Definition:

The multilayer is defined as a list of layers, each consisting of a list of properties. That is, individual layers and the multilayer are defined as

$$\text{Layer1}=\{h, E, v, CTE, \{T_{ref}, T_{ref}\}, \epsilon_g, k, \{T_{cur}, T_{cur}\}\}$$

$$\text{Layer2}=\{h, E, v, CTE, \{T_{ref}, T_{ref}\}, \epsilon_g, k, \{T_{cur}, T_{cur}\}\}$$

$$\text{Multilayer}=\{\text{Layer1, Layer2, ..., LayerN}\}$$

where $h$ is the layer thickness, $E$ is the elastic modulus, $v$ is the Poisson's ratio, $CTE$ is the coefficient of thermal expansion, and $T_{ref}$ is the reference temperature for that layer. It is repeated so that multilayer definitions from *LSB* can be copied: *LS-FEA* does not allow for temperature gradients to be imposed. $T_{cur}$ is the current temperature at which the analysis is conducted. $\epsilon_g$ is a growth strain (uniform in the layer), while $k$ is not used but represents the coefficient of thermal expansion (see Chapter 15). The layers are numbered from top to bottom, as shown in Figure 16.6(a).

### Geometry Definition:

The penetrating crack geometry shown in Figure 16.6 requires the following input:

$$\text{Geometry}=\{\{l_c, f\}, w_1, w_2\}$$

where $l_c$ is the layer number containing the crack, $0 < f < 1$ is the fractional distance from the bottom of that layer to the crack tip. Note that the crack tip should *not* be placed at an interface ($f = 0$ or $f = 1$), as this creates ambiguity regarding the energy release rate (see Chapter 8). The length of the penetrating crack is computed internally and provided as an output. The parameter $w_1$ defines the distance from the crack tip to the internal rollers. Depending on the loading, $w_1$ may not be used explicitly but still must be prescribed to complete the geometry. The parameter $w_2$ defines the distance from the internal rollers to the outer right edge of the specimen. Hence, the total specimen width is $W = w_1 + w_2$.

### Meshing Parameters:

The parameters used by the automatic meshing algorithm in `PenetrateFEA` are given by

$$\text{Mesh}=\{N_{arc}, N_{ray}, \{N_v, N_h\}\}$$

where $N_{arc}$ defines the number of elements along a circumference surrounding the crack tip. The crack tip mesh is a radial fan mesh that is broken into four arc segments (i.e., a pie slice every $45°$), each with $N_{arc}$ around the circumference; since the total number of elements surrounding a crack tip is $4N_{arc}$, this parameter typically can be small (8 or less). Similarly, $N_{ray}$ is the number of elements in the radial direction of the fan mesh. The parameter $N_v$ defines the number of elements used to fill in the vertical direction; in the layer with the crack, this is the number of elements spanning from the radial fan mesh box to the outer edge of that layer. For layers without cracks, $N_v$ specifies the total number of elements through the thickness of the layer. $N_h$ specifies the number of elements used to fill out the mesh in the horizontal direction.

Table 16.1 provides recommendations for meshing parameters, based on exhaustive convergence studies described in Chapter 17.

The coherent meshing procedure implies the total number of elements in the horizontal direction (outside the fan mesh region) is given by $N_{h,tot} = 2N_h + N_{ray}$. Referencing Figure 16.6(a) and starting from the left edge, this total reflects the elements the focused region, the number of elements from the focused region to the inner rollers (the first $N_h$), and the number of elements from the inner rollers to the outer right edge of the specimen (the second $N_h$).

The routine MeshMultilayerP (which is used internally in the module PenetrateFEA) produces a list of model information used to impose boundary conditions and loading. The output of the MeshMultilayerP module is the following list:

{finaljob, {lefttop, leftbot, bot, right, top}, iprops, bendpts}

where finaljob is a compound list of the final nodal locations and the element definitions. The second entry in the above list provides a set of coordinates that can be used to impose boundary conditions on various edges of the model. For instance, lefttop is a pair of $(x, y)$ points that define the left edge of the specimen above the delamination crack, leftbot is a list of end points on the left edge of the geometry below the delamination crack and so forth. The list *iprops* contains the elastic properties above and below the crack, which are used to compute crack tip parameters. The list bendpts contains the six nodes at the 'control' locations shown in Figure 16.6(a). The settings to impose various loading conditions are described next; one can easily adapt them, using the output of the meshing routine MeshMultilayerD to create additional modules (such as tensile forces applied to the multilayer).

## Loading and Boundary Conditions:

The penetration case is selected from those shown in Figure 16.6 using the input

$$load = \{case, value\}$$

where *case* indicates which of the three loading conditions is analyzed, and *value* indicates the value of the applied forces.

For the free bending case shown in Figure 16.6(b), *case* = 0: the bottom left corner of the specimen is fixed in place, while the bottom right corner is free to expand or contract in the horizontal direction. For *case* = 0, *value* is irrelevant but still must be specified.

For four-point bending shown in Figure 16.6(c), *case* = 4: the left edge of the specimen is constrained against horizontal motion, corresponding to a symmetry condition along that plane. The inner roller on the bottom is fixed to have zero displacement, while an applied force, with magnitude equal to *value*, is specified on the upper right corner of the multilayer.

For three-point bending shown in Figure 16.6(d), *case* = 3: the left edge of the specimen is constrained against horizontal motion, corresponding to a symmetry condition along that plane. The bottom of the left edge is pinned to have zero displacement, while an applied force, with magnitude equal to *value*, is specified on the upper right corner of the multilayer.

**Crack Tip Numerical Parameters:**
The numerical parameters used to compute the energy release rate are specified in the input as

$$\texttt{jkparam} = \{\tilde{r}_S, \Delta\tilde{a}, \tilde{K}^B\}$$

where $\tilde{r}_S$ is the normalized size of the shift zone given by $r_S/L_e$, where $L_e$ is the element size in the focused radial fan mesh surrounding the crack tip. $L_e$ is computed within the module, such that the user in effect specifies the number of elements included in the shifted zone. Similarly, $\Delta\tilde{a}$ is the normalized shift in crack length given by $\Delta a/L_e$. (The normalized crack advance cannot be larger than the elements in the focused region, or the elements will invert.) $\tilde{K}^B$ is not used in `PenetrateFEA`, as $K_2 = 0$ by definition because the geometry is symmetric. Note $K_1 = \sqrt{\bar{E}_i G}$, where $\bar{E}_i$ is the plane strain modulus of the layer containing the crack tip.

Appropriate values for these parameters are listed in Table 16.1 and discussed in detail in Chapter 17.

**Summary of Complete Input:**
Thus, the complete analysis of a bilayer with a penetrating crack half-way through the top layer is given by

```
Layer1= {2,200.,0.33,0.x10⁻⁶,{1020., 1020.}, 0, 25.,{20.,20.}}
Layer2= {1,200.,0.33,40.x10⁻⁶,{1020., 1020.}, 0, 2.,{20.,20.}}
MultilayerEx={Layer1,Layer2}
GeometryEx={{2, 0.5},2.,2.}
MeshparamEx = {2, 3, {5, 5}}
LoadEx={0,0}
JKparamEx = {0.5, 0.001, 0.1}
PenetrateFEA[MultilayerEx, GeometryEx, MeshparamEx, LoadEx,
  JKparamsEx]
```

Aside from graphical images, the last command produces $G_p$, the energy release rate for crack penetration; $G_c$, the energy release rate for crack channeling into the plane of the paper; $\Pi_c$, the total potential energy in the model in the cracked state; and $\Pi_i$, the total potential energy in the model in the intact state.

**Example: Penetrating Cracks in an Environmental Barrier Coating System**
In Section 15.1, the software *LayerSlayer* was utilized to study a broad range of delamination possibilities in *semi-infinite* multilayered environmental barrier coating, with Figure 15.5 showing the energy release rates for all possible delamination positions within the stack. Here, *LS-FEA* is used to compute the energy release rates for all possible *penetrating* cracks; this can be done with the module `Penetrate4pt` simply by

**Table 16.4** List of Properties Used in the Study of Penetrating Cracks in an Environmental Barrier Coating (EBC) System, Conducted with `Penetrate4pt []`. The Discussion References the Order Shown, with the YbMS Layer Referred to as the Top Layer

| Layer | Thickness | $E$, GPa | $v$ | $\alpha$, ppm/$^\circ$C |
|---|---|---|---|---|
| YbMS, $Yb_2Si_2O_5$ | 12.5, 25, 50, 75, 100 μm | 172 | 0.27 | 7.1 |
| YbDS, $Yb_2Si_2O_7$ | 100 μm | 180 | 0.27 | 3.1 |
| Si | 50 μm | 163 | 0.22 | 4.1 |
| SiC | 2 mm | 300 | 0.16 | 4.67 |

performing multiple function calls, each with a different crack location specifier. Figure 16.7 shows the crack driving force in this system for several different combinations of thickness of the top two layers, with a complete set of properties listed in Table 16.4. The temperature change is equal to 1000°C.

Note that the CTE mismatches imply that the top layer will be in tension, while the layer underneath will be in compression. Hence, the energy release rate for penetration decreases once the tip of the penetrating crack passes the interface, due to the

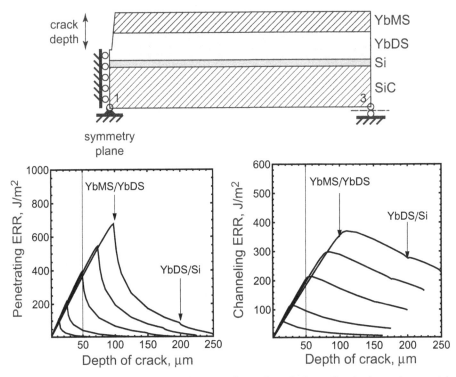

**Figure 16.7** Energy release rates for crack penetration and crack channeling in the environmental barrier coating system described in Table 16.4, as a function of crack depth; various top coat (YbMS) thicknesses are shown. The second layer from the top (YbDS) is in compression, such that the driving force decreases with crack length for crack tips within this layer.

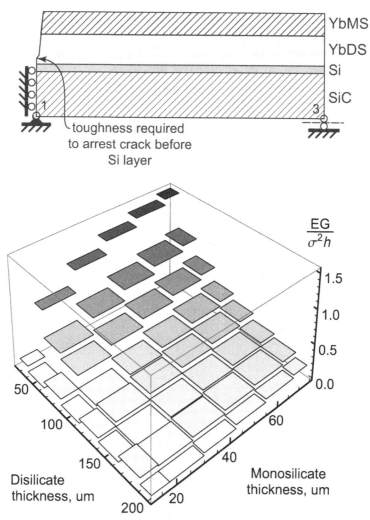

**Figure 16.8** Cracking parameter that defines whether or not penetrating cracks occur across the top two layers in the environmental barrier coating system described in Table 16.4. For values of $EG_c/\sigma^2 h$ greater than those indicated, the penetrating crack will arrest prior to penetrating both layers. The top layer (ytterbium monosilicate) is in tension, while the layer underneath (ytterbium disilicate) is in compression.

compression in the second layer. The same holds true for the energy release rate for channeling cracks that grow in the out-of-plane direction, as described in Chapter 6.

The results in Figure 16.7 can be used to infer the stability of penetrating cracks. Note that the energy release rate for penetration increases with increasing penetration into the top layer, and decreases once the tip of the crack as reached the second layer. At the interface, there is somewhat complicated and nuanced behavior, as the energy release rate strictly drops to zero as the interface is approached from the top layer. However, this behavior occurs over exceedingly small distances (crack lengths), on the order of

50 nm. Hence, it is reasonable to assume a crack penetrating the top layer will jump the interface.

The energy release rate for penetration decreases as the crack tip penetrates the second layer due to the compressive stresses in that layer. A key practical question is whether or not the crack arrests prior to the third layer, which would imply the system would exhibit greater resistance to oxidation. The answer to this central question of whether the penetrating crack arrests in the second layer is a function of the relative thickness of those layers. This is clearly illustrated in Figure 16.8, which plots a normalized energy release rate for a penetrating crack whose tip is just above the YbDS/Si interface as a function of the thickness of the two top layers. In the normalization, $h$ refers to the top layer thickness and $\sigma$ to the stress in the top layer assuming plane strain deformation without bending in the layers.

One can interpret the vertical axis value as the toughness required of the YbDS layer to arrest the penetrating crack prior to exposure of the Si layer, as a function of the stress in the top layer at its thickness. Very thick top coats of YbMS high tensile stress require either very large toughness values (if the YbDS layer is very thin) or very thick YbDS layers (if the toughness is very small). Conversely, if the YbMS layer is thin, and the YbDS layer is thick, very low toughness values are needed to prevent penetration of the top two layers.

# 17  Convergence and Benchmarks with *LayerSlayer FEA*

As described in the previous chapter, the computation of energy release rates and stress intensity factors using the finite element method requires specification of various numerical parameters. These parameters relate to both the FE model itself (basically, the mesh) and the calculations performed during postprocessing of the FE results (basically, the calculation of the stiffness derivitive). In this chapter, the powerful automation of the *LS-FEA* framework is exploited to conduct extensive parametric studies, which illustrate the impact of these parameters on computational accuracy. The outcomes of these studies are used as the basis to establish recommended practices and default values for the associated numerical parameters, such as those in Table 16.1.

## 17.1  Convergence of Crack Tip Parameters

Numerical convergence in the present context is conceptually very simple; when the values of interest no longer change upon increases in mesh density (i.e., decreases in element size, often referred to as refinement), the model is said to have converged. In practice and especially for fracture problems, convergence can be more nuanced, as the behavior of different outputs will converge at different levels of mesh density. For example, when considering the potential energy of the system, and in many cases the energy release rate, convergence is achieved for much coarser meshes than when considering the crack tip stress intensity factors.

Such differences can create headaches if the spatial distribution of elements changes with refinement. For example, this would occur if the number of elements inside the focused fan region were increased, without changing the number of elements outside the focused region. Strictly speaking, mesh refinement to test convergence should involve a uniform increase in mesh density, such as would be achieved by splitting every element in the model into four smaller elements. This can be prohibitively expensive, particularly for components that require dense meshes in multiple locations: for example, bending of a multilayer with one particularly thin layer.

In this chapter, we consider mesh refinements associated with increasing the number of elements around the crack tip in the focused region and in the fill regions. Practically speaking, the crack tip parameters become independent of the mesh upon increase of all or even some of these parameters. However, the above comments should be kept in mind when attempting to impose stringent convergence criteria – that is, very very small changes in numerical values. If highly accurate solutions are sought (with differences much smaller than 1%), one should be aware that such small differences can arise from element distributions in a convergence studies that do not impose uniform increases in mesh density.

As a practical matter, convergence is defined here by identifying meshing parameters for which the crack tip parameters do not change by more than 1–3% upon further mesh refinements. This threshold is motivated by the small variability caused by nonuniform mesh refinements, and by variability that is associated with round-off errors during post-processing that can require a great deal of sophisticated code to eliminate. It is worth noting that units should be chosen such that the dimensions of the layers are on the order of unity, such that meshing headaches associated with small element sizes are avoided. (See discussion in Chapter 16.) As will be illustrated, convergence to within 1% is generally easily achieved with the present implementation.

In addition to mesh refinement, the numerical parameters used to compute crack parameters also play a role in convergence. As described in Section 16.2, these are the magnitude of the virtual crack extension ($\Delta a$), the size of the mesh region shifted in the extension ($r_S$) and the magnitude of the imposed stress intensity factor ($K_1^B$). Extensive numerical studies show that the results are completely insensitive to $K_1^B$ when $0.01 < K_1^B < 1$, and as such, this parameter will not be discussed further. With either $\Delta a$ or $r_S$, there are both lower and upper limits to consider: one cannot simply decrease their magnitudes and expect convergence. As will be illustrated, the errors associated with computing crack tip parameters are far more sensitive to these parameters than to the number of elements, provided the mesh is not extremely coarse.

Recognizing that the stiffness matrix is a function of crack length, $\Delta a$ must be kept small enough such that one accurately computes the derivative of the stiffness matrix for a given crack length. Further, it must be small enough such that shifting the nodes in a fixed region does not dramatically distort the ring of elements that define the boundary of the shift. This creates a bit of a complication, since it implies that the upper limit on $\Delta a$ is tied to the details of the mesh (specifically, $N_{ray}$ the number of elements along a ray in a focused fan mesh, and the physical size of the focused mesh). On the other hand, if $\Delta a$ is *too* small, round-off errors in the computation of the derivative of the stiffness matrix will introduce variability into the results.

These effects are illustrated in the sections that follow, using three benchmark cases. For each case, the influence of crack tip parameters $r_S$ and $\Delta a$ is shown for three different meshes.

## 17.2     Benchmark: Delamination during Four-Point Bending of a Bilayer

This section considers delamination in a bilayer of equal thickness and with no elastic or thermal mismatch, subject to four-point bending. Results are presented for the two different crack lengths and specimen sizes, shown to scale at the top of Figure 17.1. For the long crack, $a/h = 30$, where $a$ is the crack length $h$ is the half-thickness of the specimen; for the short crack, $a/h = 10$. The parameters defining the model input for

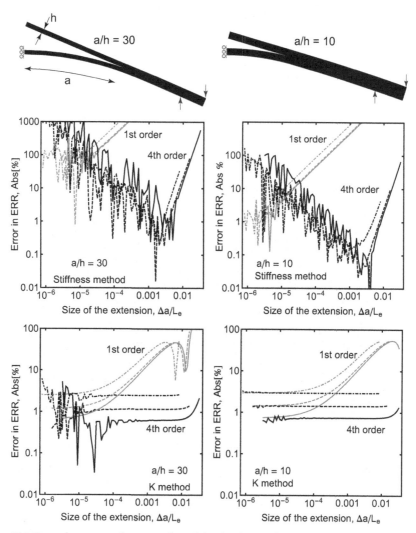

**Figure 17.1** Errors in energy release rate for a delamination crack in a uniform bilayer subjected to four-point bending, as a function of numerical parameters used in the calculation. Results from three meshes are shown: coarse (dot-dashed), medium (dashed) and fine (solid). The first-order method refers to a first-order estimate of the stiffness matrix derivative; the fourth-order method refers to a fourth-order estimate of the stiffness matrix derivative. The 'K method' refers to computing the energy release rate from the computed stress intensity factors, which minimizes round-off errors.

the module DelamFEA are provided in Table 17.1. In many respects, this problem is one of the more challenging of the benchmarks considered in this chapter, because the large aspect ratio of the specimen creates a stringent meshing challenge.

Figures 17.1 and 17.2 show errors in the predicted energy release rate and stress intensity factors, respectively, as a function of several numerical parameters used in their computation. The errors are computed using the analytical result given as (4.22) in

**Table 17.1** Parameters for the Convergence Study of a Bilayer with Equal Thickness Layers Subjected to Four-Point Bending, without Thermal or Elastic Mismatch

| Layer | $h$ | $E$ | $v$ | $\alpha$ | $T_{ref}$ | $T$ |
|---|---|---|---|---|---|---|
| 1 (bottom) | 10 | $200 \times 10^3$ | 0.33 | 0 | 1020 | 20 |
| 2 (top) | 10 | $200 \times 10^3$ | 0.33 | 0 | 1020 | 20 |

| Crack length, $a$ | Interior span, $w_1$ | Exterior span, $w_2$ |
|---|---|---|
| 300 ($a/h = 30$) | 125 | 50 |
| 100 ($a/h = 10$) | 125 | 50 |

| Mesh | $N_{arc}$ | $N_{ray}$ | $N_v$ | $N_h$ | $L_e$ |
|---|---|---|---|---|---|
| coarse | 4* | 4 | 2 | 4 | 1.25 |
| medium | 4* | 8 | 4 | 8 | 0.625 |
| fine | 4* | 16 | 8 | 16 | 0.3125 |

* Total number of elements around circumference of crack is $4N_{arc}$.
Results for the study are shown in Figures 17.1 and 17.2.

Chapter 4, which provides the energy release rate and the mode mix. The phase angle is $\psi = 40.89°$, which corresponds to $K_2/K_1 = 0.866$.

In Figure 17.1, one can see that the size of the extension $\Delta a$ plays a pivotal role; for very small extensions ($\Delta a/L_e < 10^{-3}$), round-off errors dominate the response. The first-order method refers to the use of the first-order estimate for the stiffness matrix derivative given as (16.8). The fourth-order method uses the higher-order estimate for this same derivative given as (16.9). For large extensions, the implicit error in the derivative computation dominates the response. One can see that it can be quite difficult to identify an extension that produces acceptable errors. The results shown in Figure 17.1 tell a cautionary tale; using very small crack extensions appears to be just fine for short cracks (with errors on the order of 1%) but is clearly problematic for longer cracks.

A second method, apparently less susceptible to round-off errors, is also shown in Figure 17.1; here, the 'K method' refers to computing the energy release rate from the stress intensity factors. That is, the energy release rate is estimated as before, but then recomputed once the stress intensity factors have been extracted. Presumably, the process of adding the pure mode 1 (or mode 2) crack tip fields to that computed for the given problem washes out round-off errors present in the ERR computation. It is worth noting, however, that the large aspect ratio of the long crack specimen still makes round-off errors worse (as compared to the short crack specimen).

Figure 17.2 illustrates stress intensity factors as a function of crack extension. The results appear less susceptible to round-off error for small extensions, with the implicit error of the derivative dominating for large extensions. While the fourth-order stiffness derivative estimate produces more consistent results for large extensions, the first-order

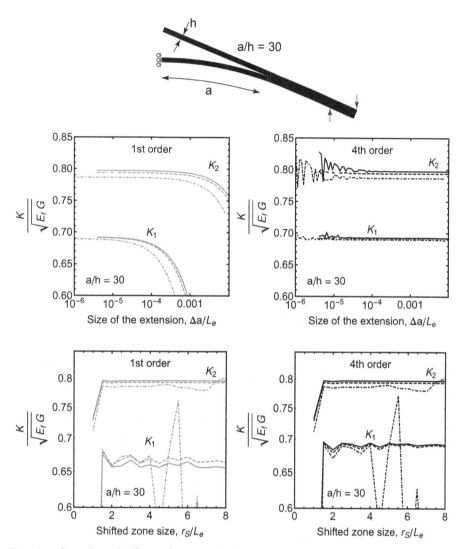

**Figure 17.2** Stress intensity factors for a delamination crack in a uniform bilayer subjected to four-point bending, as a function of numerical parameters used in the calculation. Results are shown for a long crack (top) and a short crack (bottom). In each case, three meshes are shown: coarse (dot-dashed), medium (dashed) and fine (solid). The first-order method refers to a first-order estimate of the stiffness matrix derivative; the fourth-order method refers to a fourth-order estimate of the stiffness matrix derivative.

method is much faster (requiring only one computation of the stiffness matrix associated with a single advance). The results shown in Figure 17.2 show that the choice of the size of the shifted zone, $r_S$, is relatively immaterial, *provided that the shifted zone includes at least two elements ($r_S/L_e > 2$) and does not extend past the region with the focused fan mesh.* Note that in the bottom two panels of Figure 17.2, the fan region of coarse mesh (represented by dot-dashed lines) has only four elements, such that problems arise for shifted zone sizes larger than $r_S/L_e \geq 4$.

**Table 17.2** Parameters for the Convergence Study of Delamination of a Thin Film on a Thick Substrate

| Layer | $h$ | $E$ | $v$ | $\alpha$ | $T_{ref}$ | $T$ |
|---|---|---|---|---|---|---|
| 1 (bottom) | 1000 | $200 \times 10^3$ | 0.33 | 0 | 1020 | 20 |
| 2 (top) | 10 | $20 \times 10^3$ | 0.33 | $10 \times 10^{-6}$ | 1020 | 20 |

| Crack length, $a$ | Interior half-span, $w_1$ | Exterior span, $w_2$ |
|---|---|---|
| 300 ($a/h = 30$) | 500 | 500 |

| Mesh | $N_{arc}$ | $N_{ray}$ | $N_v$ | $N_h$ | $L_e$ |
|---|---|---|---|---|---|
| coarse | 4* | 4 | 2 | 4 | 1.25 |
| medium | 4* | 8 | 4 | 8 | 0.625 |
| fine | 4* | 16 | 8 | 16 | 0.3125 |

* Total number of elements around circumference of crack is $4N_{arc}$. Results from the study are shown in Figures 17.3 and 17.4.

## 17.3 Benchmark: Delamination of Thermally Stressed Thin Film on a Thick Substrate

In this section, we examine the numerical behaviors of a thin film delaminating from a thick substrate, with both thermal and elastic mismatch and no mechanical loads. The parameters used to define the model are listed in Table 17.2. Though not shown, results for short or longer cracks show similar behaviors; the results presented here are essentially those associated with steady-state delamination of a thin film from a semi-infinite substrate. For the present problem, $\alpha_D = -0.818$ and $\beta_D = \alpha_D/4 = -0.205$ since $v_1 = v_2 = 1/3$. The relevant analytical solution is given in Section 4.3; the stress intensity factors and energy release rate are given by (4.11) with $P = \sigma h$ and $M = 0$, where $\sigma = E_1(\alpha_2 - \alpha_1)\Delta T/(1 - v_1)$.

Figure 17.3 plots the error in the computed energy release rate as a function of crack tip parameters, using the solution for the semi-infinite substrate in Section 4.3 as reference. Interestingly, the presence of thermal stresses seems to mitigate round-off errors associated with small crack extensions; for example, compare Figure 17.3 with the similar plots in Figure 17.1. Based on a wide range of problems analyzed with *LS-FEA*, this appears to be a general result: thermal loading is much less susceptible to round-off errors. The reason for this is not entirely clear, but it is likely related to the fact that thermal stresses introduce the equivalent of body force terms at all nodes in the mesh, such that round-off errors associated with nominally zero entries in the stiffness matrix are mitigated. Figure 17.3 illustrates that direct computation of the ERR and indirect computation of the ERR through stress intensity factors produce very similar results, particularly for refined meshes.

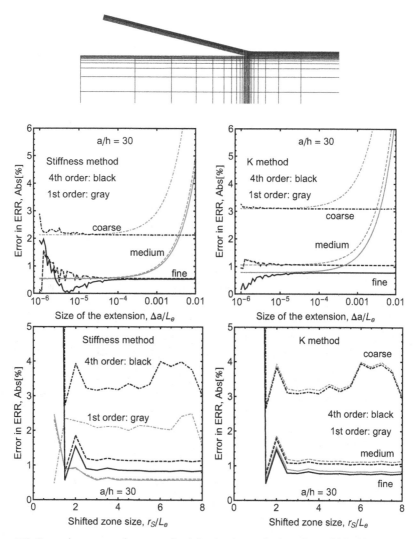

**Figure 17.3** Errors in energy release rate for delamination at the interface of thin film on a thick substrate subjected to thermal and elastic mismatch, as a function of numerical parameters used in the calculation. Results are shown for a long crack near the steady-state limit. In each case, three meshes are shown: coarse (dot-dashed), medium (dashed) and fine (solid). The first-order method refers to a first-order estimate of the stiffness matrix derivative; the fourth-order method refers to a fourth-order estimate of the stiffness matrix derivative.

The results for stress intensity factors in Figure 17.4 illustrate that robust results are obtained for sufficiently small crack extensions and shifted zone sizes that encompass more than two elements and are restricted to fall within the focused fan region. For reference, the solution from Chapter 4 for the semi-infinite substrate (with $\omega \approx 56°$) predicts $K_1/\sqrt{E_1 G} = 0.988$ and $K_2/\sqrt{E_1 G} = 1.070$ for the properties listed in Table 17.2. Clearly, the values computed here are in excellent quantitative agreement for the finest mesh.

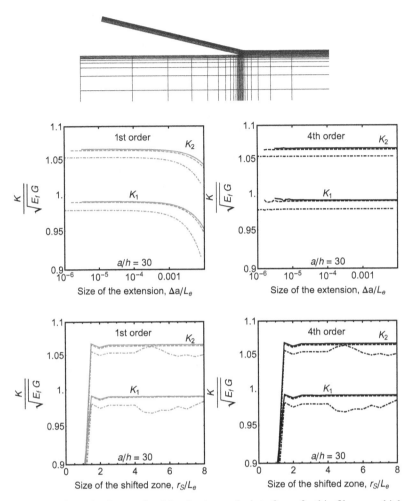

**Figure 17.4** Stress intensity factors for delamination at the interface of a thin film on a thick substrate subject to thermal and elastic mismatch, as a function of numerical parameters used in the calculation. Results are shown for a long crack (top) and a short crack (bottom). In each case, three meshes are shown: coarse (dot-dashed), medium (dashed) and fine (solid). The first-order method refers to a first-order estimate of the stiffness matrix derivative; the fourth-order method refers to a fourth-order estimate of the stiffness matrix derivative.

## 17.4  Benchmark: Penetrating Crack in a Homogeneous Specimen Subject to Three-Point Bending

In the final example of benchmarking and convergence, we consider a penetrating edge crack in a homogeneous specimen subjected to three-point bending, as shown in Figure 17.5. The specimen's aspect ratios is $s/b = 4$, where $s$ is the distance between the outer load points and $b$ is the total specimen height. The parameters that define the model are listed in Table 17.3. Results are shown for both a shallow crack ($\simeq 1\%$ of the total specimen height) and a deep crack (50% of the total specimen height). The

**Table 17.3** Parameters for the Study of a Penetrating Crack in a Homogenous Three-Point Bend Specimen. The Distance Between the Outer Load Points Divided by the Total Specimen Height is $s/h_{tot} = 4$

| Layer | $h$ | $E$ | $v$ | $\alpha$ | $T_{ref}$ | $T$ |
|---|---|---|---|---|---|---|
| 1 (bottom) | 100 | $100 \times 10^3$ | 0.33 | $3 \times 10^{-6}$ | 1020 | 20 |
| 2 (top) | 1, 100 | $100 \times 10^3$ | 0.33 | $3 \times 10^{-6}$ | 1020 | 20 |

| Crack depth, $d$ | Interior half span, $w_1$ | Exterior span, $w_2$ |
|---|---|---|
| 1 ($d/h_{tot} = 0.01$) | 100 | 100 |
| 100 ($d/h_{tot} = 0.5$) | 100 | 100 |

| Mesh | $N_{arc}$ | $N_{ray}$ | $N_v$ | $N_h$ | $L_e$ |
|---|---|---|---|---|---|
| coarse | 4* | 8 | 4 | 4 | 1.25 |
| medium | 4* | 12 | 8 | 8 | 0.625 |
| fine | 4* | 16 | 12 | 16 | 0.3125 |

* Total number of elements around circumference of crack is $2N_{arc}$. The results of the study are shown in Figures 17.5 and 17.6.

basis for computing the error in the FEA results is the formula provided by Tada et al. (2000).

The results shown in Figures 17.5 and 17.6 illustrate that these cases produced the smallest errors with very little evidence of round-off errors, with the possible exception of the fourth-order stiffness derivative method applied to the shallow crack. One observes very systematic convergence, with the results asymptotically approaching a mesh-independent result for fine meshes. The reason for the lack of round-off error is likely tied to the fact that the specimens' aspect ratios are relatively modest; this implies that the size of the focused fan mesh around the crack is a significant fraction of the entire model, especially for the deep crack. By contrast, in the four-point bending specimen shown in Figure 17.1 for $a/h = 30$, the focused fan mesh surrounding the crack tip is a very small fraction of the whole model. Generally speaking, round-off errors are most worrisome for specimens in which the crack tip mesh is a small fraction of the specimen.

## 17.5 Recommended Numerical Practices

Systematic convergence studies are always good practice, and *LS-FEA* is set up specifically to make such studies extremely easy to set up and fast to execute. It is worth noting the entire set of computations in the numerical studies in Figures 17.1 through 17.6 involves about $10^4$ finite element analyses, yet requires computing times of only

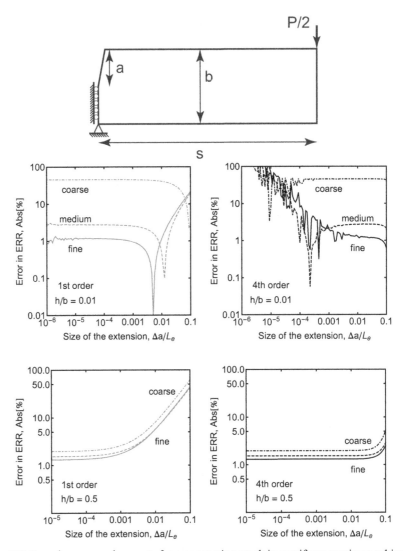

**Figure 17.5** Error in energy release rate for a penetrating crack in a uniform specimen subject to three-point bending, as a function of the crack extension used to compute the derivative of the stiffness matrix. Results are shown for a short crack (top) and deep crack (bottom). In each case, three meshes are shown: coarse (dot-dashed), medium (dashed) and fine (solid). The first-order method refers to a first-order estimate of the stiffness matrix derivative; the fourth-order method refers to a fourth-order estimate of the stiffness matrix derivative.

1–2 hours on a Macintosh MacBook Pro laptop, circa 2016. As such, this is a very small barrier to performing the type of systematic studies covered in Sections 17.2 through 17.4.

Nevertheless, it is worthwhile to summarize a 'best practice', which provides guidance for novice users and provides reasonable default parameters that in most cases, are sufficient to ensure convergence. A summary of recommended defaults is included in

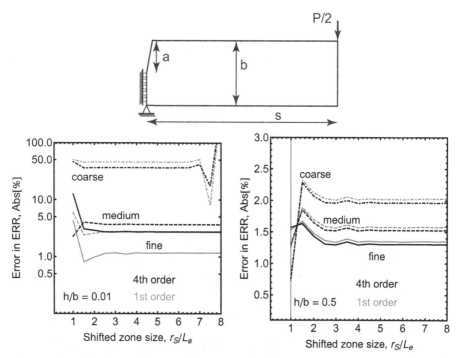

**Figure 17.6** Error in energy release rate for a penetrating crack in a uniform specimen subject to three-point bending, as a function of the size of the shifted zone surrounding the crack tip. Results are shown for a short crack (top) and deep crack (bottom). In each case, three meshes are shown: coarse (dot-dashed), medium (dashed) and fine (solid). The first-order method refers to a first-order estimate of the stiffness matrix derivative; the fourth-order method refers to a fourth-order estimate of the stiffness matrix derivative.

Table 17.4, with rationales based on Figures 17.1 through 17.6 and articulated below. It should be noted that Table 17.4 is identical to Table 16.1, which is repeated here for convenience.

**Size of the focused fan region:** The provided modules described in Sections 16.4 and 16.5 set the size of the region to be one-half of smallest of the following two quantities: the smallest layer thickness that is involved in the crack, or the crack length. That is, if the crack length is smaller than the thinnest layer participating in the cracking, the focused region is set to be one-half of the crack length. For very long cracks (such as occur in delamination studies), the focused fan region is set to be one-half of the thinnest layer participating in the cracking. This strategy allows for a single meshing strategy regardless of the number of layers, the relative thickness of the layers and the crack length.

**Mesh density:** In the modules described in Section 16.4 and 16.5, the mesh is dictated by the number of elements in a circumferential direction around the crack tip ($N_{arc}$), the number of elements along a ray emanating from the crack time ($N_{ray}$), and the number of elements in the vertical and horiontal directions that fill in the remaining mesh ($N_h$ and $N_v$). Choosing $N_{arc} = 4$ for the number of elements surrounding the crack tip

**Table 17.4** Recommendations for Numerical Parameters for Delamination and Penetrating Cracks. In the Prepackaged Modules, $L_e = \ell/N_{ray}$, Where $\ell$ is the Minimum of Either the Crack Length or the Minimum Thickness of the Layers Participating in the Crack

| Feature | Parameter | Suggested Minimum | Suggested Default | Suggested Maximum |
|---|---|---|---|---|
| Thickness | $h$ | $O[0.1]$ | $O[1]$ | $O[10^3]$ |
| Circumferential elements | $N_{arc}$ | 2 | 4* | 8 |
| Ray elements | $N_{ray}$ | 4 | 12 | 16 |
| Vertical fill elements | $N_v$ | 5 | 10 | $20^+$ |
| Horizontal fill elements | $N_h$ | 5 | 10 | $20^+$ |
| Crack extension | $\Delta a$ | $10^{-6}L_e$ | $10^{-5}L_e$ | $10^{-3}L_e$ |
| Shift zone size | $r_S$ | $2L_e$ | $4L_e$ | $(N_{ray} - 1)L_e$ |

* Total number of elements around circumference of delamination crack is $8N_{arc}$, while that around the penetrating crack is $4N_{arc}$.

$^+$ These can and should be adjusted based on specimen aspect ratio; to date, the only aspect strained by using large numbers is one's patience with computation times. The basis for these recommendations is discussed in detail in Chapter 17.

provides good convergence while avoiding complications associated with high aspect ratio elements at the crack tip (associated with more elements in circumferential direction) that can cause problems if taken to the extreme. Similarly, choosing $N_{ray} = 8$–16 typically provides adequate convergence and controls balances the aspect ratio for elements in the focused radial fan region. The number of elements in the horizontal and vertical 'fill' directions should be between $10 < N_{v,h} < 20$, with either parameter increased if the specimen has a large aspect ratio in the relevant direction. Setting these mesh parameters in this range typically leads to models with ~2,500 elements and ~7,500 nodes, or ~$10^4$ degrees of freedom. For such models, computation times are typically on the order of a couple seconds, which makes trouble shooting and/or parameter studies tolerable.

**Crack tip parameters:** As described in Section 16.2, the parameters to be specified include the size of the virtual crack extension ($\Delta a$), the size of the region to be shifted ($r_S$) and the magnitude of the superimposed stress intensity factor ($K_1^B$). The convergence studies in Sections 17.2–17.4 strongly support using the first-order approximation for stiffness matrix derivatives, with the 'K method' for delamination problems that estimate the ERR using $K_1$ and $K_2$. Recall that this method for computing the ERR mitigates round errors that are often present when computing the ERR directly using the stiffness derivative method. Choosing $\Delta a = 10^{-5}L_e$, where $L_e$ is the size of the element in the focused region, typically provides convergent results (i.e., those independent of $\Delta a$) and minimal round-off error.

A very effective check that the results are acceptable is to compare the ERR estimates from the 'K method' and the direct computation using the stiffness derivative method; that they are in close agreement indicates that round-off errors have been avoided and the answers are independent of numerical parameters. It should be noted that this comparison is not possible using the module provided for penetrating cracks, as the symmetric mesh does not allow application of the default module to compute $K_1$ (recall that $K_2 = 0$).

The size of the shifted region, $r_S$, can then be set to be one-half of size of the focused fan region: with the element defaults suggested above, this implies the shifted region will involve the fourth to eighth ring of elements around the crack tip. As mentioned earlier, the magnitude of the superimposed $K$ field, $K_1^B$ or $K_2^B$, has almost no effect when it is in the range $0.01 < K_1^B < 1$; all computations to date suggest this parameter can be set 0.1 without any effect on the results.

# Appendix: Asymptotic Crack Tip Displacement Fields for an Interface Crack

The complete asymptotic displacement fields near the crack tip are useful for numerical computation of the mode mix (see Section 16.2). These displacements are written below in terms of polar coordinates $(r, \theta)$, but expressed in terms of Cartesian displacements in the $(x_1, x_2)$ directions (as is needed in the finite element calculation):

$$G_i = \frac{E_i}{2(1 + v_i)}$$

$$\mu_i = 3 - 4v_i$$

$$\epsilon = \frac{1}{2\pi} \log \frac{\frac{1}{G_1} + \frac{1}{G_2}}{\frac{\mu_2}{G_2} + \frac{\mu_1}{G_1}} \tag{A.1}$$

$$\Psi = \epsilon \log r + \frac{\theta}{2}$$

$$d_1 = e^{-\epsilon(\pi - \theta)}; \qquad d_2 = e^{\epsilon(\pi + \theta)} \tag{A.2}$$

$$\gamma_i = \mu_i d_i - \frac{1}{d_i}; \qquad \gamma_i' = \mu_i d_i + \frac{1}{d_i} \tag{A.3}$$

$$\beta = \frac{[\cos(\epsilon \log r) + 2\epsilon \sin(\epsilon \log r)]}{\frac{1}{2} + 2\epsilon^2}$$

$$\beta' = \frac{[\cos(\epsilon \log r) - 2\epsilon \sin(\epsilon \log r)]}{\frac{1}{2} + 2\epsilon^2} \tag{A.4}$$

$$D_i = \beta \gamma_i \cos \frac{\theta}{2} + \beta' \gamma_i' \sin \frac{\theta}{2}$$

$$C_i = \beta' \gamma_i \cos \frac{\theta}{2} - \beta \gamma_i' \sin \frac{\theta}{2} \tag{A.5}$$

$$(f_{I1})_k = D_k + 2d_k \sin \theta \sin \Psi$$

$$(f_{I2})_k = -C_k - 2d_k \sin \theta \cos \Psi$$

$$(f_{III})_k = -C_k + 2d_k \sin \theta \cos \Psi$$

$$(f_{II2})_k = -D_k + 2d_k \sin \theta \sin \Psi \tag{A.6}$$

Using the above definitions, the mode I displacements (with $(u, v)$ being the displacements in the $(x, y)$ directions) are given by

$$u_I^{(k)} = \frac{K_1}{2G_1} \sqrt{\frac{r}{2\pi}} \frac{e^{\pi\epsilon}}{1 + e^{2\pi\epsilon}} (f_{I1})_k$$

$$v_I^{(k)} = \frac{K_1}{2G_1} \sqrt{\frac{r}{2\pi}} \frac{e^{\pi\epsilon}}{1 + e^{2\pi\epsilon}} (f_{I2})_k$$

(A.7)

where $k$ indicates the material in which the displacements are sought: $k = 1$ for $0 \leq \theta \leq \pi$ (the top material) and $k = 2$ for $0 \geq \theta \geq -\pi$ (the bottom material). Similarly, the mode II displacements are given by

$$u_{II}^{(k)} = \frac{K_2}{2G_1} \sqrt{\frac{r}{2\pi}} \frac{e^{\pi\epsilon}}{1 + e^{2\pi\epsilon}} (f_{II1})_k$$

$$v_{II}^{(k)} = \frac{K_2}{2G_1} \sqrt{\frac{r}{2\pi}} \frac{e^{\pi\epsilon}}{1 + e^{2\pi\epsilon}} (f_{II2})_k$$

(A.8)

# References

Ambrico, J. M. & Begley, M. R. (2002) 'The role of initial flaw size, elastic compliance and plasticity in channel cracking of thin films', *Thin Solid Films* **419**(1), 144–153.

Ambrico, J. M. & Begley, M. R. (2003) 'The role of flaw geometry in thin film delamination from two-dimensional interface flaws along free edges', *Engineering Fracture Mechanics* **70**(13), 1721–1736.

Anderson, T. L. & Anderson, T. L. (2005) *Fracture Mechanics: Fundamentals and Applications*, CRC Press.

Audoly, B. (1999) 'Stability of straight delamination blisters', *Physical Review Letters* **83**(20): 4124.

Bagchi, A. & Evans, A. G. (1996) 'Measurements of the debond energy for thin metallization lines on dielectrics', *Thin Solid Films* **286**(1), 203–212.

Balint, D. S. & Hutchinson, J. W. (2001) 'Mode 2 edge delamination of compressed thin films', *Journal of Applied Mechanics* **68**(5), 725–730.

Bank-Sills, L., Travitzky, N., Ashkenazi, D. & Eliasi, R. (1999) 'A methodology for measuring interface fracture properties of composite materials', *International Journal of Fracture* **99**(3), 143–161.

Banks-Sills, L., Freed, Y., Eliasi, R. & Fourman, V. (2006) 'Fracture toughness of the $+45/-45$ interface of a laminate composite', *International Journal of Fracture* **141**(1–2), 195–210.

Bathe, K.-J. (1997) *Finite Element Procedures*, Prentice-Hall, Upper Saddle River, NJ.

Begley, M. R. & Ambrico, J. M. (2001) 'Delamination of thin films from two-dimensional interface flaws at corners and edges', *International Journal of Fracture* **112**(3), 205–222.

Begley, M. R. & Bart-Smith, H. (2005) 'The electro-mechanical response of highly compliant substrates and thin stiff films with periodic cracks', *International Journal of Solids and Structures* **42**(18), 5259–5273.

Begley, M. R., Collino, R. R., Israelachvili, J. N. & McMeeking, R. M. (2014) 'Peeling of a tape with frictional sliding and large deformations', *Journal of the Mechanics and Physics of Solids* **61**, 1265–1279.

Begley, M. R., Mumm, D. R., Evans, A. G. & Hutchinson, J. W. (2000) 'Analysis of a wedge impression test for measuring the interface toughness between films/coatings and ductile substrates', *Acta Materialia* **48**(12), 3211–3220.

Beuth, J. L. (1992) 'Cracking of thin bonded films in residual tension', *International Journal of Solids and Structures* **29**, 1657–1675.

Beuth, J. L. & Klingbeil, N. W. (1996) 'Cracking of thin films bonded to elastic-plastic substrates', *Journal of the Mechanics and Physics of Solids* **44**, 1411–1428.

Bilby, B. A., Cardew, G. E. & Howard, I. C. (1977) 'Stress intensity factors at the tips of kinked and forked cracks', *Fracture* **3**, 197–200.

Bower, A. F. (2010) *Applied Mechanics of Solids*, CRC Press, Boca Raton, FL.

Broek, D. (2012) *Elementary Engineering Fracture Mechanics*, Springer Science & Business Media, Martinus Nijhoff, The Hague.

Buchanan, G. (1994) *Schaum's Outline of Finite Element Analysis*, McGraw-Hill Education, New York.

Cao, H. C. & Evans, A. G. (1989) 'An experimental study of the fracture resistance of bimaterial interfaces', *Mechanics of Materials* **7**(4), 295–304.

Charalambides, P. G., Cao, H. C., Lund, J. & Evans, A. G. (1990) 'Development of a test method for measuring the mixed mode fracture resistance of bimaterial interfaces', *Mechanics of Materials* **8**(4), 269–283.

Charalambides, P. G., Lund, J., Evans, A. G. & McMeeking, R. M. (1989) 'A test specimen for determining the fracture resistance of bimaterial interfaces', *Journal of Applied Mechanics* **56**(1), 77–82.

Chen, J. & Bull, S. J. (2010) 'Approaches to investigate delamination and interfacial toughness in coated systems: An overview', *Journal of Physics D: Applied Physics* **44**(3), 034001.

Cherepanov, G. P. (1962) 'The stress state in a heterogeneous plate with slits', *Izvestia AN SSSR, OTN, Mekhan. i Mashin* **1**, 131–137.

Collino, R. R., Philips, N., Rossol, M., McMeeking, R. & Begley, M. R. (2014) 'Detachment of compliant thin films adhered to stiff substrates via Van der Waals interactions', *Proceedings of the Royal Society: Interface* **11**, 20140453.

Comninou, M. (1977) 'The interface crack', *Journal of Applied Mechanics* **44**(4), 631–636.

Cotterell, B. & Chen, Z. (2000) 'Buckling and cracking of thin films on compliant substrates under compression', *International Journal of Fracture* **104**(2), 169–179.

Cotterell, B. & Rice, J. R. (1980) 'Slightly curved or kinked cracks', *International Journal of Fracture* **16**(2), 155–169.

Dauskardt, R. H., Lane, M., Ma, Q. & Krishna, N. (1998) 'Adhesion and debonding of multi-layer thin film structures', *Engineering Fracture Mechanics* **61**(1), 141–162.

Delannay, F. & Warren, P. (1991) 'On crack interaction and crack density in strain-induced cracking of brittle films on ductile substrates', *Acta Metallurgica et Materialia* **39**(6), 1061–1072.

Douville, N. J., Li, Z., Takayama, S. & Thouless, M. D. (2011) 'Fracture of metal coated elastomers', *Soft Matter* **7**(14), 6493–6500.

Dundurs, J. (1969) 'Discussion: Edge-bonded dissimilar orthogonal elastic wedges under normal and shear loading (Bogy, D.B., 1968, ASME J. Appl. Mech., 35, pp. 460–466)', *Journal of Applied Mechanics* **36**(3), 650–652.

Dvorak, G. J. & Laws, N. (1986) 'Analysis of first ply failure in composite laminates', *Engineering Fracture Mechanics* **25**(5), 763–770.

Eberl, C., Wang, X., Gianola, D. S., Nguyen, T. D., He, M. Y., Evans, A. G. & Hemker, K. J. (2011) 'In situ measurement of the toughness of the interface between a thermal barrier coating and a Ni alloy', *Journal of the American Ceramic Society* **94**(s1), 120–127.

England, A. H. (1965) 'A crack between dissimilar media', *Journal of Applied Mechanics* **32**(2), 400–402.

Erdogan, F. O. (1965) 'Stress distribution in bonded dissimilar materials with cracks', *Journal of Applied Mechanics* **32**(2), 403–410.

Erdogan, F. & Sih, G. (1963) 'On the crack extension in plates under plane loading and transverse shear', *Journal of Basic Engineering* **85**(4), 519–525.

Evans, A. G., Drory, M. D. & Hu, M. S. (1988) 'The cracking and decohesion of thin films', *Journal of Materials Research* **3**(05), 1043–1049.

Evans, A. G. & Hutchinson, J. W. (1989) 'Effects of non-planarity on the mixed mode fracture resistance of bimaterial interfaces', *Acta Metallurgica* **37**(3), 909–916.

Evans, A. G., Rühle, M., Dalgleish, B. J. & Charalambides, P. G. (1990) 'The fracture energy of bimaterial interfaces', *Metallurgical Transactions A* **21**(9), 2419–2429.

Evans, A. & Hutchinson, J. (2007) 'The mechanics of coating delamination in thermal gradients', *Surface and Coatings Technology* **201**(18), 7905–7916.

Faou, J.-Y., Parry, G., Grachev, S. & Barthel, E. (2012) 'How does adhesion induce the formation of telephone cord buckles?', *Physical Review Letters* **108**(11), 116102.

Faou, J.-Y., Parry, G., Grachev, S. & Barthel, E. (2015) 'Telephone cord buckles: A relation between wavelength and adhesion', *Journal of the Mechanics and Physics of Solids* **75**, 93–103.

Florence, A. L. & Goodier, J. N. (1960) 'Thermal stresses due to disturbance of uniform heat flow by an insulated ovaloid hole', *Journal of Applied Mechanics* **27**(4), 635–639.

Freund, L. B. & Suresh, S. (2003) *Thin Film Materials*, Cambridge University Press, Cambridge.

Gille, G. (1985) 'Strength of thin films and coatings', in E. Kaldis, ed., *Current Topics in Materials Science*, Vol. 12, North Holland, Amsterdam.

Guo, S., Mumm, D., Karlsson, A. M. & Kagawa, Y. (2005) 'Measurement of interfacial shear mechanical properties in thermal barrier coating systems by a barb pullout method', *Scripta Materialia* **53**(9), 1043–1048.

Hayashi, K. & Nemat-Nasser, S. (1981) 'Energy-release rate and crack kinking under combined loading', *Journal of Applied Mechanics* **48**(3), 520–524.

He, M.-Y., Bartlett, A., Evans, A. G. & Hutchinson, J. W. (1991) 'Kinking of a crack out of an interface: Role of in-plane stress', *Journal of the American Ceramic Society* **74**(4), 767–771.

He, M. Y., Evans, A. G. & Hutchinson, J. W. (1994) 'Crack deflection at an interface between dissimilar elastic materials: Role of residual stresses', *International Journal of Solids and Structures* **31**(24), 3443–3455.

He, M.-Y. & Hutchinson, J. W. (1989a) 'Crack deflection at an interface between dissimilar elastic materials', *International Journal of Solids and Structures* **25**(9), 1053–1067.

He, M.-Y. & Hutchinson, J. W. (1989b) 'Kinking of a crack out of an interface', *Journal of Applied Mechanics* **56**(2), 270–278.

Ho, S. & Suo, Z. (1993) 'Tunneling cracks in constrained layers', *Journal of Applied Mechanics* **60**(4), 890–894.

Hofinger, I., Oechsner, M., Bahr, H.-A. & Swain, M. V. (1998) 'Modified four-point bending specimen for determining the interface fracture energy for thin, brittle layers', *International Journal of Fracture* **92**(3), 213–220.

Hohlfelder, R. J., Luo, H., Vlassak, J. J., Chidsey, C. E. D. & Nix, W. D. (1996) 'Measuring interfacial fracture toughness with the blister test', in *MRS Proceedings*, Vol. 436, Cambridge University Press, Cambridge, p. 115.

Hutchinson, J. W. & Suo, Z. (1992) 'Mixed mode cracking in layered materials', *Advances in Applied Mechanics* **29**, 63–191.

Hutchinson, J. W., Thouless, M. D. & Liniger, E. G. (1992) 'Growth and configurational stability of circular, buckling-driven film delaminations', *Acta Metallurgica et Materialia* **40**(2), 295–308.

Hutchinson, R. G. & Hutchinson, J. W. (2011) 'Lifetime assessment for thermal barrier coatings: Tests for measuring mixed mode delamination toughness', *Journal of the American Ceramic Society* **94**(s1), 85–95.

Jackson, R. W. & Begley, M. R. (2014) 'Critical cooling rates to avoid transient-driven cracking in thermal barrier coating (TBC) systems', *International Journal of Solids and Structures* **51**(6), 1364–1374.

Jackson, R. W., Zaleski, E. M., Poerschke, D. L., Hazel, B. T., Begley, M. R. & Levi, C. G. (2015) 'Interaction of molten silicates with thermal barrier coatings under temperature gradients', *Acta Materialia* **89**, 396–407.

Jensen, H. M. (1991) 'The blister test for interface toughness measurement', *Engineering Fracture Mechanics* **40**(3), 475–486.

Jensen, H. M., Hutchinson, J. W. & Kim, K.-S. (1990) 'Decohesion of a cut prestressed film on a substrate', *International Journal of Solids and Structures* **26**(9), 1099–1114.

Jensen, H. M. & Thouless, M. D. (1993) 'Effects of residual stresses in the blister test', *International Journal of Solids and Structures* **30**(6), 779–795.

Jorgensen, D. J., Pollock, T. M. & Begley, M. R. (2015) 'Dynamic response of thin films on substrates subjected to femtosecond laser pulses', *Acta Materialia* **84**, 136–144.

Kanninen, M. F. & Popelar, C. L. (1985) *Advanced Engineering Fracture Mechanics*, Oxford University Press, Oxford.

Kardomateas, G. A., Berggreen, C. & Carlsson, L. A. (2013) 'Energy-release rate and mode mixity of face/core debonds in sandwich beams', *AIAA Journal* **51**(4), 885–892.

Knott, J. F. (1973) *Fundamentals of Fracture Mechanics*, Gruppo Italiano Frattura, Butterworth, London.

Kuo, A.-Y. (1990) 'Effects of crack surface heat conductance on stress intensity factors', *Journal of Applied Mechanics* **57**(2), 354–358.

Liechti, K. M. & Chai, Y.-S. (1991) 'Biaxial loading experiments for determining interfacial fracture toughness', *Journal of Applied Mechanics* **58**(3), 680–687.

Liechti, K. M. & Chai, Y.-S. (1992) 'Asymmetric shielding in interfacial fracture under in-plane shear', *Journal of Applied Mechanics* **59**(2), 295–304.

Malyshev, B. M. & Salganik, R. L. (1965) 'The strength of adhesive joints using the theory of cracks', *International Journal of Fracture Mechanics* **1**(2), 114–128.

Marthelot, J. (2014) Delamination of thin films, PhD thesis, ESPCI-PSL, University Pierre and Marie Curie, Paris, France.

Marthelot, J., Bico, J., Melo, F. & Roman, B. (2015) 'A new failure mechanism in thin film by collaborative fracture and delamination: Interacting duos of cracks', *Journal of the Mechanics and Physics of Solids* **84**, 214–229.

Matos, P. L., McMeeking, R. M., Charlambides, G., P. & Drory, M. D. (1989) 'A method for calculating stress intensities in bimaterial fracture', *International Journal of Fracture* **40**, 235–244.

Mei, H., Landis, C. M. & Huang, R. (2011) 'Concomitant wrinkling and buckle-delamination of elastic thin films on compliant substrates', *Mechanics of Materials* **43**(11), 627–642.

Moon, M.-W., Lee, K.-R., Oh, K.-H. & Hutchinson, J. W. (2004) 'Buckle delamination on patterned substrates', *Acta Materialia* **52**(10), 3151–3159.

Noijen, S. P. M., van der Sluis, O., Timmermans, P. H. M. & Zhang, G. Q. (2012) 'A semi-analytic method for crack kinking analysis at isotropic bi-material interfaces', *Engineering Fracture Mechanics* **83**, 8–25.

Parks, D. M. (1974) 'A stiffness derivative method finite element technique for determination of crack tip stress intensity factors', *International Journal of Fracture* **10**, 487–502.

Parry, G., Colin, J., Coupeau, C., Foucher, F., Cimetière, A. & Grilhé, J. (2005) 'Effect of substrate compliance on the global unilateral post-buckling of coatings: AFM observations and finite element calculations', *Acta Materialia* **53**(2), 441–447.

Rice, J. R. (1988) 'Elastic fracture mechanics concepts for interfacial cracks', *Journal of Applied Mechanics* **55**(1), 98–103.

Rice, J. R. & Sih, G. C. (1965) 'Plane problems of cracks in dissimilar media', *Journal of Applied Mechanics* **32**(2), 418–423.

Rice, J. R., Suo, Z. & Wang, J.-S. (1990) 'Mechanics and thermodynamics of brittle interfacial failure in bimaterial systems', in M. Rühle, A. G. Evans, M. F. Ashby & J. P. Hirth, eds., *Metal-Ceramic Interfaces*, Pergamon Press, Oxford, pp. 269–294.

Shield, T. W. & Kim, K. S. (1992) 'Beam theory models for thin film segments cohesively bonded to an elastic half space', *International Journal of Solids and Structures* **29**(9), 1085–1103.

Sih, G. C. (1962) 'On the singular character of thermal stresses near a crack tip', *Journal of Applied Mechanics* **29**(3), 587–589.

Sofla, A., Seker, E., Landers, J. P. & Begley, M. R. (2010) 'PDMS-glass interface adhesion energy determined via comprehensive solutions for thin film bulge/blister tests', *Journal of Applied Mechanics* **77**(3), 031007.

Sørensen, B. F., Jørgensen, K., Jacobsen, T. K. & Østergaard, R. C. (2006) 'Dcb-specimen loaded with uneven bending moments', *International Journal of Fracture* **141**(1–2), 163–176.

Stringfellow, R. G. & Freund, L. B. (1993) 'The effect of interfacial friction on the buckle-driven spontaneous delamination of a compressed thin film', *International Journal of Solids and Structures* **30**(10), 1379–1395.

Suga, T., Elssner, G. & Schmauder, S. (1988) 'Composite parameters and mechanical compatibility of material joints', *Journal of Composite Materials* **22**(10), 917–934.

Sundaram, S., Lipkin, D., Johnson, C. & Hutchinson, J. (2013) 'The influence of transient thermal gradients and substrate constraint on delamination of thermal barrier coatings', *Journal of Applied Mechanics* **80**(1), 011002.

Suo, Z. & Hutchinson, J. W. (1989) 'Sandwich test specimens for measuring interface crack toughness', *Materials Science and Engineering: A* **107**, 135–143.

Suo, Z. & Hutchinson, J. W. (1990) 'Interface crack between two elastic layers', *International Journal of Fracture* **43**(1), 1–18.

Tada, H., Paris, P. C. & Irwin, G. R. (2000) *The Stress Analysis of Cracks Handbook* 3rd rev. ed., ASME Press, New York.

Théry, P.-Y., Poulain, M., Dupeux, M. & Braccini, M. (2009) 'Spallation of two thermal barrier coating systems: Experimental study of adhesion and energetic approach to lifetime during cyclic oxidation', *Journal of Materials Science* **44**(7), 1726–1733.

Thouless, M. D. (1990) 'Crack spacing in brittle films on elastic substrates', *Journal of the American Ceramic Society* **73**(7), 2144–2146.

Thouless, M. D., Li, Z., Douville, N. J. & Takayama, S. (2011) 'Periodic cracking of films supported on compliant substrates', *Journal of the Mechanics and Physics of Solids* **59**(9), 1927–1937.

Tvergaard, V. & Hutchinson, J. W. (1993) 'The influence of plasticity on mixed mode interface toughness', *Journal of the Mechanics and Physics of Solids* **41**(6), 1119–1135.

Tvergaard, V., Xia, Z. C. & Hutchinson, J. W. (1993) 'Cracking due to localized hot shock', *Journal of the American Ceramic Society* **76**(3), 729–736.

Vasinonta, A. & Beuth, J. L. (2001) 'Measurement of interfacial toughness in thermal barrier coating systems by indentation', *Engineering Fracture Mechanics* **68**(7), 843–860.

Vaunois, J. R. (2014) Modèlisation de la durèe de vie des barrières thermiques, par le dèveloppement et l'exploitation d'essais d'adhèrence, PhD thesis, Materials, Universitè de Grenoble.

Vaunois, J.-R., Poulain, M., Kanoute, P. & Chaboche, J.-L. (2016) 'Development of bending tests for near shear mode interfacial toughness measurement of eb-pvd thermal barrier coatings', *Engineering Fracture Mechanics*, **to appear**.

Wang, J.-S. & Suo, Z. (1990) 'Experimental determination of interfacial toughness curves using Brazil-nut-sandwiches', *Acta Metallurgica et Materialia* **38**(7), 1279–1290.

Williams, M. L. (1959) 'The stresses around a fault or crack in dissimilar media', *Bulletin of the Seismological Society of America* **49**(2), 199–204.

Wolfram Research (2016) *Mathematica* v. 11. Wolfram Research Inc.

Xia, Z. C. & Hutchinson, J. W. (2000) 'Crack patterns in thin films', *Journal of the Mechanics and Physics of Solids* **48**(6), 1107–1131.

Xue, Z., Evans, A. & Hutchinson, J. (2009) 'Delamination susceptibility of coatings under high thermal flux', *Journal of Applied Mechanics* **76**(4), 041008.

Ye, T., Suo, Z. & Evans, A. G. (1992) 'Thin film cracking and the roles of substrate and interface', *International Journal of Solids and Structures* **29**(21), 2639–2648.

Yu, H.-H., He, M.-Y. & Hutchinson, J. W. (2001) 'Edge effects in thin film delamination', *Acta Materialia* **49**(1), 93–107.

Yu, H.-H. & Hutchinson, J. W. (2002) 'Influence of substrate compliance on buckling delamination of thin films', *International Journal of Fracture* **113**(1), 39–55.

Yu, H.-H. & Hutchinson, J. W. (2003) 'Delamination of thin film strips', *Thin Solid Films* **423**(1), 54–63.

Zak, A. R. & Williams, M. L. (1963) 'Crack point stress singularities at a bi-material interface', *Journal of Applied Mechanics* **30**(1), 142–143.

Zheng, J. & Sitaraman, S. K. (2007) 'Fixtureless superlayer-driven delamination test for nanoscale thin-film interfaces', *Thin Solid Films* **515**(11), 4709–4716.

Zhuk, A. V., Evans, A. G., Hutchinson, J. W. & Whitesides, G. M. (1998) 'The adhesion energy between polymer thin films and self-assembled monolayers', *Journal of Materials Research* **13**(12), 3555–3564.

# Index